Advanced Series in Nonlinear Dynamics - Volume 10

RENORMALIZATION AND GEOMETRY IN ONE-DIMENSIONAL AND COMPLEX DYNAMICS

ADVANCED SERIES IN NONLINEAR DYNAMICS*

Editor-in-Chief: R. S. MacKay *(Univ. Warwick)*

Published

Vol. 12 Localization and Solitary Waves in Solid Mechanics
A. R. Champneys, G. W. Hunt & J. M. T. Thompson

Vol. 13 Time Reversibility, Computer Simulation, Algorithms, Chaos (2nd Edition)
W. G. Hoover & C. G. Hoover

Vol. 14 Topics in Nonlinear Time Series Analysis – With Implications for EEG Analysis
A. Galka

Vol. 15 Methods in Equivariant Bifurcations and Dynamical Systems
P. Chossat & R. Lauterbach

Vol. 16 Positive Transfer Operators and Decay of Correlations
V. Baladi

Vol. 17 Smooth Dynamical Systems
M. C. Irwin

Vol. 18 Symplectic Twist Maps
C. Gole

Vol. 19 Integrability and Nonintegrability of Dynamical Systems
A. Goriely

Vol. 20 The Mathematical Theory of Permanent Progressive Water-Waves
H. Okamoto & M. Shoji

Vol. 21 Spatio-Temporal Chaos & Vacuum Fluctuations of Quantized Fields
C. Beck

Vol. 22 Energy Localisation and Transfer
eds. T. Dauxois, A. Litvak-Hinenzon, R. MacKay & A. Spanoudaki

Vol. 23 Geometrical Theory of Dynamical Systems and Fluid Flows (Revised Edition)
T. Kambe

Vol. 24 Microscopic Chaos, Fractals and Transport in Nonequilibrium Statistical Mechanics
R. Klages

Vol. 25 Smooth Particle Applied Mechanics – The State of the Art
W. G. Hoover

Vol. 26 Geometry of Nonholonomically Constrained Systems
by R. H. Cushman, J. Śniatycki & H. Duistermaat

Vol. 27 Simulation and Control of Chaotic Nonequilibrium Systems
by W. G. Hoover & C. G. Hoover

*For the complete list of titles in this series, please visit
http://www.worldscientific.com/series/asnd

Advanced Series in Nonlinear Dynamics - Volume 10

RENORMALIZATION AND GEOMETRY IN ONE-DIMENSIONAL AND COMPLEX DYNAMICS

Yunping Jiang

(City University of New York)

World Scientific

NEW JERSEY • LONDON • SINGAPORE • BEIJING • SHANGHAI • HONG KONG • TAIPEI • CHENNAI • TOKYO

Published by

World Scientific Publishing Co. Pte. Ltd.
5 Toh Tuck Link, Singapore 596224
USA office: 27 Warren Street, Suite 401-402, Hackensack, NJ 07601
UK office: 57 Shelton Street, Covent Garden, London WC2H 9HE

British Library Cataloguing-in-Publication Data
A catalogue record for this book is available from the British Library.

Advanced Series in Nonlinear Dynamics — Vol. 10
RENORMALIZATION AND GEOMETRY IN ONE-DIMENSIONAL AND COMPLEX DYNAMICS

ISBN-13 978-981-02-2326-7
ISBN-10 981-02-2326-9

To my Mother and Father

To Bin and Jeffrey

Preface

This monograph summarizes my research in dynamical systems during the past eight years. Included too are many facts, techniques, and results which I have learned from others and which have greatly enhanced my own work.

In September 1985, I arrived in the United States to pursue a doctoral degree in mathematics. One and a half years later, I wrote to Dennis Sullivan and asked if I could study under his supervision. I became a Ph.D. student of Sullivan and so began this research. During my years as his student, Sullivan gave classes and ran seminars on Tuesday and Thursday at the Graduate Center of the City University of New York, where he held an Einstein Chair. These sessions sometimes lasted all day.

My first problem in this area of research was suggested by Sullivan in one of his classes. The problem was first, to understand the asymptotic geometry of Cantor sets generated by a family of dynamical systems involving a singular point and second, to investigate the geometric property of conjugacy between two such dynamical systems. This work is included in Chapter Two of the present monograph as an application of the Koebe distortion principle. One recognized program is to "fill in" the dictionary between the theory of one-dimensional dynamical systems and the theory of Kleinian groups. Seeking to advance this program and to generalize my first research in this area, I studied the space of geometrically finite one-dimensional maps and the classification of these maps up to conjugacy by quasisymmetric homeomorphisms and up to conjugacy by diffeomorphisms. This work is described in Chapter Three. Part of it comes from my Ph.D. thesis. Frederick Gardiner, Charles Tresser, and Richard Sacksteder gave me their help during this study and Sullivan provided his own insightful suggestions.

In 1987, Peter Veerman gave me a paper of Robert MacKay concerning Denjoy's theorem for circle diffeomorphisms. In this paper, MacKay applies the renormalization method to an old theorem. I became interested in this approach. It is described in Chapter One as an introduction to renormalization theory and as an application of the Denjoy distortion principle. That same year, Welington de Melo and Sebastian van Strien visited the Einstein Chair and presented their results on one-dimensional dynamical systems. Grzegorz Świątek also paid a visit and presented his results on critical circle mappings. Interestingly, they had independently developed a technique to estimate the distortion of a one-dimensional map having a critical point. This technique was later generalized to a larger class of one-dimensional maps by Sullivan; it is called the Koebe distortion principle because of its similarity to Koebe's distortion theorem in one complex variable discovered some eighty years ago. Chapter Two contains several versions of the distortion principle.

A universal rule governs the transition from simple motion to chaos in a one-parameter family of dynamical systems with a unique quadratic critical point. Mitchell Feigenbaum discovered this in the 1970s. The rule can be explained by means of a family of one-dimensional dynamical systems like those generated by quadratic polynomials. Feigenbaum calculated period doubling bifurcations for such a family and showed that the limit of these period doubling bifurcations is a chaotic dynamical system in the family. Furthermore, the appearance of the chaotic dynamical system follows a universal pattern which is described by the so-called Feigenbaum universal number. Oscar Lanford III gave the first proof of this discovery with some computer help. For the chaotic dynamical system, the interesting object is its attractor. The attractor is uncountable, perfect, and totally disconnected: a Cantor set. Feigenbaum, and independently, Pierre Coullet and Charles Tresser, discovered in the 1970s that the geometry of this Cantor set is universal, meaning that it does not depend on the specific family being studied. This discovery is similar to Mostow's rigidity theorem, which says that in the class of closed hyperbolic three-manifolds, topology determines geometry. During my years as a Ph.D. student, some work of Sullivan led to an important mathematical understanding of this discovery. Chapter Four contains part of this work based on my class notes.

During my time as his Ph.D. student, Sullivan showed me how to deform a Feigenbaum-like map. I began to think about this topic and also to study the spectrum of the period doubling operator. Meanwhile, Takehiko Morita visited the Einstein Chair. I told him what I was working on and he showed me a general strategy to study the spectrum of a transfer operator in thermodynamical formalism. I applied this strategy to the study of the tangent map of the period doubling operator by connecting it with a transfer operator. This led eventually to a conceptual proof of the existence of the Feigenbaum universal number in a joint paper with Morita and Sullivan. This is the origin of Chapter Six. During this study, conversations with David Ruelle and Henri Epstein helped me to better understand the spectrum of the period doubling operator and other related topics. In the summer of 1993, Viviane Baladi lectured on thermodynamical formalism at a workshop held in Hillerød, Denmark. After her lectures, I asked her about generalizing some of the results in her lectures to the Zygmund continuous vector space. She showed me some calculations that suggested this possibility. In the summer of 1994, I visited the Forschungsinstitut für Mathematik at the Eidgenössische Technische Hochschule in Zürich to work with Baladi and Lanford on this problem. The fruit of this work is described in a paper written by Baladi, Lanford, and myself. A special case of our result is presented in Chapter Six for the purpose of studying the spectrum of the renormalization operator.

After completing my Ph.D. study in May 1990, I went to the Institute for Mathematical Sciences at Stony Brook in September 1990. There I was influenced by John Milnor's investigations in complex dynamics; I began to work on some problems in this field.

Yakov Sinai constructed Markov partitions for Anosov dynamical systems in the 1960s. Rufus Bowen generalized this method to Axiom A dynamical systems. This method became very important in the study of hyperbolic dynamical systems. During the academic year of 1991, Mitsuhiro Shishikura visited the Institute for Mathematical Sciences at Stony Brook and presented his work in complex dynamics. In his lectures, he introduced me to the result of Jean-Christophe Yoccoz on the local connectivity of the Julia set of a non-renormalizable quadratic polynomial and to the technique called Yoccoz puzzles. This technique was first used by Bodil Branner and John Hubbard in their study of certain cubic polynomials and was successfully used by Yoccoz in his study of non-renormalizable quadratic polynomials. The technique is a little different from that of the Markov partitions but is motivated by the same philosophy. By learning this technique, I was able to apply it, along with my knowledge of infinitely renormalizable maps, to the study of infinitely renormalizable quadratic polynomials. I proved that some conditions on an infinitely renormalizable quadratic polynomial are sufficient to ensure that its Julia set is locally connected. This is described in Chapter Five.

The first time I applied Yoccoz puzzles in my research was in the study of bounded and bounded nearby geometry of certain infinitely renormalizable folding maps and in the quasisymmetric classification of these maps. In this study, I combined the technique of Yoccoz puzzles with Markov partitions and with my previous work on geometrically finite one-dimensional maps. This research is described in Chapter Four.

After completing my contract with the State University of New York at Stony Brook, I started work, in September 1992, at Queens College of the City University of New York. I began to regularly attend Sullivan's seminars at the Graduate Center of the City University of New York. Jun Hu told me there that Sullivan had completed his work about the *a priori* complex bounds for the Feigenbaum quadratic polynomial. This caught my attention because the *a priori* complex bounds is a sufficient condition that the Julia set of a real infinitely renormalizable quadratic polynomial is locally connected. After Hu explained to me the idea of Sullivan's proof of the *a priori* complex bounds for the Feigenbaum quadratic polynomial, I went on to write a joint paper with him about the local connectivity of the Julia set of the Feigenbaum quadratic polynomial. Later I realized that unless I had a complete proof of Sullivan's result, my understanding of the local connectivity of the Julia set of the Feigenbaum polynomial was incomplete. I began a serious study of Sullivan's result, which appears in Chapter Five. During this study, several conversations with Sullivan, along with the thesis of Edson de Faria, provided a lot help. During my research into infinitely renormalizable quadratic polynomials, communication with Curt McMullen via e-mail was very helpful. Several statements were made more precise because of his comments. Moreover, they led me to combine several of my papers in this direction into one self-contained paper, which is the origin of Chapter Five.

After I explained to Sullivan my research about the local connectivity of the Mandelbrot set at certain infinitely renormalizable points, he suggested that I might be able to reduce the computational aspect of many proofs by addressing the argument from a topological perspective. This is also presented in Chapter Five. In this study, many conversations with Tan Lei helped me to better understand the topological structure of the Mandelbrot set.

The survey articles written by E. B. Vul, Ya. G. Sinai, and K. M. Khanin [VSK], by J. Milnor [MI2,MI3], and by J. Hubbard [HUB], the book written by P. Collet and J.-P. Eckmann [COE] and the book edited by P. Cvitanović [CVI] provided valuable guidance, not only at the beginning, but also throughout the whole period of my research.

An invitation from the Advanced Series in Nonlinear Dynamics gave me a chance to work on this research monograph. Its editor, Robert MacKay, not only suggested this monograph but also encouraged me to complete it. Frank Isaacs gave me advice on English and on mathematical presentation.

In addition to the names of colleagues mentioned above, Benjamin Bielefeld, Elise Cawley, Hsinta Frank Cheng, Guizhen Cui, Jack Diamond, Jozef Dodziuk, Lisa Goldberg, Sen Hu, Huyi Hu, Weihua Jiang, Jeremy Kahn, Linda Keen, Ravi Kulkarni, Genadi Levin, Shantao Liao, Arthur Lopes, Feng Luo, Jiaqi Luo, Mikhail Lyubich, Michael Maller, Jürgen Moser, Waldemar Pałuba, Alberto Pinto, Feliks Przytycki, David Rand, Michael Shub, Meiyu Su, Scott Sutherland, Folkert Tangerman, David Tischler, He Wu, Zhihong (Jeff) Xia, Shing-Tung Yau, Lai-Sang Young have each given valuable counsel and help. During the publication of my results, the referees provided invaluable comments. I offer thanks to all.

During the past eight years, while I visited the following institutes, my research has been supported by the Institut des Hautes Études Scientifiques in France, by the Nonlinear Systems Laboratory in the Mathematics Institute at the University of Warwick in England, by the Forschungsinstitut für Mathematik at the Eidgenössische Technische Hochschule in Switzerland, by the Max-Planck-Institut für Mathematik in Germany, and by the Mathematical Sciences Research Institute in the United States. Valuable research time was granted by the Department of Mathematics at Queens College of the City University of New York. This research was also funded in part by the National Science Foundation, by the Professional Staff Congress-City University of New York Research Award Program, by the City University of New York Collaborative Incentive Research Grant Program, and by the New York State Graduate Research Initiative Grant Program. I am deeply grateful for their support.

Queens, New York — YUNPING JIANG

September 1995

Contents

Chapter One

The Denjoy Distortion Principle and Renormalization

This first chapter introduces an old technique to the study of one-dimensional dynamics; then it introduces the renormalization of circle mappings.

Let $f(x)$ be a self-map of either an interval I or the circle $\mathbf{T}^1$. The dynamical system generated by f is the semigroup $\{f^{\circ n}\}_{n=0}^{\infty}$ of iterations of f. The simplest dynamical systems are those generated by linear mappings of the real line. An important analytic characterization of these linear dynamical systems is that $(f^{\circ n})'(x)/(f^{\circ n})'(y) = 1$ for x and y both real and n a non-negative integer. For a dynamical system $\{f^{\circ n}\}_{n=0}^{\infty}$, we call the set of ratios $\{\log |(f^{\circ n})'(x)/(f^{\circ n})'(y)|\}$ the distortion. The distortion of a linear dynamical system contains only zero. Indeed, estimating the distortion of a dynamical system is useful and important. An old technique, going back to Denjoy [DEN], estimates the distortion of a one-dimensional dynamical system generated by a self-map without critical point of either an interval or the circle. The technique was discovered in the study of circle diffeomorphisms with irrational rotation numbers, and was developed and widely used in the study of one (and even higher) dimensional dynamical systems. The original Denjoy's theorem says that a C^2 irrational circle diffeomorphism is topologically conjugate to the rigid rotation with the same rotation number. A well-known version of Denjoy's theorem applies to irrational circle diffeomorphisms whose derivatives have bounded variation; it is proved in §1.4 to §1.11 by means of the renormalization method worked out by MacKay [MAC].

Cantor sets which are the non-escaping sets of degree two expanding interval maps, and the scaling functions defined for mathematics by Sullivan [SU3] associated with these Cantor sets are discussed in §1.2 and §1.3.

1.1. Naive Distortion Lemmas

Let f be a function defined on a set U of the real line $\mathbf{R}$. It is said to be C^1 (or $C^{1+\alpha}$ for $0 < \alpha \leq 1$ or C^{1+bv}) if it can be extended to a differentiable function defined on an open set containing U and if the derivative of the extension is continuous (or is α-Hölder continuous or is of bounded variation).

Suppose f is a C^1 function on a set U of the real line $\mathbf{R}$ and $X = \{x_i\}_{i=1}^n$ and $Y = \{y_i\}_{i=1}^n$ are two sequences of points in U. The number

$$\log\left|\prod_{i=1}^{n} \frac{f'(x_i)}{f'(y_i)}\right|$$

is called the distortion of f along X and Y.

Lemma 1.1. *Suppose $\kappa = \inf_{x\in U} |f'(x)| > 0$. Then the distortion of f along X and Y can be estimated as*

$$\left|\log\left|\prod_{i=1}^{n} \frac{f'(x_i)}{f'(y_i)}\right|\right| \leq \frac{1}{\kappa}\sum_{i=1}^{n} |f'(x_i) - f'(y_i)|. \tag{1.1}$$

Proof. The proof of this lemma is easy because

$$\begin{aligned}\left|\log\left|\prod_{i=1}^{n} \frac{f'(x_i)}{f'(y_i)}\right|\right| &\leq \sum_{i=1}^{n} \big|\log|f'(x_i)| - \log|f'(y_i)|\big| \\ &\leq \frac{1}{\kappa}\sum_{i=1}^{n} |f'(x_i) - f'(y_i)|.\end{aligned}$$

■

The next two lemmas are easily derived from Lemma 1.1.

Lemma 1.2 ($C^{1+\alpha}$-Denjoy distortion lemma). *Suppose f is $C^{1+\alpha}$ for some $0 < \alpha \leq 1$ and $\kappa = \inf_{x\in U} |f'(x)| > 0$. Let $\iota = \sup_{x\neq y\in U} \big(|f'(x) - f'(y)|/|x-y|^\alpha\big) < \infty$. Then the distortion of f along X and Y is bounded by $(\iota/\kappa)\sum_{i=1}^{n} |x_i - y_i|^\alpha$, that is,*

$$\left|\log\left|\prod_{i=1}^{n} \frac{f'(x_i)}{f'(y_i)}\right|\right| \leq \frac{\iota}{\kappa}\sum_{i=1}^{n} |x_i - y_i|^\alpha. \tag{1.2}$$

Proof. Since $|f'(x_i) - f'(y_i)| \leq \iota|x_i - y_i|^\alpha$, it follows directly from Lemma 1.1. ■

Lemma 1.3 (C^{1+bv}-Denjoy distortion lemma). *Suppose f is C^{1+bv}. Then there is a constant $C > 0$ so that the distortion of f along X and Y is bounded by C, that is,*

$$\left|\log\left|\prod_{i=1}^{n} \frac{f'(x_i)}{f'(y_i)}\right|\right| \leq C, \tag{1.3}$$

whenever the open intervals I_i, bounded by x_i and y_i, for $i = 1, \ldots, n$, are pairwise disjoint.

Proof. Let V be the total variation of f on U. Then $\sum_{i=1}^{n} |f'(x_i) - f'(y_i)|$ is bounded by V for $\{I_i\}_{i=1}^{n}$ are pairwise disjoint. We can take $C = V/\kappa$. ■

1.2. Hyperbolic Cantor Sets

Suppose $I = [0,1]$ is the unit interval and $I_0 = [0,a]$ and $I_1 = [b,1]$ are two subintervals where $0 < a < b < 1$. A C^1 map f defined on $I_0 \cup I_1$ is said to be degree two if $f|I_i$ from I_i to I is bijective for $i = 0, 1$ (see Fig. 1.1); f is said to be expanding if there are constants $C > 0$ and $\lambda > 1$ so that $|(f^{\circ n})'(x)| \geq C\lambda^n$ whenever $f^{\circ i}(x)$ are in $I_0 \cup I_1$ for all $i = 0, 1, \ldots, n-1$.

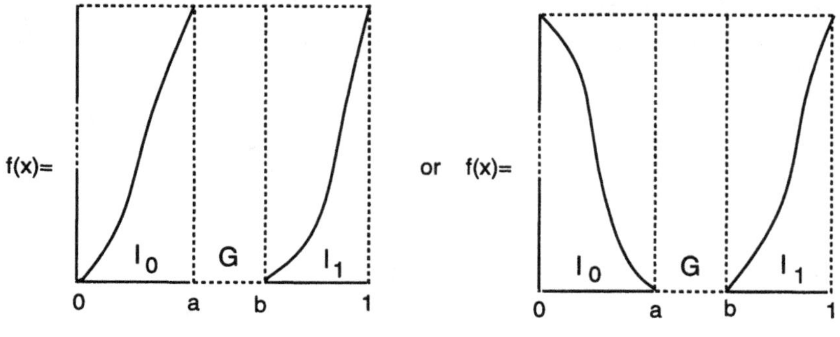

Fig. 1.1

Suppose $f : I_0 \cup I_1 \to I$ is a degree two expanding map. Let $G = (a,b)$ be the complement of $I_0 \cup I_1$ in I. A number x in I is said to be escaping to G if $f^{\circ k}(x)$ is in G for some integer $k \geq 0$ (where $f^{\circ 0}$ is the identity). The set $\Omega \subseteq I$ of escaping points is an open subset of the real line. The complement Λ of Ω in I is called the non-escaping set under f. It is a compact (closed and bounded) subset of the real line $\mathbf{R}$.

Example 1.1 ($\frac{1}{3}$-Cantor set). *Suppose $a = 1/3$ and $b = 2/3$. Define*

$$f(x) = \begin{cases} 3x, & \text{if } 0 \leq x \leq a; \\ 3x - 2, & \text{if } b \leq x \leq 1. \end{cases}$$

Then f is a degree two expanding map for which the non-escaping set Λ under f is the famous $\frac{1}{3}$-Cantor set.

Topologically, a Cantor subset of the real line $\mathbf{R}$ is compact, uncountable, totally disconnected, and perfect.

Theorem 1.1 ($C^{1+\alpha}$-hyperbolic Cantor set). *If $f : I_0 \cup I_1 \to I$ is a $C^{1+\alpha}$ degree two expanding map for some $0 < \alpha \leq 1$. Then the non-escaping set Λ under f is a Cantor set whose Lebesgue measure is zero.*

Proof. Let f_i be the restriction of the function f to I_i, and $g_i = f_i^{-1} : I \to I_i$ be the inverse of f_i for $i = 0$ or 1. We can consider compositions $g_{w_n} = g_{i_0} \circ g_{i_1} \circ \cdots \circ g_{i_n}$ for all strings $w_n = i_0 i_1 \ldots i_n$ of $0's$ and $1's$. These compositions are contracting; this means that there are constants $C > 0$ and $0 < \mu < 1$ so that $|g'_{w_n}(x)| < C\mu^n$ for all x in I.

Suppose $w_n = i_0 i_1 \ldots i_n$ is a string of $0's$ and $1's$. Let $I_{w_n} = g_{w_n}(I)$ be the image of I under g_{w_n}, and let $G_{w_n} = g_{w_n}(G)$ be the set of all the points escaping to G under $g_{w_n}^{-1}$. The union $\cup_{w_n} I_{w_n}$ is the set of all points not escaping to G under the iterates $f^{\circ k}$ for $k = 0, 1, \ldots, n$, where w_n runs over all the strings of $0's$ and $1's$ of length $n+1$. The set $\{I_{w_n}\}$ is a collection of pairwise disjoint closed intervals and one to one correspondence with the set $\{w_n\}$ of all the strings of $0's$ and $1's$ of length $n+1$. Hence $\Lambda = \cap_{n=0}^{\infty} \cup_{w_n} I_{w_n}$, where w_n runs over all the strings of $0's$ and $1's$ of length $n+1$.

Let us first prove that Λ is uncountable. For a string $w_n = i_0 i_1 \ldots i_n$ of $0's$ and $1's$ and a digit $i_{n+1} = 0$ or 1, $I_{w_n i_{n+1}} \subseteq I_{w_n}$ since $I_{i_{n+1}} \subseteq I$. This implies that

$$\cdots \subseteq I_{i_0 i_1 \ldots i_n} \subseteq \cdots \subseteq I_{i_0 i_1} \subseteq I_{i_0}$$

and that $I_w = \cap_{n=0}^{\infty} I_{i_0 i_1 \ldots i_n}$ is a non-empty closed subset for any infinite string $w = i_0 i_1 \ldots$ of $0's$ and $1's$. Hence the set $\{I_w\}$ is a collection of pairwise disjoint non-empty closed subsets and is in one to one correspondence with the uncountable set $\{w = i_0 i_1 \ldots\}$ of all infinite strings of $0's$ and $1's$. Hence the set $\{I_w\}$ is uncountable. So too is the set Λ because $\Lambda = \cup_w I_w$ where $w = i_0 i_1 \ldots$ runs over all infinite strings of $0's$ and $1's$.

Since g_{w_n} is contracting, the length of I_{w_n} is less than $C\mu^n$ for any string $w_n = i_0 i_1 \ldots i_n$ of $0's$ and $1's$ of length $n+1$. This implies that I_w contains a single number x_w, and the map $\pi(w) = x_w$ from $\{w\}$ to Λ is bijective. We use this to prove that Λ is totally disconnected, that is, every (connected) component Π of Λ contains only one number. Suppose there is a component Π of Λ which contains two different numbers x_w and $x_{w'}$ where $w = i_0 i_1 \ldots i_n i_{n+1} \ldots$ and $w' = i_0 i_1 \ldots i_n i'_{n+1} \ldots$ where $i_{n+1} \neq i'_{n+1}$. Both x_w and $x_{w'}$ are in I_{w_n} where $w_n = i_0 i_1 \ldots i_n$. The set I_{w_n} is the union of an open interval G_{w_n} and two closed intervals $I_{w_n i_{n+1}}$ and $I_{w_n i'_{n+1}}$ which are on different sides of G_{w_n}. The numbers x_w and $x_{w'}$ are in $I_{w_n i_{n+1}}$ and $I_{w_n i'_{n+1}}$, respectively. Take a point z in G_{w_n}. Then

$$\Pi = \Big(\Pi \cap (-\infty, z)\Big) \cup \Big(\Pi \cap (z, \infty)\Big).$$

This contradicts the statement that Π is a component of Λ and proves that Λ is totally disconnected.

Since Λ is closed, the set Λ' of limit points of Λ is contained in Λ. To prove that Λ is a perfect set, we only need to show that Λ is contained in Λ'. Let x_w be a number in Λ and $w = i_0 i_1 \ldots i_n i_{n+1} \ldots$. Let $r(i) = i+1 \pmod 2$; $r(i)$ is 1 for $i = 0$ and 0 for $i = 1$. Take $w^{(n)} = i_0 \ldots i_{n-1} r(i_n) i_{n+1} \ldots$; $w^{(n)}$ differs from w at $(n+1)^{th}$ position. Then $x_{w^{(n)}} \neq x_w$ and both of them are in $I_{i_0 \ldots i_{n-1}}$. Since the length of $I_{i_0 \ldots i_{n-1}}$ tends to zero as n goes to infinity, $x_{w^{(n)}}$ tends to x_w as n goes to infinity. This says that x_w is a limit point of Λ. So Λ is contained in Λ'. Hence Λ is a Cantor set.

Now let us prove that the Lebesgue measure of the Cantor set Λ is zero. Let $m(\cdot)$ mean the Lebesgue measure and let $|J|$ mean the length of an interval. An inequality which can be easily obtained is

$$m(\Lambda) \leq \sum_{w_n} |I_{w_n}| < C 2^{n+1} \mu^n,$$

where w_n runs over all the strings of $0's$ and $1's$ of length $n+1$. This inequality is true because $\{I_{w_n}\}$ is a cover of Λ and the total number of the strings of $0's$ and $1's$ of length $n+1$ is 2^{n+1}. If $\mu < 1/2$, it is much easier to see $m(\Lambda) = 0$. However, to prove that the Lebesgue measure of Λ is zero for any $0 < \mu < 1$, we need help from Lemma 1.2.

Suppose $w_n = i_0 \ldots i_n$ is a string of $0's$ and $1's$ of length $n+1$. The map f^{n+1} from I_{w_n} to I is a monotone function and its inverse is g_{w_n}. For any two numbers x and y in I_{w_n}, let $x_i = f^{\circ(n-i+1)}(x)$ and $y_i = f^{\circ(n-i+1)}(y)$ for $i = 0, 1, \ldots, n+1$. By the mean value theorem and the chain rule, $|x_i - y_i| < C\mu^i$ and $\sum_{i=0}^{n+1} |x_i - y_i|^\alpha < C/(1-\mu^\alpha)$. According to Lemma 1.2, the distortion of f along $X = \{x_i\}$ and $Y = \{y_i\}$ is bounded by the constant $C' = (\iota/\kappa)(C/(1-\mu^\alpha))$, that is,

$$\left| \log \left| \frac{\left(f^{\circ(n+1)}\right)'(x)}{\left(f^{\circ(n+1)}\right)'(y)} \right| \right| \leq C',$$

where ι is the Hölder constant of f' on $I_0 \cup I_1$ and $\kappa = \inf_{x \in I_0 \cup I_1} |f'(x)|$. This implies that

$$\frac{|G_{w_n}|}{|I_{w_n}|} \geq c = e^{-C'} |G|,$$

since $G = f^{\circ(n+1)}(G_{w_n})$ and $I = f^{\circ(n+1)}(I_{w_n})$. Now we have that

$$|I_{w_n 0}| + |I_{w_n 1}| \leq (1-c)|I_{w_n}|$$

because $I_{w_n} = I_{w_n 0} \cup G_{w_n} \cup I_{w_n 1}$; moreover,

$$m(\Lambda) \leq \sum_{w_{n+1}} |I_{w_{n+1}}| = \sum_{w_n} (|I_{w_n 0}| + |I_{w_n 1}|)$$
$$\leq (1-c) \sum_{w_n} |I_{w_n}| \leq \cdots \leq (1-c)^{n+1}$$

for all positive integers n. Hence the Lebesgue measure of Λ is zero. ■

Remark 1.1. From the proof, one can see that the non-escaping set Λ of a C^1 degree two expanding map is a Cantor set in the real line. There is a Cantor set with positive Lebesgue measure. An interesting problem is to construct a C^1 degree two expanding map whose non-escaping set is a Cantor set with positive Lebesgue measure (see [BO3]).

1.3. Scaling Functions of Hyperbolic Cantor Sets

Suppose $f : I_0 \cup I_1 \to I$ is a $C^{1+\alpha}$ degree two expanding map for some $0 < \alpha \leq 1$ and Λ is the non-escaping set of f. We use the same notation as in §1.2.

In the proof of Theorem 1.1, we read every string w_n of $0's$ and $1's$ in ascending order, that is, $w_n = i_0 i_1 \ldots i_n$, and obtained a bijective map π, from $\Sigma = \{w = i_0 i_1 \ldots\}$, the set of all infinite strings of $0's$ and $1's$, to Λ, where $\pi(w) = \cap_{n=0}^{\infty} I_{w_n}$ and $w = w_n \ldots$. Let σ be the shift map; $\sigma : i_0 i_1 \ldots \mapsto i_1 \ldots$. Then $\pi \circ \sigma = f \circ \pi$. This says that if we consider Σ equipped with the product topology, then (f, Λ) and (σ, Σ) are topologically conjugate. Here (σ, Σ) is called a symbolic dynamical system. It is a topological model of all dynamical systems generated by degree two expanding maps. Now let us begin to read every finite string w_n of $0's$ and $1's$ in descending order, that is, $w_n = j_n \ldots j_1 j_0$. Let $\sigma^*(w_n) = j_n \ldots j_1$; σ^* knocks off the first digit of w_n on the right. One can check that I_{w_n} is a subinterval of $I_{\sigma^*(w_n)}$ (see Fig. 1.2).

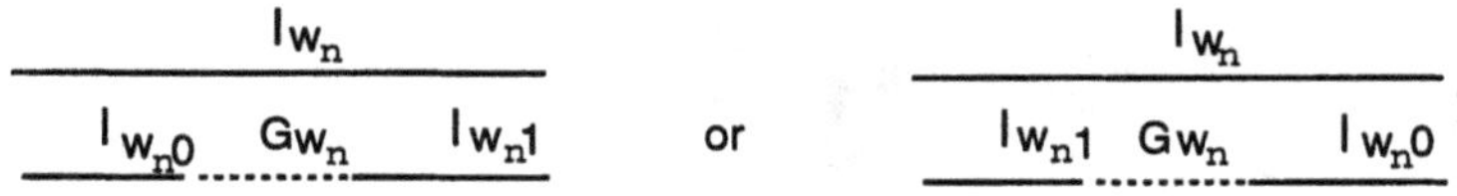

Fig. 1.2

Let $|J|$ mean the length of an interval J. Define the scaling of f at w_n as the ratio

$$s(w_n) = \frac{|I_{w_n}|}{|I_{\sigma^*(w_n)}|}. \tag{1.4}$$

Lemma 1.4. *For any infinite string* $w = \ldots j_n j_{n-1} \ldots j_1 j_0$ *of* $0's$ *and* $1's$, *let* $w_n = j_n \ldots j_1 j_0$. *The limit*

$$s(w) = \lim_{n \to \infty} s(w_n) \tag{1.5}$$

exists.

Proof. For any $n > m$, $f^{\circ(n-m)}$ sends I_{w_n} and $I_{\sigma^*(w_n)}$ onto I_{w_m} and $I_{\sigma^*(w_m)}$, and $f^{\circ m}$ sends I_{w_m} and $I_{\sigma^*(w_m)}$ onto I_{j_0} and $I = [0, 1]$. So

$$s(w_n) = \frac{\left(f^{\circ(n-m)}\right)'(\xi)}{\left(f^{\circ(n-m)}\right)'(\eta)} \cdot s(w_m)$$

for some ξ and η in $I_{\sigma^*(w_n)}$, and

$$s(w_m) = \frac{(f^{\circ m})'(\xi')}{(f^{\circ m})'(\eta')} \cdot s(j_0)$$

for some ξ' and η' in $I_{\sigma^*(w_m)}$. From Lemma 1.2 (refer to the proof of Theorem 1.1), there is a constant $C > 0$ such that

$$\log\left|\frac{(f^{\circ(n-m)})'(\xi)}{(f^{\circ(n-m)})'(\eta)}\right| \leq C|I_{w_m}|^\alpha$$

and

$$\log\left|\frac{(f^{\circ m})'(\xi')}{(f^{\circ m})'(\eta')}\right| \leq C.$$

Hence there is a positive constant which we still denote as C such that

$$\begin{aligned}|s(w_n) - s(w_m)| &= \left|\frac{(f^{\circ(n-m)})'(\xi)}{(f^{\circ(n-m)})'(\eta)} - 1\right| \cdot \left|\frac{(f^{\circ m})'(\xi')}{(f^{\circ m})'(\eta')}\right| \cdot s(j_0) \\ &\leq C|I_{w_m}|^\alpha.\end{aligned}$$

This implies that $\{s(w_n)\}_{n=0}^\infty$ is a Cauchy sequence. ∎

Definition 1.1. *The space* $\Sigma^* = \{w = \ldots j_n \ldots j_1 j_0 \mid j_n = 0$ or 1, $n = 0, 1, \ldots\}$ *with the product topology is called the dual symbolic space of an degree two expanding map* $f : I_0 \cup I_1 \to I$.

From Lemma 1.4, $s(w)$ defines a function on Σ^*.

Definition 1.2. *The function* s *defined on* Σ^* *is called the scaling function of* (f, Λ).

Remark 1.2. The scaling function of (f, Λ) is defined by Sullivan [SU3] for mathematics and is, in some sense, a generalization of the thickness of a Cantor set defined by Newhouse [NEW].

An object is called a C^1-invariant of a dynamical system $\{f^{\circ n}\}_{n=0}^\infty$ if it is the same for f and for $h \circ f \circ h^{-1}$ whenever h is a C^1 diffeomorphism.

Lemma 1.5. *The scaling function* s *of* (f, Λ) *is a* C^1*-invariant.*

Proof. Consider degree two expanding maps f and g with scaling functions s_f and s_g. Further, suppose f and g are C^1 conjugate; there is a C^1 diffeomorphism h of I such that $h \circ f = g \circ h$.

For every $w = \dots j_n \dots j_1 j_0$, let $w_n = j_n \dots j_1 j_0$. Then

$$s_g(w_n) = \left|\frac{h'(\xi_n)}{h'(\eta_n)}\right| \cdot s_f(w_n)$$

for some ξ_n and η_n in $I_{\sigma^*(w_n)}$. Hence

$$s_g(w) = s_f(w)$$

because $\lim_{n\to\infty} |h'(\xi_n)|/|h'(\eta_n)| = 1$. ∎

Definition 1.3. *A function s defined on Σ^* is said to be Hölder if there are constants $C > 0$ and $0 < \mu < 1$ such that*

$$|s(w) - s(w')| \le C\mu^m$$

whenever the first m digits of w and w' from right are the same.

Theorem 1.2 [SU3]. *Suppose f is a $C^{1+\alpha}$ degree two expanding map and Λ is its non-escaping Cantor set. Then the scaling function s of (f, Λ) is a C^1-invariant Hölder function on Σ^*.*

Proof. It remains to prove only that s is Hölder. Let w and w' be two points in Σ^* whose first m digits are the same. Then $w = \dots j_{m+1} w_m$ and $w' = \dots j'_{m+1} w_m$. For every $n > m$,

$$\begin{aligned}|s(w_n) - s(w'_n)| &= \left|\frac{\left(f^{\circ(n-m)}\right)'(\xi)}{\left(f^{\circ(n-m)}\right)'(\eta)} - \frac{\left(f^{\circ(n-m)}\right)'(\xi')}{\left(f^{\circ(n-m)}\right)'(\eta')}\right| \cdot s(w_m) \\ &= \left|\frac{\left(f^{\circ(n-m)}\right)'(\xi)}{\left(f^{\circ(n-m)}\right)'(\eta)} - \frac{\left(f^{\circ(n-m)}\right)'(\xi')}{\left(f^{\circ(n-m)}\right)'(\eta')}\right| \cdot \left|\frac{\left(f^{\circ m}\right)'(\xi'')}{\left(f^{\circ m}\right)'(\eta'')}\right| \cdot s(j_0),\end{aligned}$$

where $w_n = j_n \dots j_{m+1} w_m$ and $w'_n = j'_n \dots j'_{m+1} w_m$ and where $\xi, \eta \in I_{\sigma^*(w_n)}$, $\xi', \eta' \in I_{\sigma^*(w'_n)}$, and $\xi'', \eta'' \in I_{\sigma^*(w_m)}$. According to Lemma 1.2 (refer also to the proofs of Theorem 1.1 and Lemma 1.4), there are constants $C > 0$ and $0 < \mu < 1$ so that

$$|s(w_n) - s(w'_n)| \le C\mu^m.$$

Now by taking limits

$$|s(w) - s(w')| \le C\mu^m.$$

∎

Remark 1.3. Sullivan [SU3] also proved that the scaling function of (f, Λ) is a complete C^1-invariant; meaning that if the scaling functions of (f, Λ_f) and (g, Λ_g) are the same, then they are C^1 conjugate, that is, there is a C^1 diffeomorphism h of I such that $f \circ h = h \circ g$ on Λ_g (see the paper of Bedford and Fisher [BEF] as well). We discuss this kind of property in Chapter Three for geometrically finite one-dimensional maps.

1.4. Circle Mappings

Suppose $\mathbf{R}$ is the real line and $\mathbf{T}^1 = \mathbf{R}/\mathbf{Z}$, the real line modulo of the set of integers, is the circle. For a real number x, let $p = [x]$ denote the equivalent class $\{y \mid y - x \in \mathbf{Z}\}$ of x in $\mathbf{R}$ and let $x \pmod 1$ denote the unique number in $[x]$ and in $[0, 1)$. Let $\mathcal{S}$ be the set of strictly increasing continuous functions f on $\mathbf{R}$ satisfying

(i) $f(x+1) = f(x) + 1$ for all real numbers x and

(ii) $0 \leq f(0) < 1$.

For a function f in $\mathcal{S}$, we define an orientation-preserving homeomorphism of the circle $\mathbf{T}^1$:

$$F([x]) = [f(x)]. \tag{1.6}$$

Actually, there is a one-to-one correspondence between $\mathcal{S}$ and the set of orientation-preserving homeomorphisms of the circle $\mathbf{T}^1$ (one can do it as an exercise). We call a function f in $\mathcal{S}$ a circle mapping. Then F is called the corresponding homeomorphism of the circle $\mathbf{T}^1$.

Lemma 1.6 (Poincaré rotation number). *Suppose f is a circle mapping. Then the limit*

$$\rho_f = \lim_{|n| \to +\infty} \frac{f^{\circ n}(x)}{n} \tag{1.7}$$

exists and does not depend on x.

Proof. First we prove that $\lim_{|n|\to+\infty} f^{\circ n}(0)/n$ exists. For any integer m, there is an integer k_m so that

$$k_m \leq f^{\circ m}(0) < k_m + 1.$$

From the condition **(i)**,

$$nk_m \leq f^{\circ nm}(0) < n(k_m + 1) \qquad \text{or} \qquad n(k_m+1) < f^{\circ nm}(0) \leq nk_m$$

for any integer n. Then for any integers n and m,

$$\left| \frac{f^{\circ nm}(0)}{nm} - \frac{f^{\circ m}(0)}{m} \right| < \frac{1}{|m|}.$$

By exchanging the positions of n and m,

$$\left| \frac{f^{\circ nm}(0)}{nm} - \frac{f^{\circ n}(0)}{n} \right| < \frac{1}{|n|}.$$

Thus

$$\left|\frac{f^{\circ n}(0)}{n} - \frac{f^{\circ m}(0)}{m}\right| < \frac{1}{|n|} + \frac{1}{|m|}.$$

This says that $\{f^{\circ n}(0)/n\}$ is a Cauchy sequence.

Now for any real number x, there is an integer k so that $k \leq x < k+1$. From the condition **(i)**,

$$f^{\circ n}(0) + k \leq f^{\circ n}(x) \leq f^{\circ n}(0) + k + 1.$$

This implies that

$$\lim_{|n|\to+\infty} \frac{f^{\circ n}(x)}{n} = \lim_{|n|\to+\infty} \frac{f^{\circ n}(0)}{n}.$$

Hence $\rho_f = \lim_{|n|\to+\infty} f^{\circ n}(x)/n$ exists and does not depend on x. ∎

Definition 1.4. *The number ρ_f is called the rotation number of a circle mapping f. A circle mapping f is called irrational if its rotation number ρ_f is an irrational number. Otherwise, it is called rational.*

Example 1.2 (rigid rotation). *The function $R_\rho(x) = x + \rho$ for $0 \leq \rho < 1$ is a circle mapping and the rotation number of R_ρ is ρ. It is called the rigid rotation by rotation number ρ.*

1.5. Commuting Pairs

For a circle mapping f, we define an induced pair of functions (l_0, L_0) and (r_0, R_0) (with L_0 or R_0 possibly reducing to a point) as follows: let $L_0 = [f(0) - 1, 0]$, $R_0 = [0, f(0)]$, $T_0 = L_0 \cup R_0$, and

$$l_0 = f|L_0 : L_0 \to T_0$$

and

$$r_0 = f|R_0 - 1 : R_0 \to T_0.$$

It is easy to check that

$$l_0(r_0(0)) = r_0(l_0(0)).$$

Definition 1.5. *A commuting pair* $\{(l, L); (r, R)\}$ *is two intervals* L *and* R *(each may reduce to a point) on the left and right of* 0*, joining at* 0*, and two increasing functions* l *and* r *from* L *and* R *into* $T = L \cup R$ *with* $L = [r(0), 0]$*,* $R = [0, l(0)]$*, and the commuting condition* $l \circ r(0) = r \circ l(0)$ *(see Fig. 1.3).*

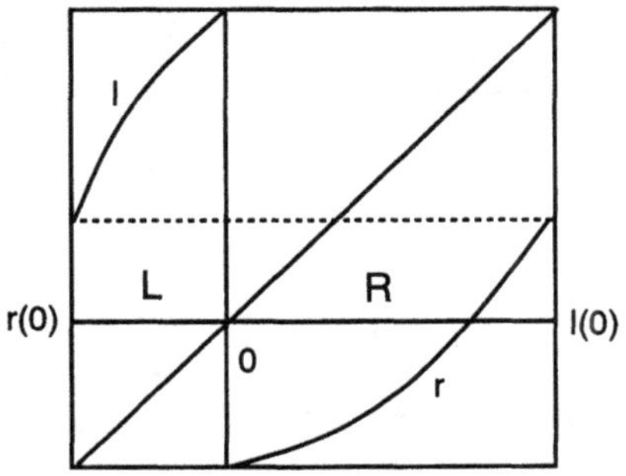

Fig. 1.3

The induced pair of functions $\{(l_0, L_0); (r_0, R_0)\}$ from a circle mapping f is a commuting pair.

Remark 1.4 . A commuting pair is called degenerate if L or R is a point. For a circle mapping f, the induced commuting pair is degenerate if and only if $f(0) = 0$.

1.6. The Renormalization of a Commuting Pair

For a commuting pair $\{(l,\ L);\ (r,\ R)\}$, one can define a new commuting pair $\{(\tilde{l},\ \tilde{L});\ (\tilde{r},\ \tilde{R})\}$ as follows (refer to Fig. 1.4):

Case $(+)$ $(l \circ r(0) > 0)$**:** Let $\tilde{L} = L$ and $\tilde{R} = [0,\ l \circ r(0)]$, and let $\tilde{l} = r \circ l$ and $\tilde{r} = r|\tilde{R}$.

Case $(-)$ $(l \circ r(0) \leq 0)$**:** Let $\tilde{L} = [l \circ r(0),\ 0]$ and $\tilde{R} = R$, and let $\tilde{l} = l|\tilde{L}$ and $\tilde{r} = l \circ r$.

One can check that in each case, $\tilde{l} \circ \tilde{r}(0) = \tilde{r} \circ \tilde{l}(0)$.

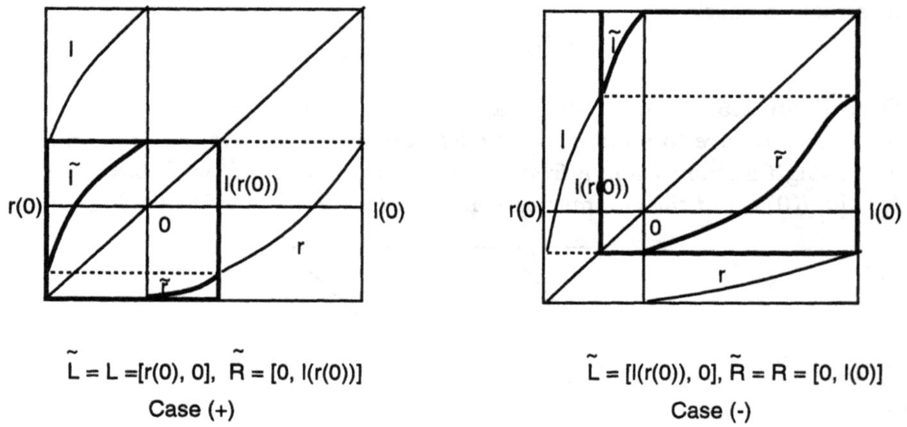

Fig. 1.4

Definition 1.6. *Suppose* $\{(l, L);\ (r, R)\}$ *is a commuting pair. Then the commuting pair*

$$\{(\tilde{l}, \tilde{L}); (\tilde{r}, \tilde{R})\}$$

defined either in Case $(+)$ *or in Case* $(-)$ *is called the renormalization of* $\{(l, L);\ (r, R)\}$.

Remark 1.5. If $\{(l, L);\ (r, R)\}$ is a degenerate commuting pair, then the renormalization

$$\{(\tilde{l}, \tilde{L}); (\tilde{r}, \tilde{R})\}$$

is also degenerate.

1.7. Rotation Numbers Defined through Renormalization

Suppose f is a circle mapping and $\{(l_0, L_0); (r_0, R_0)\}$ is the induced commuting pair from f. One can generate a sequence of commuting pairs $\{(l_n, L_n); (r_n, R_n)\}$ so that $\{(l_n, L_n); (r_n, R_n)\}$ is the renormalization of $\{(l_{n-1}, L_{n-1}); (r_{n-1}, R_{n-1})\}$ for $n \geq 1$, and a corresponding sequence $\sigma = \{\sigma_n\}_{n=0}^{\infty}$ of signs $\pm$ where σ_n is $\pm$ if $\{(l_n, L_n); (r_n, R_n)\}$ is in Case ($\pm$).

Let $\Pi_0 = [0/1, 1/1]$, and define a nested sequence of intervals $\Pi_n = [p_n/q_n, P_n/Q_n]$ by substituting the median $(p_n + P_n)/(q_n + Q_n)$ for p_n/q_n if $\{(l_n, L_n); (r_n, R_n)\}$ is in Case (+) and for P_n/Q_n if $\{(l_n, L_n); (r_n, R_n)\}$ is in Case (−). So in Case (+)

$$\frac{p_{n+1}}{q_{n+1}} = \frac{p_n + P_n}{q_n + Q_n} \qquad \text{and} \qquad \frac{P_{n+1}}{Q_{n+1}} = \frac{P_n}{Q_n}$$

and in Case (−)

$$\frac{p_{n+1}}{q_{n+1}} = \frac{p_n}{q_n} \qquad \text{and} \qquad \frac{P_{n+1}}{Q_{n+1}} = \frac{p_n + P_n}{q_n + Q_n}.$$

One can check by induction that

$$P_n q_n - p_n Q_n = 1,$$

$$L_n = [f^{\circ Q_n}(0) - P_n, 0], \qquad R_n = [0, f^{\circ q_n}(0) - p_n], \tag{1.8}$$

and

$$l_n = f^{\circ q_n} - p_n, \qquad r_n = f^{\circ Q_n} - P_n. \tag{1.9}$$

Then $\{\Pi_n = [p_n/q_n, P_n/Q_n]\}$ is a nested sequence of intervals and at least one of q_n and Q_n tends to infinity as n goes to infinity. Thus the intersection $\cap_{n=0}^{\infty}\Pi_n$ of the sequence of intervals Π_n is a single number. It is the rotation number ρ_f of f because $f^{\circ q_n}(0) = p_n + l_n(0)$ and $f^{\circ Q_n}(0) = Q_n + r_n(0)$, and $l_n(0)$ and $r_n(0)$ are bounded, so

$$\lim_{q_n \to +\infty} f^{\circ q_n}(0)/q_n = \lim_{q_n \to +\infty} p_n/q_n = \rho_f$$

or

$$\lim_{Q_n \to +\infty} f^{\circ Q_n}(0)/Q_n = \lim_{Q_n \to +\infty} P_n/Q_n = \rho_f$$

whenever q_n or Q_n goes to infinity.

Remark 1.6. Suppose one of the commuting pairs in the sequence of renormalizations is degenerate; let the first degenerate one be

$$\{(l_m, L_m); (r_m, R_m)\}.$$

Then if $m = 0$, then $f(0) = 0$, and if $m \geq 1$, then $l_m(0)$ or $r_m(0)$ equals 0. For $f(0) = 0$, it is easy to see the rotation number ρ_f is zero. For $m \geq 1$, suppose $r_m(0) = 0$ (for $l_m(0) = 0$ the argument is similar), then $f^{\circ Q_m}(0) = P_m$ and, also, $f^{\circ nQ_m}(0) = nP_m$. Hence the rotation number $\rho_f = P_m/Q_m$ is rational and 0 is a periodic point, of period Q_m, of the corresponding homeomorphism F of the circle $\mathbf{T}^1$.

Remark 1.7. In general, suppose f is a circle mapping. Let $f_t(x) = f(x+t) - t$. If there is a $0 \leq t < 1$ such that $f_t(0) \leq 0$ or $f_t(0) \geq 1$. Then the rotation number ρ_f is zero or one. Suppose the rotation number ρ_f is neither 0 nor 1. Then f_t is a circle mapping for every $0 \leq t < 1$. Let

$$\Big\{\{(l_{n,t}, L_{n,t}); (r_{n,t}, R_{n,t})\}\Big\}_{n=0}^{\infty}$$

be the corresponding sequence of renormalizations to f_t. Then the rotation number $\rho_f \neq 0, 1$ is rational if and only if there is a $0 \leq t < 1$ so that the commuting pairs $\{(l_{n,t}, L_{n,t}); (r_{n,t}, R_{n,t})\}$ are all degenerate except for finitely many.

1.8. Irrational Circle Mappings

Suppose f is an irrational circle mapping and F is the corresponding homeomorphism of the circle $\mathbf{T}^1$. A positive orbit $\{F^{\circ n}(p)\}_{n=0}^{\infty}$ is said to be recurrent if there is a subsequence $F^{\circ n_i}(p)$ tending to p in $\mathbf{T}^1$ as n_i goes to infinity.

Lemma 1.7 (recurrence implies dense orbit). *Suppose f is an irrational circle mapping and F is the corresponding homeomorphism of the circle $\mathbf{T}^1$. If every positive orbit $\{F^{\circ n}(p)\}_{n=0}^{\infty}$ is recurrent, then every positive orbit $\{F^{\circ n}(p)\}_{n=0}^{\infty}$ as well as every negative orbit $\{F^{\circ n}(p)\}_{-\infty}^{n=0}$ is dense in $\mathbf{T}^1$.*

Proof. Given any x and $y \neq 0$ in $[0,1)$ and any small $\epsilon > 0$ such that $(y-\epsilon, y+\epsilon) \subset (0,1)$, we must find a positive integer n such that

$$|f^{\circ n}(x)\ (\mathrm{mod} 1) - y| < \epsilon.$$

From the assumption that every positive orbit is recurrent, one can find an integer $n_0 > 0$ such that

$$|f^{\circ n_0}(y)\ (\mathrm{mod} 1) - y| < \epsilon.$$

Let J be the interval bounded by $f^{\circ n_0}(y)\ (\mathrm{mod} 1)$ and y. Then the intervals

$$F^{-jn_0}([J]) = [f^{-jn_0}(J)]$$

for $j = 0, 1, \ldots$, are neighboring and cover the circle $\mathbf{T}^1$. (If they do not cover $\mathbf{T}^1$, then $f^{-jn_0}(y)\ (\mathrm{mod} 1)$ is a monotone sequence in $[0,1)$ and accumulates at some z with $f^{\circ n_0}(z) = z + m_0$ for some integer m_0, and furthermore, $f^{\circ nn_0}(z) = z + nm_0$ and $\rho_f = m_0/n_0$, contradicting the assumption of irrational rotation number.) There is an integer $j > 0$ such that x is in $f^{-jn_0}(J)\ (\mathrm{mod} 1)$. Let $n = jn_0 > 0$. Then $f^{\circ n}(x)\ (\mathrm{mod} 1)$ is in J, that is, $|f^{\circ n}(x)\ (\mathrm{mod} 1) - y| < \epsilon$. This says that the positive orbit $\{F^{\circ n}(p)\}_{n=0}^{\infty}$ for $p = [x]$ is dense in $\mathbf{T}^1$. One can prove that the negative orbit $\{F^{\circ n}(p)\}_{-\infty}^{n=0}$ is dense in $\mathbf{T}^1$ by similar arguments. ∎

Lemma 1.8. *Suppose f is an irrational circle mapping. Let*

$$\left\{\{(l_n, L_n); (r_n, R_n)\}\right\}_{n=0}^{\infty}$$

be the corresponding sequence of renormalizations (none of which is degenerate) to f and $[a_L, a_R] = \cap_{n=0}^{\infty}(L_n \cup R_n)$. Then for any positive integer n, $r_n(a_R) \leq a_L$ and $l_n(a_L) \geq a_R$.

Proof. Let us prove that both $r_n(a_R) > a_L$ and $l_n(a_L) < a_R$ are impossible. First, suppose $r_n(a_R) > a_L$. Since for any x, $r_k(x)$ is an increasing sequence, then $r_k(a_R) > a_L$ for every $k \geq n$. Because $l_k(0) \geq a_R$ and r_k is an increasing function, then $r_k \circ l_k(0) \geq r_k(a_R) > a_L$ for $k \geq n$. Since $r_k \circ l_k(0) = l_k \circ r_k(0)$ is an endpoint of $L_{k+1} \cup R_{k+1} \supset [a_L, a_R]$, it must be the right endpoint and $r_k \circ l_k(0) > 0$ for $k \geq n$. This implies that the commuting pair $\{(l_k, L_k); (r_k, R_k)\}$ is in Case (+) for every $k \geq n$. By the definition of rotation number through renormalization, the rotation number ρ_f is P_n/Q_n, contradicting the assumption that f is an irrational circle mapping. A similar argument holds if $l_n(a_L) < a_R$. ■

Let f be an irrational circle mapping f. For a number x in $\mathbf{R}$, let $y(x)$ be the unique number in $[f(0)-1, f(0))$ such that $y(x) = x + m$ for some integer m. Let $\chi : x \mapsto y(x)$. In the next two lemmas, we identify every interval I, whose length is less than or equal to 1, with $\chi(I)$.

Lemma 1.9. *Suppose f is an irrational circle mapping. Let*

$$\left\{\{(l_n, L_n); (r_n, R_n)\}\right\}_{n=0}^{\infty}$$

be the corresponding sequence of renormalizations to f. Then for any integer $n > 0$, the interiors of $f^{\circ i}(L_n)$ and $f^{\circ j}(R_n)$ (in $[f(0)-1, f(0)]$) are disjoint from $T_n = L_n \cup R_n$ for any $0 < i < q_n$ and any $0 < j < Q_n$.

Proof. The proof is by induction. For $n = 1$, in Case (+), we have

$$L_1 = [r_0(0), 0], \qquad R_1 = [0, l_0 \circ r_0(0)], \qquad f(L_1) = [l_0 \circ r_0(0), l_0(0)],$$

and $q_1 = 2$, $Q_1 = 1$; in Case (−), we have

$$L_1 = [l_0 \circ r_0(0), 0], \qquad R_1 = [0, l_0(0)], \qquad f(R_1) = [r_0(0), l_0 \circ r_0(0)],$$

and $q_1 = 1$, $Q_n = 2$. So the lemma is true for $n = 1$.

Suppose the lemma is true for $n-1 \geq 1$. We must prove that it is true for n. We offer a proof when $\{(l_{n-1}, L_{n-1}); (r_{n-1}, R_{n-1})\}$ is in Case (+) (the argument is similar when it is in Case (−)). In this case,

$$L_n = [r_{n-1}(0), 0] = L_{n-1}, \qquad R_n = [0, l_{n-1} \circ r_{n-1}(0)] \subset R_{n-1},$$

and $q_n = q_{n-1} + Q_{n-1}$, $Q_n = Q_{n-1}$. The interiors of $f^{\circ i}(L_n)$ and $f^{\circ j}(R_n)$ are disjoint from $T_{n-1} \supset T_n$ for $0 < i < q_{n-1}$ and $0 < j < Q_n$. Since

$$f^{\circ q_{n-1}}(L_n) = [l_{n-1} \circ r_{n-1}(0), l_{n-1}(0)],$$

the interior of $f^{\circ q_{n-1}}(L_n)$ is disjoint from $T_n = [r_n(0), l_n(0)]$. Let $i = q_{n-1} + i'$ with $0 < i' < Q_{n-1}$. Since $f^{\circ q_{n-1}}(L_n) \subset R_{n-1}$ and $f^{\circ i}(L_n) = f^{\circ i'}(f^{\circ q_{n-1}}(L_n))$, the interior of $f^{\circ i}(L_n)$ is disjoint from $T_{n-1} \supset T_n$ for any $q_{n-1} < i < q_n$. ■

Lemma 1.10. *For any integer $n > 0$, the interiors of $f^{\circ i}(L_n)$ and $f^{\circ j}(L_n)$ (in $[f(0)-1, f(0)]$) are disjoint for any $0 \leq i \neq j < q_n$ and the interiors of $f^{\circ i}(R_n)$ and $f^{\circ j}(R_n)$ (in $[f(0)-1, f(0)]$) are disjoint for any $0 \leq i \neq j < Q_n$.*

Proof. Suppose there are $0 < i < j < q_n$ such that the interiors of $f^{\circ i}(L_n)$ and $f^{\circ j}(L_n)$ overlap. Then the interiors of $f^{\circ(j-i)}(L_n)$ and L_n overlap, contradicting Lemma 1.9. Similar arguments can be used to prove that the interiors of $f^{\circ i}(R_n)$ and $f^{\circ j}(R_n)$ are disjoint for $0 < i < j < Q_n$. ■

1.9. The Distortion of an Irrational Circle Mapping

A (non-degenerate) commuting pair $\{(l,\ L);\ (r,\ R)\}$ is said to be C^1 if both l and r are C^1 functions. The distortion of a C^1 commuting pair $\{(l,\ L);\ (r,\ R)\}$ is the set of numbers

$$\left\{\log\left(\frac{l'(x)}{l'(y)}\right)\right\}_{x,y\in L}\cup\left\{\log\left(\frac{r'(x)}{r'(y)}\right)\right\}_{x,y\in R}.$$

Suppose f is a C^1 irrational circle mapping. Define $f_t(x)=f(x+t)-t$ for $0\leq t<1$. Then f_t is also a C^1 irrational circle mapping since $t<f(t)<1+t$ (otherwise, there would be a number $0<\xi<1$ such that $f(\xi)=\xi$ or $f(\xi)=\xi+1$, which would imply that the rotation number of f is 0 or 1.) Let

$$\Big\{\{(l_{n,t},L_{n,t});(r_{n,t},R_{n,t})\}\Big\}_{n=0}^{\infty}$$

be the corresponding sequence of renormalizations to f_t.

Definition 1.7. *A C^1 irrational circle mapping f is said to have bounded distortion if there is a constant $C>0$ such that for all integers $n\geq 0$ and all $0\leq t<1$,*

$$\left|\log\left(\frac{l'_{n,t}(x)}{l'_{n,t}(y)}\right)\right|\leq C$$

for all x and y in $L_{n,t}$, and

$$\left|\log\left(\frac{r'_{n,t}(x)}{r'_{n,t}(y)}\right)\right|\leq C$$

for all x and y in $R_{n,t}$.

Lemma 1.11 (recurrence). *Suppose f is a C^1 irrational circle mapping having bounded distortion. Let F be the corresponding homeomorphism of the circle $\mathbf{T}^1$. Then every positive orbit $\{F^{\circ n}(p)\}_{n=0}^{\infty}$ is recurrent.*

Proof. We prove it by contradiction. Suppose there is a point $p=[t]$ in $\mathbf{T}^1$ whose orbit under F is not recurrent. Without loss of generality, we assume that $p=[0]$ (otherwise, consider $f_t(x)=f(x+t)-t$). Let $\Big\{\{(l_n,L_n);(r_n,R_n)\}\Big\}_{n=0}^{\infty}$ be the corresponding sequence of renormalizations to f, and $[a_L,a_R]=\cap_{n=0}^{\infty}(L_n\cup R_n)$. Then $a_L<0<a_R$ (see Fig. 1.5).

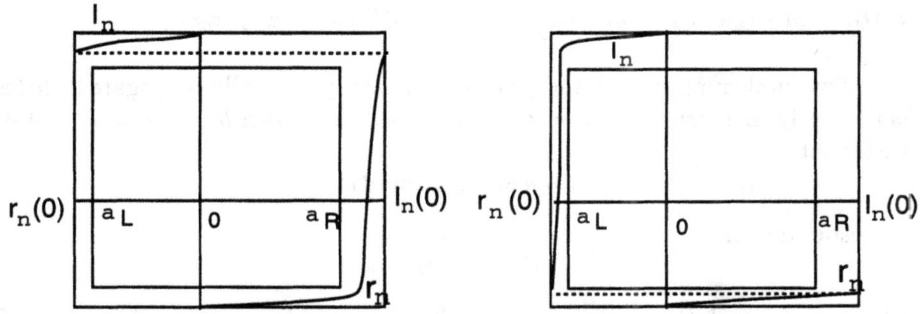

Fig. 1.5

For every integer $n > 0$, the interval $T_n = L_n \cup R_n = [r_n(0), l_n(0)]$ contains the interval $[a_L, a_R]$. From Lemma 1.8, $l_n([a_L, 0])$ and $r_n([0, a_R])$ are contained in $[a_R, l_n(0)]$ and $[r_n(0), a_L]$, respectively. These imply that both $\min_{x\in L_n}\{l'_n(x)\}$ and $\min_{x\in R_n}\{r'_n(x)\}$ tend to zero as n goes to infinity. Moreover, since f has bounded distortion, both $\max_{x\in L_n}\{l'_n(x)\}$ and $\max_{x\in R_n}\{r'_n(x)\}$ tend to zero as n goes to infinity. By the commuting condition that $l_n \circ r_n(0) = r_n \circ l_n(0)$, at least one of the intervals, $l_n([r_n(0), a_L])$ or $r_n([a_R, l_n(0)])$, contains $[a_L, a_R]$ (see Fig. 1.5). So one of the sequences $\max_{x\in R_n}\{r'_n(x)\}$ and $\max_{x\in L_n}\{l'_n(x)\}$ has to tend to infinity as n goes to infinity. This is a contradiction. ■

1.10. Conjugacies between Irrational Circle Mappings

Two circle mappings f and g are said to be topologically conjugate if there is a strictly increasing function h from $\mathbf{R}$ onto itself with $h(x+1) = h(x)+1$ such that

$$f \circ h = h \circ g \,(\mathrm{mod}\, 1),$$

correspondingly,

$$F \circ H = H \circ G$$

where F, G, and H are corresponding homeomorphisms of $\mathbf{T}^1$. A version of Denjoy's theorem is

Theorem 1.3. *Every C^1 irrational circle mapping f having bounded distortion, whose rotation number is ρ_f, is topologically conjugate to the rigid rotation $R(x) = x + \rho_f$ by rotation number ρ_f.*

Proof. Let $\rho = \rho_f$. For any number $q\rho - p$ where (p, q) is a pair of integers, define $h(q\rho - p) = f^{\circ q}(0) - p$. Then $h(x+1) = h(x) + 1$ where defined. For any $x = q\rho - p > 0$, let us prove that $h(x)$ is positive. Suppose $h(x) = f^{\circ q}(0) - p < 0$. Since f is strictly increasing,

$$f^{\circ nq}(0) - np < f^{\circ (n-1)q}(0) - (n-1)p.$$

Therefore, $f^{\circ nq}(0) - np < 0$ for all $n > 0$. This implies that $q\rho - p \leq 0$, contradicting $q\rho - p > 0$.

For any $x_1 = q_1\rho - p_1 > x_2 = q_2\rho - p_2$, let $x = (q_1 - q_2)\rho - (p_1 - p_2) = q\rho - p > 0$. From $h(x) = f^{\circ q}(0) - p > 0$, we have

$$h(x_1) = f^{\circ q_1}(0) - p_1 > h(x_2) = f^{\circ q_2}(0) - p_2.$$

Hence h is strictly increasing.

Following Lemma 1.11 and Lemma 1.7, the sets $\{q\rho - p = R^{\circ q}(0) - p\}_{p,q\in\mathbf{Z}}$ and $\{f^{\circ q}(0) - p\}_{p,q\in\mathbf{Z}}$ are both dense in the real line $\mathbf{R}$. Hence h extends uniquely to a continuous function of the whole real line $\mathbf{R}$. The extension is a strictly increasing function with $h(x+1) = h(x) + 1$ and

$$f \circ h = h \circ R \,(\mathrm{mod}\, 1)$$

on $\mathbf{R}$. Hence f is topologically conjugate to R. ∎

1.11. C^{1+bv}-Irrational Circle Diffeomorphisms

A circle mapping f is said to be a C^{1+bv} circle diffeomorphism if

$$\iota = \min_{x\in\mathbf{R}} f'(x) > 0$$

and the total variation of f' on $[0,1]$ is bounded, that is,

$$V(f) = \sup_{0\leq x_1<x_2<\cdots<x_n\leq 1} \sum_{i=1}^{n} |f'(x_i) - f'(x_{i+1})| < \infty.$$

Note that the total variation of f on any interval $[t, t+1]$ equals $V(f)$.

Lemma 1.12. *Every C^{1+bv} irrational circle diffeomorphism f has bounded distortion.*

Proof. Let $\left\{\{(l_{n,t}, L_{n,t}); (r_{n,t}, R_{n,t})\}\right\}_{n=0}^{\infty}$ be the corresponding sequence of renormalizations to $f_t(x) = f(x+t) - t$ where $0 \leq t < 1$. Then

$$l_{n,t}(x) = f_t^{\circ q_n}(x) - p_n$$

for x in $L_{n,t}$ and

$$r_{n,t}(x) = f_t^{\circ Q_n}(x) - P_n$$

for x in $R_{n,t}$. From Lemma 1.10,

$$\left|\log\left(\frac{l'_{n,t}(x)}{l'_{n,t}(y)}\right)\right| \leq \frac{1}{\iota}\sum_{i=0}^{q_n-1} |f'_t(f^{\circ i}(x)) - f'_t(f^{\circ i}(y))| \leq \frac{V_f}{\iota}$$

for x and y in $L_{n,t}$ and $0 \leq t < 1$. Similarly,

$$\left|\log\left(\frac{r'_{n,t}(x)}{r'_{n,t}(y)}\right)\right| \leq \frac{1}{\iota}\sum_{i=0}^{Q_n-1} |f'_t(f^{\circ i}(x)) - f'_t(f^{\circ i}(y))| \leq \frac{V_f}{\iota}$$

for any x and y in $R_{n,t}$ and $0 \leq t < 1$. This says that f has bounded distortion. ∎

A well-known version of Denjoy's theorem is

Theorem 1.4. *Every C^{1+bv} irrational circle diffeomorphism f, whose rotation number is ρ_f, is topologically conjugate to the rigid rotation $R(x) = x+\rho_f$ by rotation number ρ_f.*

Proof. This follows from Lemma 1.12 and Theorem 1.3. ∎

Chapter Two

The Koebe Distortion Principle

A new technique in the study of the dynamics of one-dimensional maps with critical points is introduced in this chapter. This technique was discovered and developed in the papers of, among others, Herman [HER], Van Strien [STR], Yoccoz [YO3], Świątek [SW1], De Melo and Van Strien [MV1], Guckenheimer and Johnson [GUJ], and Sullivan [SU4]. The technique is similar in many respects to Koebe's distortion theorem for holomorphic maps of one complex variable. We discuss several versions of the technique. In §2.1, a version for C^3-diffeomorphisms of an interval is proved using the theory of differential equations. In §2.4, we discuss a version of this technique which applies to C^{1+Z}-diffeomorphisms of an interval and which was worked out by Sullivan [SU4]. Koebe's distortion theorem in one complex variable is discussed in §2.6. Some elements of hyperbolic geometry are discussed in §2.2 and in §2.6. In §2.7, we prove the geometric distortion theorem, which applies to finitely generated regular $C^{1+\alpha}$ contracting semigroups (see [JI8]). It generalizes Koebe's distortion theorem.

A Cantor system is defined in §2.3. The geometry of a certain family of Cantor systems is studied in the same section. Such a family arises in dynamical systems as hyperbolicity is created. We prove that the bridge geometry of a Cantor system in such a family is uniformly bounded and that the gap geometry is regulated by the size of the leading gap. We discuss a version of Denjoy's theorem which applies to C^{1+Z}-circle diffeomorphisms in §2.5. In §2.8, we discuss the "Julia set", which is a Jordan curve, of a small perturbation $f_\epsilon(z) = z^{\frac{p+2}{2}}(\overline{z})^{\frac{p-2}{2}} + b\overline{z} + c$ of the quadratic polynomial $P_0(z) = z^2$, where $p > 1$ is real, where c and b are complex, and where $\epsilon = (p-2, b, c)$.

2.1. Nonlinearity and the Schwarzian Derivative

Let $\mathcal{C}^r$ be the space of C^r diffeomorphisms $f : I \to J$ where I and J are intervals of the real line $\mathbf{R}$ and where $r \geq 1$ is an integer. Let C^0 be the space of continuous functions defined on intervals of the real line $\mathbf{R}$. The derivative is an operator $D : \mathcal{C}^1 \to C^0$ defined as

$$D(f) = \log |f'|. \tag{2.1}$$

We have the chain rule: for any two maps $g : K \to I$ and $f : I \to J$ in $\mathcal{C}^1$,

$$D(f \circ g) = g^*\big(D(f)\big) + D(g), \tag{2.2}$$

where $g^*(h) = h \circ g$ for a function h.

The nonlinearity is an operator $N : \mathcal{C}^2 \to C^0$ defined as

$$N(f) = \frac{f''}{f'}. \tag{2.3}$$

We have the chain rule: for any two maps $g : K \to I$ and $f : I \to J$ in $\mathcal{C}^2$,

$$N(f \circ g) = g^{**}\big(N(f)\big) + N(g), \tag{2.4}$$

where $g^{**}(h) = (h \circ g) \cdot g'$ for a function h. The kernel $Ker(N) = \{f \in \mathcal{C}^2 \mid N(f) = 0\}$ consists of all linear maps $f(x) = ax + b : I \to J$ with $a \neq 0$. It is easy to see that

$$N(f) = \big(D(f)\big)'. \tag{2.5}$$

Remark 2.1. One can think the nonlinearity as a one-form

$$n(f) = \frac{f''}{f'}\, dx.$$

Lemma 2.1. *Suppose $f : I \to J$ is a map in $\mathcal{C}^2$. Then for any x and y in I,*

$$\log\Big(\frac{f'(y)}{f'(x)}\Big) = \int_x^y N(f)(\xi)\, d\xi.$$

Moreover, if $\sup_{\xi \in I} \big|N(f)(\xi)\big| \leq C$, then

$$\Big|\log\Big(\frac{f'(y)}{f'(x)}\Big)\Big| \leq C|x - y|.$$

Proof. This is calculus. ∎

Let $S : \mathcal{C}^3 \to \mathcal{C}^0$ be an operator defined as

$$S(f) = \frac{f'''}{f'} - \frac{3}{2}\Big(\frac{f''}{f'}\Big)^2. \tag{2.6}$$

It is called the Schwarzian derivative. We have the chain rule: for any two maps $g : K \to I$ and $f : I \to J$ in $\mathcal{C}^3$,

$$S(f \circ g) = g^{***}(S(f)) + S(g), \tag{2.7}$$

where $g^{***}(h) = (h \circ g) \cdot (g')^2$ for a function h. It is easy to see that

$$S(f) = (N(f))' - \frac{1}{2}(N(f))^2. \tag{2.8}$$

Lemma 2.2. *The kernel $Ker(S) = \{f \in \mathcal{C}^3 \mid S(f) = 0\}$ consists of all Möbius transformations*

$$f(x) = \frac{ax + b}{cx + d}$$

from I onto J where a, b, c, and d are real numbers satisfying $ad - bc \neq 0$ and where I and J are intervals.

Proof. If $f(x) = (ax + b)/(cx + d) : I \to J$, then $f'(x) = (ad - bc)/(cx + b)^2$, $f''(x) = -2c(ad-bc)/(cx+b)^3$ and $f'''(x) = 6c^2(ad-bc)/(cx+b)^4$. Hence $S(f) = 0$.

Suppose f is in $Ker(S)$. Then $S(f) = 0$. From Eq. (2.8),

$$(N(f))' = \frac{1}{2}(N(f))^2.$$

Solve this and then $\big(D(f)\big)' = N(f)$. We get $f(x) = (ax + b)/(cx + d)$. ∎

Remark 2.2. One can think the Schwarzian derivative as a two-form

$$s(f) = \Big(\frac{f'''}{f'} - \frac{3}{2}\Big(\frac{f''}{f'}\Big)^2\Big)\, dx^2.$$

Remark 2.3. For any pair of triple points in the real line $\mathbf{R}$, say $\{x_1 < x_2 < x_3\}$ and $\{y_1 < y_2 < y_3\}$, there is a unique map h in $Ker(S)$ so that $h(x_i) = y_i$ (or y_{4-i}) for $i = 1$, 2, and 3.

A map $f : I \to J$ is said to have non-negative (or non-positive or positive or negative) Schwarzian derivative if $S(f)(x) \geq 0$ (or $S(f)(x) \leq 0$ or $S(f)(x) > 0$ or $S(f)(x) < 0$) for all x in I. This is an invariant property of functions under composition.

Lemma 2.3. *Suppose $g : K \to I$ and $f : I \to J$ are maps in $\mathcal{C}^3$ having non-negative (or non-positive or positive or negative) Schwarzian derivatives. Then $f \circ g : K \to J$ has non-negative (or non-positive or positive or negative) Schwarzian derivative too.*

Proof. It is easy to see from the chain rule $S(f\circ g) = (g')^2 \cdot S(f)\circ g + S(g)$. ∎

Example 2.1. *Every power function $p(x) = -|x|^\gamma$ for $\gamma > 1$ has negative Schwarzian derivative on $(-\infty, 0)$ and on $(0, \infty)$. Every root function $r(x) = -|x|^{\frac{1}{\gamma}}$ for $\gamma > 1$ has positive Schwarzian derivative on $(-\infty, 0)$ and on $(0, \infty)$.*

Remark 2.4. Suppose $f : I \to J$ is a map in $\mathcal{C}^3$. Then $g = f^{-1} : J \to I$ is also a map in $\mathcal{C}^3$ and

$$S(g) = -\frac{S(f)\circ g}{(f'\circ g)^2}.$$

Hence g has positive (or negative) Schwarzian derivative if and only if f has negative (or positive) Schwarzian derivative.

Remark 2.5. Suppose $f : I \to J$ is a map in $\mathcal{C}^3$ and h is a map in $Ker(S)$. Then from the chain rule,

$$S(h\circ f) = S(f),$$
$$S(f\circ h) = (h')^2 \cdot S(f)\circ h.$$

Hence f has non-negative (or non-positive or positive or negative) Schwarzian derivative if and only if $h_1 \circ f \circ h_2$ has non-negative (or non-positive or positive or negative) Schwarzian derivative for every pair of maps h_1 and h_2 in $Ker(S)$.

Remark 2.6. Suppose $f : I \to J$ is a map in $\mathcal{C}^3$ having non-negative Schwarzian derivative and $I = [a, b]$. Then $g = 1/\sqrt{|f'|}$ is a concave downward function on I since $g'' = -(1/2)gS(f)$. This implies that

$$g(ta + (1-t)b) \geq tg(a) + (1-t)g(b)$$

for $0 \leq t \leq 1$, and that

$$|f'(x)| \leq \max\{|f'(a)|, |f'(b)|\}$$

for x in I. If $J = I$ and $f(a) = a$ and $f(b) = b$, then f can be in only one of the following cases (see Fig. 2.1):

(1) $f(x) > x$ for all $a < x < b$;

(2) $f(x) < x$ for all $a < x < b$; and

(3) there are $a \le c \le d \le b$ so that $f(x) > x$ for $a < x < c$, $f(x) = x$ for $c \le x \le d$, and $f(x) < x$ for $c < x < b$.

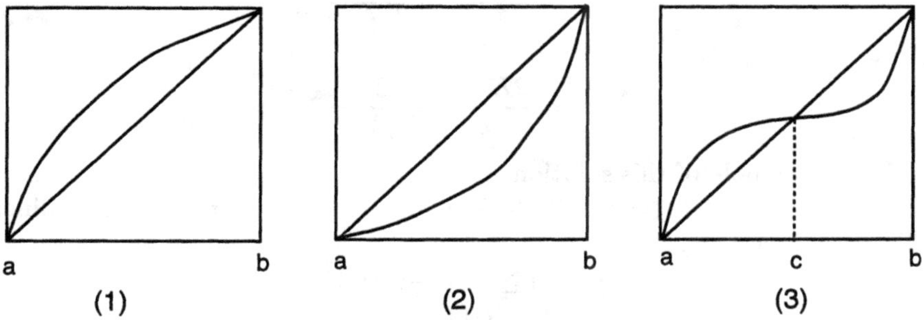

Fig. 2.1

Lemma 2.4 (C^3-Koebe distortion lemma). *Let $f : I \to J$ be a map in C^3 where $I = (a, b)$ is an open interval of the real line* **R**. *Suppose there is a constant $K \ge 0$ such that $S(f)(x) \ge -K$ for all x in I. Then, for any $\epsilon > 0$,*

$$|N(f)(x)| \le \max\left\{\sqrt{2K+\epsilon},\ \frac{2K+\epsilon}{\epsilon}\frac{2}{d(x,\partial I)}\right\}$$

for any x in I, where $d(x, \partial I) = \min\{|x-a|, |x-b|\}$ is the distance between x and the boundary of I. In particular, when f has non-negative Schwarzian derivative,

$$|N(f)(x)| \le \frac{2}{d(x,\partial I)}$$

for any x in I

Proof. Let x be any point in I. If $|N(f)(x)| \le \sqrt{2K+\epsilon}$, we have nothing to prove. Suppose $|N(f)(x)| > \sqrt{2K+\epsilon}$ and suppose $N(f)(x) > 0$ (the argument is similar when $N(f)(x) < 0$). For any $0 \le t < b - x$, if $N(f)(x+t) \ge \sqrt{2K+\epsilon}$,

$$N'(f)(x+t) = \frac{1}{2}\Big(N(f)(x+t)\Big)^2 + S(f)(x+t) > \frac{1}{2}(2K+\epsilon) - K = \frac{\epsilon}{2} > 0.$$

So $N(f)$ is increasing in a small neighborhood about t whenever $N(f)(x+t) \ge \sqrt{2K+\epsilon}$. This implies that $N(f)(x+t) \ge \sqrt{2K+\epsilon}$ for all $t \in [0, b-x)$. Consider the ordinary differential equation,

$$\begin{cases} y'(t) = \frac{1}{2}\frac{\epsilon}{2K+\epsilon}\big(y(t)\big)^2; \\ y(0) = N(f)(x) \end{cases}$$

for t in $[0, b-x)$. The unique solution of this equation is the function (see Fig. 2.2)

$$y_0(t) = \frac{1}{-\frac{1}{2}\frac{\epsilon}{2K+\epsilon}t + \frac{1}{N(f)(x)}}$$

for $t \in [0, b-x)$. Let

$$t_1 = \frac{2K+\epsilon}{\epsilon}\frac{2}{N(f)(x)}$$

be the unique pole of this solution.

The function $y_1(t) = N(f)(t+x)$ on $[0, b-x)$ is a solution of the ordinary differential inequality

$$\begin{cases} y'(t) \geq \frac{1}{2}\frac{\epsilon}{2K+\epsilon}\big(y(t)\big)^2; \\ y(0) = N(f)(x) \end{cases}$$

for $t \in [0, b-x)$ since $S(f)(x+t) \geq -K$ and $N(f)(x+t) \geq \sqrt{2K+\epsilon}$ for $t \in [0, b-x)$. Then $y_1(t)$ is greater than or equal to $y_0(t)$ for $t \in [0, b-x)$ (see Fig. 2.2) because of the comparison theorem for ordinary differential equations (see [ARN]). This implies that $y_0(t)$ is continuous on $[0, b-x)$. So the pole t_1 is greater than or equal to $b-x$. Hence

$$N(f)(x) \leq \frac{2K+\epsilon}{\epsilon}\frac{2}{b-x}$$

which is less than or equal to $\big((2K+\epsilon)/\epsilon\big)(2/d(x, \partial I))$.

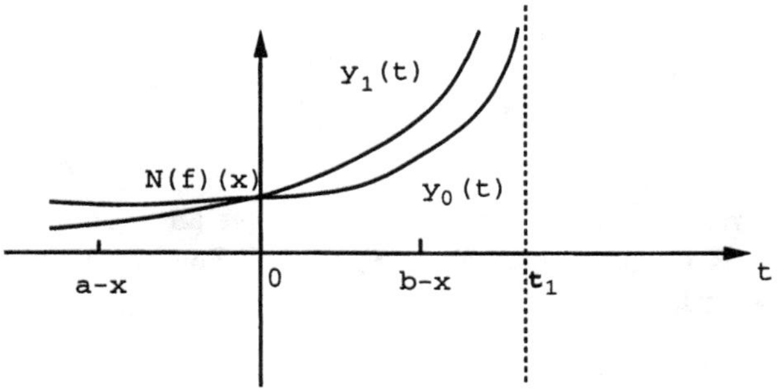

Fig. 2.2

■

2.2. The Poincaré Metric and the Schwarzian Derivative

Suppose $T = [a, d]$ is an interval of the real line $\mathbf{R}$ where a and d are real numbers. Let $M = [b, c]$ be a subinterval of T such that $a < b < c < d$. Let $L = [a, b]$ and $R = [c, d]$ be the closures of two connected components of $T \setminus M$ (see Fig. 2.3). Let $|I|$ mean the length of an interval I on the real line $\mathbf{R}$.

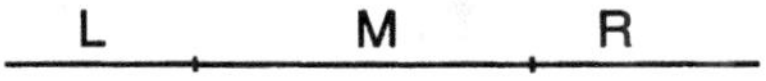

Fig. 2.3

The Poincaré length $P_T(M)$ of M in T is defined as

$$P_T(M) = \log\left(1 + \frac{|T| \cdot |M|}{|L| \cdot |R|}\right). \tag{2.9}$$

When M degenerates to a point x Eq. (2.9) becomes the Poincaré metric on T

$$dy = \frac{(d-a)}{(x-a)(d-x)}\, dx. \tag{2.10}$$

One can check that

$$P_T(M) = \int_M \frac{(d-a)}{(x-a)(d-x)}\, dx. \tag{2.11}$$

Lemma 2.5. *The Poincaré length $P_T(M)$ is invariant under the action of $Ker(S)$, more precisely,*

$$P_{h(T)}\big(h(M)\big) = P_T(M) \tag{2.12}$$

for any $h : I \to J$ in $Ker(S)$ and any intervals $M \subset T \subseteq I$.

Proof. It is clear if $h(x) = ex + f$ or $1/x$. In general, $h(x) = (px + q)/(rx + s)$ from Lemma 2.2 where p, q, r, and s are real numbers such that $ps - qr \neq 0$. This map can be factored as

$$h = h_3 \circ h_2 \circ h_1$$

where $h_1(x) = rx + s$, $h_2(x) = 1/x$, and $h_3(x) = ((qr - ps)/r)x + p/r$. ∎

Lemma 2.6. *Suppose $f : I \to I$ is a map in $\mathcal{C}^3$ having non-negative (or positive) Schwarzian derivative. Then f decreases (or strictly decreases) the Poincaré metric $P_T(M)$ for any $M \subset T \subseteq I$.*

Proof. Given any $M \subset T \subseteq I$, choose two maps $h_1 : [0,1] \to T$ and $h_2 : f(T) \to [0,1]$ in $Ker(S)$ such that $\tilde{f} = h_2 \circ f \circ h_1$ is a diffeomorphism of $[0,1]$ and fixes 0, 1/2 and 1 (refer to Remark 2.3) and such that $h_1^{-1}(M) = [t, 1/2]$. From Lemma 2.5,

$$P_{f(T)}(f(M)) = P_{[0,1]}(\tilde{f}([t, \frac{1}{2}])) = \log\Big(1 + \frac{2|\frac{1}{2} - \tilde{f}(t)|}{\tilde{f}(t)}\Big).$$

The graph of $\tilde{f}$ must look like Fig. 2.1 (3). This implies $\tilde{f}(t) \geq t$ (or $\tilde{f}(t) > t$); moreover,

$$P_{f(T)}(f(M)) \leq \text{ (or } <) \log\Big(1 + \frac{2|\frac{1}{2} - t|}{t}\Big) = P_{[0,1]}\Big([t, \frac{1}{2}]\Big) = P_T(M).$$

■

Lemma 2.7. *Suppose $f : I \to I$ is a map in $\mathcal{C}^3$ having non-negative Schwarzian derivative. Let $T \subseteq I$ be a closed interval and x be an interior point of T. Let L and R be the two connected components of $T \setminus \{x\}$. Then*

$$|f'(x)| \frac{|f(T)|}{|T|} \leq \frac{|f(R)|}{|R|} \frac{|f(L)|}{|L|}. \tag{2.13}$$

Proof. This is just a corollary of Lemma 2.6. ■

2.3. The Geometry of a Family of Cantor Systems

In order to construct a Cantor set in the real line **R**, we need to remove infinitely many subintervals which are called the gaps of the Cantor set. The sizes and positions of these gaps determine the geometry of the Cantor set. Let us first give a definition of an interval system which determines a Cantor set in the real line **R**. Let $\mathcal{I} = \{\mathcal{I}_n\}_{n=0}^{\infty}$ be a sequence of families of disjoint, non-empty, compact intervals. Let $\mathcal{G} = \{\mathcal{G}_n\}_{n=1}^{\infty}$ be a sequence of families of disjoint, non-empty, open intervals. Let $\mathcal{CS} = \{\mathcal{I}, \mathcal{G}\}$.

Definition 2.1. *We call $\mathcal{CS}$ a Cantor system if*

(i) *for each $0 \leq n < \infty$ and each interval $I \in \mathcal{I}_n$, there is a unique interval G in $\mathcal{G}_{n+1}$ and two intervals L and R in $\mathcal{I}_{n+1}$ which lie to the left and to the right of G such that $I = L \cup G \cup R$ (see Fig. 2.4), and*

(ii) $\Lambda = \bigcap_{n=0}^{+\infty} \bigcup_{I \in \mathcal{I}_n} I$ *is totally disconnected.*

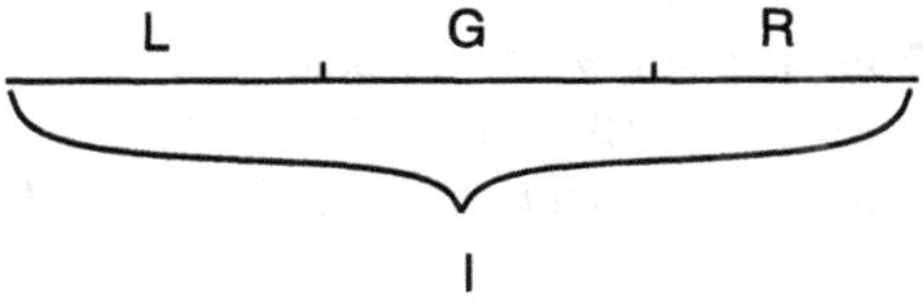

Fig. 2.4

The set Λ in Definition 2.1 is a Cantor set in the real line **R**. We call each interval I in $\mathcal{I}_n$ for $0 \leq n < \infty$ a $\mathcal{CS}$-bridge and call each interval G in $\mathcal{G}_n$ for $1 \leq n < \infty$ a $\mathcal{CS}$-gap. We also call each interval G in $\mathcal{G}_1$ a leading gap. We note that a Cantor system determines a Cantor set in the real line **R**.

Definition 2.2. *The bridge geometry $\mathcal{BR}$ of $\mathcal{CS}$ is the set of ratios*

$$\{\frac{|J|}{|I|}\ ;\ I = L \cup G \cup R \in \mathcal{I}_n, J = L \text{ or } R \in \mathcal{I}_{n+1}, G \in \mathcal{G}_{n+1}, n = 0, 1, 2, \ldots\}.$$

The gap geometry $\mathcal{GAP}$ of $\mathcal{CS}$ is the set of ratios

$$\{\frac{|G|}{|J|}\ ;\ I = L \cup G \cup R \in \mathcal{I}_n, J = L \text{ or } R \in \mathcal{I}_{n+1}, G \in \mathcal{G}_{n+1}, n = 0, 1, 2, \ldots\}.$$

Definition 2.3. *We say that the bridge geometry of $\mathcal{CS}$ is bounded if $\log x$ restricted to $\mathcal{BR}$ is bounded.*

Let $\{\mathcal{CS}_\epsilon\}_{0<\epsilon\leq\epsilon_0}$ be a family of Cantor systems where ϵ is a real parameter. Let $\mathcal{I}_\epsilon = \{\mathcal{I}_{n,\epsilon}\}_{n=0}^{\infty}$ and $\mathcal{G}_\epsilon = \{\mathcal{G}_{n,\epsilon}\}_{n=1}^{\infty}$ be the bridges and the gaps of $\mathcal{CS}_\epsilon$. Let $\mathcal{GAP}_\epsilon$ be the gap geometry of $\mathcal{CS}_\epsilon$ for each $0 < \epsilon \leq \epsilon_0$. Let $\alpha(\epsilon) > 0$ be a function of ϵ.

Definition 2.4. *We say the gap geometry of $\mathcal{CS}_\epsilon$ is regulated by $\alpha(\epsilon)$ if there is a constant $K > 0$ independent of ϵ such that*

$$\left|\log\left(\frac{g(\epsilon)}{\alpha(\epsilon)}\right)\right| \leq K$$

for all $g(\epsilon) \in \mathcal{GAP}_\epsilon$ and all $0 < \epsilon \leq \epsilon_0$.

Remark 2.7. A Cantor system $\mathcal{CS}$ is said to have bounded geometry (see [SU4]) if $\log x$ restricted to $\mathcal{BR} \cup \mathcal{GAP}$ is bounded.

Let $\epsilon_0 > 0$ and $\gamma > 1$ be real numbers. Let $\mathcal{F} = \{f_\epsilon\}_{0 \leq \epsilon \leq \epsilon_0}$ be a family of folding mappings.

Definition 2.5. *We say $\mathcal{F}$ is asymptotically non-hyperbolic (see Fig. 2.5) if*

(a) *every $f_\epsilon(x) = h_\epsilon(-|x|^\gamma)$ where $h_\epsilon : [-1, 0] \to [-1, 1+\epsilon]$ is a C^3 orientation-preserving diffeomorphism with non-positive Schwarzian derivative and where $f_\epsilon(-1) = -1$ and $f_\epsilon(0) = 1 + \epsilon$ for all $0 \leq \epsilon \leq \epsilon_0$,*

(b) *there is a constant $K > 0$ such that $\left|\left(\log h'_\epsilon(x)\right)'\right| \leq K$ for all $-1 \leq x \leq 0$ and all $0 \leq \epsilon \leq \epsilon_0$, and*

(c) *there is a constant $\lambda > 1$ such that $f'_\epsilon(-1) \geq \lambda$ for all $0 \leq \epsilon \leq \epsilon_0$.*

Let $\mathcal{F}$ be an asymptotically non-hyperbolic family. For each $0 < \epsilon \leq \epsilon_0$, let $\mathcal{I}_{n,\epsilon}$ be the set of intervals in $f_\epsilon^{-n}([-1, 1])$ and let $\mathcal{G}_{n,\epsilon}$ be the set of intervals in $f_\epsilon^{-n}((1, 1 + \epsilon))$. Let $\mathcal{I}_\epsilon = \{\mathcal{I}_{n,\epsilon}\}_{n=0}^{\infty}$, let $\mathcal{G}_\epsilon = \{\mathcal{G}_{n,\epsilon}\}_{n=1}^{\infty}$, and let $\mathcal{CS}_\epsilon = \{\mathcal{I}_\epsilon, \mathcal{G}_\epsilon\}$. Then $\mathcal{CS}_\epsilon$ is an interval system dynamically defined by f_ϵ. The leading gap of $\mathcal{CS}_\epsilon$ is $G_{*,\epsilon} = f_\epsilon^{-1}((1, 1 + \epsilon))$ and its size $lg(\epsilon)$ is

$$lg(\epsilon) = \left(\frac{\epsilon}{h'_\epsilon(\xi)}\right)^{\frac{1}{\gamma}} \tag{2.14}$$

for some $\xi \in [-1, 0]$. From (**b**) of Definition 2.5 and the fact that $h_\epsilon([-1, 0]) = [-1, 1+\epsilon]$, there is a constant $K > 0$ independent of ϵ such that $|\log h'_\epsilon(x)| \leq K$ for all $-1 \leq x \leq 0$ and all $0 < \epsilon \leq \epsilon_0$. Thus there is a constant $K > 0$ independent of ϵ such that

$$\left|\log\left(\frac{lg(\epsilon)}{\epsilon^{\frac{1}{\gamma}}}\right)\right| \leq K \tag{2.15}$$

for all $0 < \epsilon \leq \epsilon_0$.

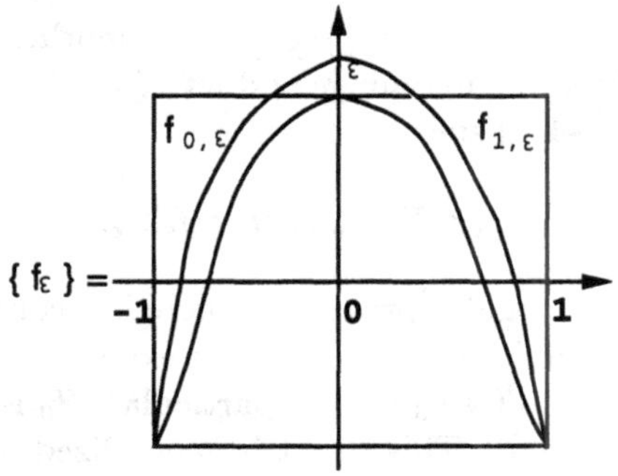

Fig. 2.5

Theorem 2.1 [JI1]. *Let $\mathcal{F}$ be an asymptotically non-hyperbolic family. For each $0 < \epsilon \leq \epsilon_0$, the dynamically defined interval system $\mathcal{CS}_\epsilon$ is a Cantor system whose bridge geometry is uniformly bounded. The gap geometry of $\mathcal{CS}_\epsilon$ is regulated by the function $\alpha : \epsilon \mapsto \epsilon^{\frac{1}{\gamma}}$.*

One of the consequences of this theorem is that

$$\Lambda_\epsilon = \cap_{n=0}^{\infty} \cup_{I \in \mathcal{I}_{n,\epsilon}} I$$

is a Cantor set on the real line $\mathbf{R}$ for each $0 < \epsilon \leq \epsilon_0$.

Remark 2.8. For $\epsilon = 0$, let $\mathcal{I}_{n,0}$ be the set of the closures of intervals in $f_0^{-n}([-1,1))$ and let $\mathcal{I} = \{\mathcal{I}_{n,0}\}_{n=0}^{\infty}$. The set $\Lambda_0 = \cap_{n=0}^{\infty} \cup_{I \in \mathcal{I}_{n,0}} I = [-1,1]$ is not a Cantor set. But from the dynamical system point of view, the interval system $\mathcal{CS}_0 = \{\mathcal{I}, \emptyset\}$ dynamically defined by f_0 can be also considered as a Cantor system with null gaps and with bridges $\mathcal{I}_0 = \{\mathcal{I}_{n,0}\}_{n=0}^{\infty}$ (refer to §3.9 and §3.10). Therefore $\mathcal{CS}_0$ can be included in the first statement of Theorem 2.1 too.

Remark 2.9. From Theorem 2.1, one can see that the Hausdorff dimension $HD(\epsilon)$ (see [FAL]) of Λ_ϵ is less than or equal to $1 - K\epsilon^{\frac{1}{\gamma}}$, where $K > 0$ is a constant independent of ϵ. For a family like $f_\epsilon(x) = 1 + \epsilon - (2+\epsilon)x^2$, we can show (see [JI3]) that

$$K^{-1}\sqrt{\epsilon} \leq 1 - HD(\epsilon) \leq K\sqrt{\epsilon}$$

for all $0 \leq \epsilon \leq \epsilon_0$, where $K > 0$ is a constant independent of ϵ. A similar result, in the theory of Fuchsian groups, is proved by Pignataro and Sullivan [PIS].

Remark 2.10. Let $\tilde{\mathcal{F}} = \{\tilde{f}_\epsilon\}_{0\leq\epsilon\leq\epsilon_0}$ be another asymptotically non-hyperbolic family of folding mappings and let $\mathcal{H} = \{H_\epsilon\}_{0\leq\epsilon\leq\epsilon_0}$ be a family of homeomorphisms of $[-1,1]$ such that

$$\tilde{f}_\epsilon \circ H_\epsilon|\Lambda_\epsilon = H_\epsilon \circ f_\epsilon|\Lambda_\epsilon.$$

From Theorem 2.1, we can also prove that there is a constant $K \geq 1$, which is independent of ϵ, such that every $H_\epsilon|C_\epsilon$ can be extended to a K-quasisymmetric homeomorphism of $[-1,1]$ (see §2.4). In particular, H_0 is a K-quasisymmetric homeomorphism (see [JI3]). This result is generalized to geometrically finite one-dimensional maps (see §3.5).

Let us use $K > 0$ to denote a constant (even though it may be different in different formulas). Let $\mathcal{F} = \{f_\epsilon\}_{0\leq\epsilon\leq\epsilon_0}$ be an asymptotically non-hyperbolic family of folding mappings. For each $0 \leq \epsilon \leq \epsilon_0$, let $f_{0,\epsilon} = f_\epsilon|[-1,0]$ and let $f_{1,\epsilon} = f_\epsilon|[0,1]$. We will omit ϵ if there can be no confusion.

Let $g_0 : [-1, 1+\epsilon] \to [-1,0]$ and $g_1 : [-1,1+\epsilon] \to [0,1]$ be the inverses of $f_{0,\epsilon}$ and $f_{1,\epsilon}$, respectively. Let $w_n = i_0 \ldots i_{n-1}$ denote a sequence of symbols 0's and 1's of length n and let

$$g_{w_n} = g_{i_0} \circ \cdots \circ g_{i_{n-1}}.$$

Let

$$I_{w_n} = g_{w_n}([-1,1]).$$

Then

$$\mathcal{I}_n = \{I_{w_n} \mid w_n = i_0 \ldots i_{n-1} \text{ is a sequence of 0's and 1's of length } n\}$$

for $n = 1, 2, \ldots$. (Note that $\mathcal{I}_0 = \{[-1,1]\}$.) Let $G_* = [-1,1] \setminus \Big(I_0 \cup I_1\Big)$ be the leading gap and let

$$G_{w_n} = g_{w_n}(G_*).$$

Then G_{w_n} is an open interval in I_{w_n},

$$\mathcal{G}_{n+1} = \{G_{w_n} \mid w_n = i_0 \ldots i_{n-1} \text{ is a sequence of 0's and 1's of length } n\}$$

for $n = 1, 2, \ldots$, and $\mathcal{G}_1 = \{G_*\}$.

Let $\{-a_\epsilon, a_\epsilon\} = f_\epsilon^{-1}(0)$ where $a_\epsilon > 0$. From (**b**) of Definition 2.5 (see the argument before Theorem 2.1), there is a constant $0 < K < 1$ such that $K \leq a_\epsilon \leq 1 - K$ for all $0 \leq \epsilon \leq \epsilon_0$. Therefore, there is a constant $K > 0$ such that $\big|\log|g_i'(x)|\big| \leq K$ for all $x \in [-1,0]$, for $i = 0$ or 1, and for all $0 \leq \epsilon \leq \epsilon_0$.

Let

$$U(\epsilon) = [-a_\epsilon, a_\epsilon]$$

and let

$$V(\varepsilon) = [-1, -a_\epsilon] \cup [a_\epsilon, 1].$$

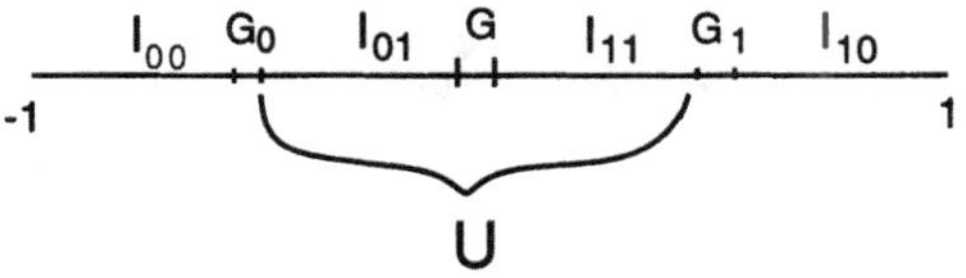

Fig. 2.6

Dividing the interval $[-1, 1]$ into two parts $U(\epsilon)$ and $V(\epsilon)$ is the key point in the proof of Theorem 2.1 (see Fig. 2.6); on $V(\epsilon)$, f_ϵ is C^3 and expanding, one can apply the naive distortion lemma (see Lemma 2.8), and on $U(\epsilon)$, although the derivative of f_ϵ may be zero, but Lemma 2.4 can take care of it (see Lemma 2.9). We also note that $f_\epsilon(U(\epsilon)) = [0, 1+\epsilon]$ and $f_\epsilon(V(\epsilon)) = [-1, 0]$.

From (**c**) of Definition 2.5, there is a constant $0 < \mu_0 < 1$, which is independent of ϵ, such that

$$|g'_{0,\epsilon}(-1)|, |g'_{1,\epsilon}(-1)| \leq \mu_0.$$

By considering $T = [-1, 1+\epsilon]$ and applying Lemma 2.7, we have that

$$|g'_{0,\epsilon}(0)|, |g'_{1,\epsilon}(0)| \leq \frac{1}{2}.$$

From Remark 2.6,

$$|g'_{0,\epsilon}(x)|, |g'_{1,\epsilon}(x)| \leq \mu = \min\{\mu_0, \frac{1}{2}\} < 1$$

for all $x \in [-1, 0]$ and all $0 \leq \epsilon \leq \epsilon_0$.

Let $w_m = j_0 j_1 \ldots j_{m-1}$ be a sequence of $0's$ and $1's$, let $g_{w_m} = g_{j_0} \circ \cdots \circ g_{j_{m-1}}$, and let $I_{w_m} = g_{w_m}([-1, 1])$. Let $w_n = i_0 \ldots i_{n-1}$ be another sequence of 0's and 1's and let $g_{w_n} = g_{i_0} \circ \cdots \circ g_{i_{n-1}}$. Let $w_{n+m} = w_n w_m = i_0 \ldots i_{n-1} j_0 \ldots j_{m-1}$ and let $g_{w_{n+m}} = g_{w_n} \circ g_{w_m}$. Then

$$I_{w_{n+m}} = g_{w_{n+m}}([-1, 1]) = g_{w_n}(I_{w_m}).$$

Let $w_{k+m} = i_{n-k} \ldots i_{n-1} w_m$ and $I_{w_{k+m}} = g_{w_{k+m}}([-1, 1])$ for $k = 1, \ldots, n$.

Lemma 2.8. *There is a constant $K > 0$ independent of ϵ such that if $I_{w_{k+m}} \subseteq V(\epsilon)$ for all $k = 1, \ldots, n$, then*

$$\left|\log\left(\frac{|g'_{w_n}(x)|}{|g'_{w_n}(y)|}\right)\right| \leq K$$

for all x and y in I_{w_m} and all $0 \leq \epsilon \leq \epsilon_0$.

Proof. Let $w_k = i_{n-k} \ldots i_{n-1}$, let $g_{w_k} = g_{i_{n-k}} \circ \cdots g_{i_{n-1}}$, and let $x_k = g_{w_k}(x)$ and $y_k = g_{w_k}(y)$ for $k = 1, \ldots, n$. Set $x_0 = x$ and $y_0 = y$. Since we have

$$\frac{|g'_{w_n}(x)|}{|g'_{w_n}(y)|} = \prod_{k=0}^{n-1} \frac{|g'_{i_{n-k-1}}(x_k)|}{|g'_{i_{n-k-1}}(y_k)|},$$

then

$$\mathcal{A} = \left|\log\left(\frac{|g'_{w_n}(x)|}{|g'_{w_n}(y)|}\right)\right| \leq \sum_{k=0}^{n-1} \Big| \log|g'_{i_{n-k-1}}(x_k)| - \log|g'_{i_{n-k-1}}(y_k)|\Big|.$$

Since $\left|\left(\log|g'_i(x)|\right)'\right| \leq K$ for all $i = 0$ and 1, all $x \in [-1, 0]$, and all $0 \leq \epsilon \leq \epsilon_0$ and since x_k and y_k are in $[-1, 0]$ for all $k = 0, \ldots, n-1$, there is a constant $K > 0$ independent of ϵ such that

$$\mathcal{A} \leq K \sum_{k=0}^{n-1} |x_k - y_k|.$$

Because g_0 and g_1 restricted to $[-1, 0]$ are contracting and because $I_{w_{k+m}} \subseteq [-1, 0]$ for all $0 \leq k \leq n-1$, we have

$$\sum_{k=0}^{n-1} |x_k - y_k| \leq \sum_{k=0}^{n-1} \mu^k \leq \frac{1}{1-\mu}.$$

Thus

$$\mathcal{A} \leq \frac{K}{1-\mu}.$$

Here $K/(1-\mu)$ is the constant we want. ∎

Lemma 2.9. *There is a constant $K > 0$ independent of ϵ such that if $I_{w_m} \subseteq U(\epsilon)$, then*

$$\left|\log\left(\frac{|g'_{w_n}(x)|}{|g'_{w_n}(y)|}\right)\right| \leq K$$

for all x and y in I_{w_m} and for all $0 \leq \epsilon \leq \epsilon_0$.

Proof. The distance $\mathrm{dist}(U(\epsilon), \partial[-1, 1+\epsilon]) = 1+\epsilon-a_\epsilon$ exceeds a positive constant K for all $0 \leq \epsilon \leq \epsilon_0$. Since g_{w_n} is a C^3-diffeomorphism defined on $(-1, 1+\epsilon)$ with non-negative Schwarzian derivative, Lemma 2.4 tells us that

$$\left|\log\left(\frac{|g'_{w_n}(x)|}{|g'_{w_n}(y)|}\right)\right| = \left|\int_x^y N(g_{w_n})(\xi)\, d\xi\right| \leq \frac{2|x-y|}{K}.$$

This implies the lemma. ∎

Proof of Theorem 2.1. For any interval I_{w_n} in $\mathcal{I}_n$, we have

$$I_{w_n} = I_{w_n0} \cup G_{w_n} \cup I_{w_n1},$$

which is (**i**) of Definition 2.1. Let

$$gg_{w_n}(\epsilon) = \frac{|G_{w_n}|}{|I_{w_n0}|} \quad \text{or} \quad \frac{|G_{w_n}|}{|I_{w_n1}|}$$

and let

$$br_{w_n}(\epsilon) = \frac{|I_{w_n0}|}{|I_{w_n}|} \quad \text{or} \quad \frac{|I_{w_n1}|}{|I_{w_n}|}.$$

Let $lg = lg(\epsilon)$ be the size of the leading gap G_*. From the argument before Theorem 2.1, there is a constant $K > 0$ such that

$$K^{-1} \leq \frac{lg(\epsilon)}{\epsilon^{\frac{1}{\gamma}}} \leq K$$

for all $0 < \epsilon \leq \epsilon_0$. Moreover, there is a constant $K > 0$ such that

$$K^{-1} \leq \frac{gg_{w_1}(\epsilon)}{\epsilon^{\frac{1}{\gamma}}} \leq K$$

and such that

$$K^{-1} \leq br_{w_1}(\epsilon) \leq K$$

for all $0 < \epsilon \leq \epsilon_0$ and for $w_1 = 0$ or 1,

Let $w_n = i_0 \dots i_{n-1}$ be any sequence of $0'$s and $1'$s of length $n > 1$. Let $w_k = i_{n-k} \dots i_{n-1}$, let $g_{w_k} = g_{i_{n-k}} \circ \cdots \circ g_{i_{n-1}}$, and let $I_k = g_{w_k}([-1,1])$ for $k = 1, \dots, n$. The sequence of intervals $\{I_k\}_{k=1}^{n}$ satisfies (**I**) $I_k \subseteq V(\epsilon)$ for all $2 \leq k \leq n$; or (**II**) there is an integer $2 \leq m \leq n$ such that $I_k \subseteq V(\epsilon)$ for all $2 \leq k < m$ and such that $I_m \subseteq U(\epsilon)$.

If the sequence $\{I_k\}_{k=1}^{n}$ satisfies (**I**), then we apply Lemma 2.8 to get a constant $K > 0$ independent of ϵ such that

$$K^{-1} \leq \frac{gg_{w_n}(\epsilon)}{\epsilon^{\frac{1}{\gamma}}} \leq K$$

and such that

$$K^{-1} \leq br_{w_n}(\varepsilon) \leq K$$

for all $0 < \epsilon \leq \epsilon_0$.

If the sequence $\{I_k\}_{k=1}^{n}$ satisfies (**II**), we have that $I_m \subseteq U(\epsilon)$ and $I_k \subseteq V(\epsilon)$ for all $2 \leq k < m$. Following the argument in (**I**), we also have in (**II**) that

$$K^{-1} \leq \frac{gg_{w_{m-1}}(\epsilon)}{\epsilon^{\frac{1}{\gamma}}} \leq K$$

and that

$$K^{-1} \leq br_{w_{m-1}}(\varepsilon) \leq K$$

for all $0 < \epsilon \leq \epsilon_0$.

From (**b**) of Definition 2.5, $f_\epsilon(x)|I_m$ is comparable with $x \mapsto (1+\epsilon) - (2+\epsilon)|x|^\gamma$ for all $0 < \epsilon \leq \epsilon_0$. There is thus a constant $K > 0$ independent of ϵ such that

$$K^{-1} \leq \frac{gg_{w_m}(\epsilon)}{gg_{w_{m-1}}} \leq K$$

and such that

$$K^{-1} \leq \frac{br_{w_m}(\varepsilon)}{br_{w_{m-1}}(\varepsilon)} \leq K$$

for all $0 < \epsilon \leq \epsilon_0$.

From Lemma 2.9, there is a constant $K > 0$ independent of ϵ such that

$$K^{-1} \leq \frac{gg_{w_n}(\epsilon)}{gg_{w_m}(\epsilon)} \leq K$$

and such that

$$K^{-1} \leq \frac{br_{w_n}(\varepsilon)}{br_{w_m}(\varepsilon)} \leq K$$

for all $0 < \epsilon \leq \epsilon_0$.

By combining the above estimates, we get that in both (**I**) and (**II**), there is a constant $K > 0$ independent of ϵ such that

$$K^{-1} \leq \frac{gg_{w_n}(\epsilon)}{\epsilon^{\frac{1}{\gamma}}} \leq K$$

and such that

$$K^{-1} \leq br_{w_n}(\varepsilon) \leq K$$

for all $0 < \epsilon \leq \epsilon_0$.

For each fixed $\epsilon > 0$, we can further get constants $0 < \nu < 1$ and $K > 0$ such that

$$|I_{w_n}| \leq K\nu^n$$

for all $I_{w_n} \in \mathcal{I}_n$ and all $0 \leq n < \infty$. This implies that

$$\Lambda_\epsilon = \cap_{n=0}^{\infty} \cup_{I \in \mathcal{I}_n} I$$

contains no interval. Therefore it is totally disconnected, and furthermore, a Cantor set (see the proof of Theorem 1.1). ■

2.4. Cross-Ratio Distortion, Quasisymmetry, and the Zygmund Condition

Let $T = [a, d]$ be an interval of the real line where a and d are real numbers. We use $M = [b, c]$ to denote a subinterval of T such that $a < b < c < d$. Let $L = [a, b]$ and $R = [c, d]$ be the closures of the connected components of $T \setminus M$ (see Fig. 2.3). Let $|I|$ denote the length of an interval I. The cross-ratio $Cr(T, M)$ is defined as

$$Cr(T, M) = \frac{|L \cup M| \cdot |M \cup R|}{|M| \cdot |T|} = 1 + \frac{|L| \cdot |R|}{|M| \cdot |T|}. \tag{2.16}$$

Suppose h is a strictly monotone function defined on an interval I. The cross-ratio distortion of h on $M \subset T \subseteq I$ is defined as

$$D(T, M; h) = \log\left(\frac{\log\left(Cr(h(T), h(M))\right)}{\log\left(Cr(T, M)\right)}\right). \tag{2.17}$$

Lemma 2.10. *The cross-ratio distortion $D(T, M; h)$ of $h : I \to J$ in $Ker(S)$ is zero for any $M \subset T \subseteq I$.*

Proof. See the proof of Lemma 2.5. ∎

Lemma 2.11. *If $h : I \to J$ is a C^3 diffeomorphism having non-negative Schwarzian derivative, then h increases the cross-ratio $Cr(T, M)$ for any $M \subset T \subseteq I$.*

Proof. Remember that the Poincaré length (see §2.2) of M in T is

$$P_T(M) = \log\left(1 + \frac{|M| \cdot |T|}{|L| \cdot |R|}\right).$$

It is easy to see that

$$Cr(T, M) = 1 + \frac{1}{e^{P_T(M)} - 1}.$$

The lemma now follows from Lemma 2.6. ∎

Definition 2.6. *A homeomorphism h from an interval I onto another interval J is called quasisymmetric if there is a constant $C > 0$ such that for any x and y in I, let $z = (x+y)/2$,*

$$C^{-1} \leq \frac{|h(x) - h(z)|}{|h(z) - h(y)|} \leq C.$$

The smallest such constant is called the quasisymmetric constant of h.

Remark 2.11. A quasisymmetric homeomorphism h is C^{α} continuous for some $0 < \alpha \leq 1$ (see [AH1]). More importantly, a quasisymmetric homeomorphism h of the real line $\mathbf{R}$ can be extended to a quasiconformal homeomorphism H of the complex plane $\mathbf{C}$ (see [AH1]).

Lemma 2.12. *Suppose h is a homeomorphism of the real line* $\mathbf{R}$ *and there is a constant* $C > 0$ *such that*

$$D(T, M; h) \geq -C$$

for any $M \subset T \subset \mathbf{R}$. *Then h is quasisymmetric.*

Proof. For any real number x, and any positive t, let $T = [x - 2t, x + t]$ and $M = [x - t, x]$. Then $Cr(T, M) = 4/3$ and

$$Cr(h(T), h(M)) \geq B = \left(\frac{4}{3}\right)^{A}$$

where $A = e^{-C}$. This implies that

$$\frac{|f(x) - f(x+t)|}{|f(x-t) - f(x)|} \geq \frac{|f(R)|}{|f(M)|}\frac{|f(L)|}{|f(T)|} \geq D^{-1} = B - 1.$$

Similarly, by taking $T = [x - t, x + 2t]$ and $M = [x, x + t]$, we have

$$\frac{|f(x-t) - f(x)|}{|f(x) - f(x+t)|} \geq D^{-1}.$$

Thus h is quasisymmetric. ∎

Lemma 2.13. *Suppose* $h : I \to J$ *is a homeomorphism where I and J are compact (closed and bounded) intervals. Suppose* $I' \subset I$ *is a closed subinterval. If there are two constants* $C > 0$ *and* $C_0 > 1$ *such that*

$$D(T, M; h) \geq -C$$

for any $M \subset T \subseteq I$ and

$$Cr(I, I') \geq C_0,$$

then $h|I'$ is quasisymmetric. The quasisymmetric constant of $h|I'$ depends on C and C_0.

Proof. Let us prove it first under the assumption that $I' = [0, 1]$. From $Cr(I, I') \geq C_0$, there is a positive number $t_0 = t_0(C_0) > 0$ such that $T = [x - 2t, x + 2t] \subseteq I$ for any $x \in I'$ and $0 < t < t_0$. From the proof of the previous lemma, there is a constant $C_1 = C_1(C) > 0$ such that

$$C_1^{-1} \leq \frac{|h(x-t) - h(x)|}{|h(x) - h(x+t)|} \leq C_1$$

for any $x \in I'$ and $0 < t < t_0$. Let

$$C_2 = \max_{|t| \geq t_0, x \in I', x \pm t \in I} \frac{|h(x-t) - h(x)|}{|h(x) - h(x+t)|} < \infty$$

and let $C_3 = \max\{C_1, C_2\}$. Then for any x and y in I',

$$C_3^{-1} \leq \frac{|h(x) - h(z)|}{|h(z) - h(y)|} \leq C_3$$

where $z = (x+y)/2$.

Now for any I', let $l : [0,1] \to I'$ be the orientation-preserving linear homeomorphism. Then $\tilde{h} = h \circ l$ is quasisymmetric when restricted on $[0, 1]$. But $h = \tilde{h} \circ l^{-1}$. So $h|I'$ is quasisymmetric. ∎

For a continuous function f, Use

$$[f]_{xy} = \frac{1}{|x-y|} \int_x^y f(\xi)\, d\xi$$

to denote the average of f on the interval $[x, y]$.

Lemma 2.14. *Suppose $h : I \to J$ is a C^1 diffeomorphism. Then the cross-ratio distortion $D(M, T; h)$ of h on any $M \subset T \subseteq I$ can be calculated as*

$$D(T, M; h) = \log|h'(\xi)| + \log|h'(\eta)| - 2\log([|h'|]_{\xi\eta})$$

for some $\xi < \eta$ in T.

Proof. Suppose $h'(x) > 0$ for all x in I (for $h'(x) < 0$, the proof is similar). For any $M = [b, c] \subset T = [a, d]$ in I, let

$$B = \{(x, y) \mid a \leq x \leq b,\ c \leq y \leq d\}$$

be a rectangle in the plane $\mathbf{R}^2$ and let

$$h^*(B) = \{(x,y) | h(a) \le x \le h(b), h(c) \le y \le h(d)\}$$

be the forward rectangle of B by h (see Fig. 2.7). Then

$$\begin{aligned}\log\Big(Cr(h(T),h(M))\Big) &= \int\int_{h^*(B)} \frac{dx\,dy}{(x-y)^2}\\ &= \int\int_B \frac{1}{(x-y)^2} \frac{h'(x)h'(y)}{\left(\frac{h(x)-h(y)}{x-y}\right)^2}\,dx\,dy\\ &= \log\Big(Cr(T,M)\Big)\left(\frac{h'(\xi)h'(\eta)}{\left(\frac{h(\xi)-h(\eta)}{\xi-\eta}\right)^2}\right)\end{aligned}$$

for some $\xi < \eta$ in T.

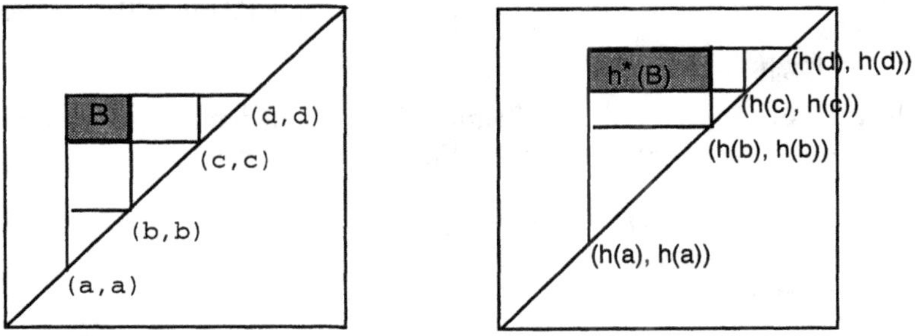

Fig. 2.7

Hence

$$\begin{aligned}D(M,T;h) &= \log\left(\frac{\log\Big(Cr(h(T),h(M))\Big)}{\log\big(Cr(T,M)\big)}\right)\\ &= \log\big(h'(\xi)\big) + \log\big(h'(\eta)\big) - 2\log\big([h']_{\xi\eta}\big).\end{aligned}$$

∎

Lemma 2.14 suggests the Zygmund condition on a function to estimate its cross-ratio distortion.

Definition 2.7. *A C^1 diffeomorphism $h: I \to J$ is C^{1+Z} if $f(x) = \log|h'(x)|$ is a Zygmund function; this means that f is continuous and that there is a constant $C > 0$ such that*

$$\Big|f(x) + f(y) - 2f\Big(\frac{x+y}{2}\Big)\Big| \le C|x-y| \qquad (2.18)$$

for all x and y in I. The smallest such C is called the Zygmund constant of f.

Remark 2.12. A Lipschitz continuous function f is a Zygmund function but there is an example of an α-Hölder continuous function which is not Zygmund for a fixed $0 \leq \alpha < 1$. Actually one can show that the modulus of continuity of a Zygmund function is $|x-y|\log|x-y|$ (see [ZYG]). Hence a Zygmund function is α-Hölder for all $0 \leq \alpha < 1$ (with different Hölder constants). So the Zygmund condition is weaker than Lipschitz but stronger than α-Hölder for any fixed $0 \leq \alpha < 1$.

From Lemma 2.14, for a C^{1+Z} diffeomorphism $h : I \to J$, if we want to control the distortion

$$D(T,M;h) = \log|h'(\xi)| + \log|h'(\eta)| - 2\log\left([|h'|]_{\xi\eta}\right),$$

we need to control

$$\left|f(\frac{\xi+\eta}{2}) - \log\left([|h'|]_{\xi\eta}\right)\right| \leq 2\left|f(\frac{\xi+\eta}{2}) - [f]_{\xi\eta}\right| + \left|[f]_{\xi\eta} - \log\Big([|h'|]_{\xi\eta}\Big)\right|,$$

where $f = \log|h'|$.

Lemma 2.15. *Suppose f is a Zygmund function defined on an interval I. Then there is a constant $C > 0$ such that for any $x < y$ in I,*

$$\left|f(\frac{x+y}{2}) - [f]_{xy}\right| \leq C|x-y|.$$

Proof. Since

$$[f]_{xy} = \frac{1}{|x-y|}\int_x^y f(\xi)\,d\xi = \lim_{n\to\infty}\frac{\sum_{i=0}^{2^n} f(x_i)}{2^n}$$

where $x_i = x + i(y-x)/2^n$, the lemma can be proved by estimating $\sum_{i=0}^{2^n} f(x_i)$. Using the Zygmund condition on f (see Inequality (2.18)),

$$\begin{aligned}\sum_{i=0}^{2^n} f(x_i) &= f(\frac{x+y}{2}) + \sum_{i=0}^{2^{n-1}-1}\Big(f(x_i) + f(x_{2^n-i})\Big)\\ &= f(\frac{x+y}{2}) + \sum_{i=0}^{2^{n-1}-1}\Big(2f(\frac{x+y}{2}) + O(|x_i - x_{2^n-i}|)\Big)\\ &= (2^n+1)f(\frac{x+y}{2}) + 2^n O(|x-y|).\end{aligned}$$

Hence

$$[f]_{xy} = f(\frac{x+y}{2}) + O(|x-y|).$$

■

Lemma 2.16. *Suppose $h : I \to J$ is a C^1 diffeomorphism from a bounded and closed interval I onto another interval J. Suppose h' is $\frac{1}{2}$-Hölder continuous. Let $f = \log|h'|$. Then there is a constant $C > 0$ such that for any $x < y$ in I,*

$$\Big|[f]_{xy} - \log\big([|h'|]_{xy}\big)\Big| \leq C|x-y|.$$

Proof. Since h' is uniformly continuous on the bounded and closed interval I and $h'(x) \neq 0$ for all x in I, there is a $\delta_0 > 0$ such that

$$\left|\frac{h'(x)}{h'(y)} - 1\right| \leq \frac{1}{2}$$

for x and y in I with $|y-x| < \delta_0$.

For any $0 < y-x < \delta_0$, let $k(\xi) = h'(\xi+x)/h'(x) - 1$ where $0 \leq \xi \leq y-x$. From the Taylor expansion of $\log(1+\eta)$ on the interval $[-1/2, 1/2]$ centered at $\eta = 0$,

$$\begin{aligned}[f]_{xy} - f(x) &= \frac{1}{|x-y|}\int_0^{y-x} \log\big(1+k(\xi)\big)\,d\xi \\ &= \frac{1}{|x-y|}\int_0^{y-x} k(\xi)\,d\xi - \frac{1}{2}\frac{1}{|x-y|}\int_0^{y-x} \big(k(\xi)\big)^2\,d\xi + \cdots,\end{aligned}$$

and

$$\begin{aligned}\log\big([|h'|]_{xy}\big) - f(x) &= \log\Big(1 + \frac{1}{|x-y|}\int_0^{y-x} k(\xi)\,d\xi\Big) \\ &= \frac{1}{|x-y|}\int_0^{y-x} k(\xi)\,d\xi - \frac{1}{2}\Big(\frac{1}{|x-y|}\int_0^{y-x} k(\xi)\,d\xi\Big)^2 + \cdots.\end{aligned}$$

This implies that

$$\Big|[f]_{xy} - \log\big([|h'|]_{xy}\big)\Big|$$

$$= \Big| -\frac{1}{2}\frac{1}{|x-y|}\int_0^{y-x} \big(k(\xi)\big)^2\,d\xi + \frac{1}{2}\Big(\frac{1}{|x-y|}\int_0^{y-x} k(\xi)\,d\xi\Big)^2 + \cdots\Big|$$

which is less than $C|x-y|$ for some constant $C > 0$ because $|k(\xi)| \leq C'\sqrt{\xi}$ for some constant C'.

Since the function

$$\frac{\Big|[f]_{xy} - \log\big([|h'|]_{xy}\big)\Big|}{|x-y|}$$

is continuous and bounded on

$$\{(x,y) \in I \times I \,\big|\, |x-y| \geq \delta_0\},$$

there is a positive constant, which we still denote as C, such that

$$\Big|[f]_{xy} - \log\Big([|h'|]_{xy}\Big)\Big| \leq C|x-y|$$

for all $x < y$ in I. ■

Lemma 2.17 (C^{1+Z}-Koebe distortion lemma [SU4]). *Suppose $h : I \to J$ is a C^{1+Z} diffeomorphism from a bounded and closed interval I onto another interval J. Then there is a constant $C > 0$, depending only on the Zygmund and 1/2-Hölder constants of $f = \log|h'|$, such that for any $M \subset T \subseteq I$, the cross-ratio distortion $D(T, M; h)$ is bounded by $C|T|$, that is,*

$$|D(T,M;h)| = \left|\log\left(\frac{\log\left(Cr(h(T),h(M))\right)}{\log\left(Cr(T,M)\right)}\right)\right| \leq C|T|.$$

Proof. It follows now from Lemmas 2.14, 2.15, and 2.16 because

$$D(T,M;h) = \log|h'(\xi)| + \log|h'(\eta)| - 2\log\left([|h'|]_{\xi\eta}\right)$$

for some $\xi < \eta$ in T. ■

2.5. The Cross-Ratio Distortion of an Irrational Circle Mapping

Suppose f is an irrational circle mapping (see §1.4). Let $f_t(x) = f(x+t)-t$ for $0 \leq t < 1$ (remember that $0 < f_t(0) < 1$), and let

$$\Big\{\{(l_{n,t}, L_{n,t}); (r_{n,t}, R_{n,t})\}\Big\}_{n=0}^{\infty}$$

be the corresponding sequence of renormalizations to f_t (see §1.7). Let $M = [b,c] \subset T = [a,d]$ be intervals such that $-\infty < a < b < c < d < \infty$.

Definition 2.8. *An irrational circle mapping f is said to have bounded cross-ratio distortion if there are constants $C > 0$ and $0 < \mu < 1$ such that for any $n \geq 0$ and $0 \leq t < 1$,*

$$|D(T, M; l_{n,t})| = \left|\log\left(\frac{\log\Big(Cr\big(l_{n,t}(T), l_{n,t}(M)\big)\Big)}{\log\big(Cr(T,M)\big)}\right)\right| \leq C$$

for any $M \subset T \subseteq (1+\mu)L_{n,t}$, and

$$|D(T, M; r_{n,t})| = \left|\log\left(\frac{\log\Big(Cr\big(r_{n,t}(T), r_{n,t}(M)\big)\Big)}{\log\big(Cr(T,M)\big)}\right)\right| \leq C$$

for any $M \subset T \subseteq (1+\mu)R_{n,t}$.

Lemma 2.18 (recurrence). *Suppose f is an irrational circle mapping having bounded cross-ratio distortion. Let F be the corresponding homeomorphism of the circle $\mathbf{T}^1$. Then every positive orbit $\{F^{\circ n}(p)\}_{n=0}^{\infty}$ is recurrent.*

Proof. The proof is by contradiction. Suppose there is a point $p = [t]$ in $\mathbf{T}^1$ whose orbit under F is not recurrent. Without loss of generality, we assume that $p = [0]$ (otherwise, consider $f_t(x) = f(x+t) - t$ and its corresponding homeomorphism F_t of the circle $\mathbf{T}^1$). Let

$$\Big\{\{(l_n, L_n); (r_n, R_n)\}\Big\}_{n=0}^{\infty}$$

be the corresponding sequence of renormalizations to f and

$$[a_L, a_R] = \cap_{n=0}^{\infty}(L_n \cup R_n).$$

Then $a_L < 0 < a_R$ (see Fig. 1.5).

For every integer $n > 0$, the interval $T_n = L_n \cup R_n = [r_n(0), l_n(0)]$ contains the interval $[a_L, a_R]$. From Lemma 1.8, $l_n([a_L, 0])$ and $r_n([0, a_R])$ are, respectively, contained in $[a_R, l_n(0)]$ and $[r_n(0), a_L]$.

Suppose $r_n \circ l_n(0) < 0$ (similar arguments work for $r_n \circ l_n(0) > 0$). Cut the interval $[r_n(0), a_L]$ into three subintervals say K_1, K_2, and K_3, such that $|K_1| = |K_2| = |K_3|$. Take a fourth interval $K_4 = [a_L, a_L + |K_1|]$ and a fifth interval $K_5 = [r_n(0) - \mu|K_1|, r_n(0)]$ where μ is the constant in Definition 2.8. Let W_1, W_2, W_3, W_4, and W_5 be the images of K_1, K_2, K_3, K_4, and K_5 under l_n (see Fig. 2.8).

Consider $T = K_1 \cup K_2 \cup K_3$ and $M = K_2$. Since the cross-ratio distortion of l_n is bounded uniformly and $Cr(T, M) = \log(4/3)$,

$$\begin{aligned}&\log\Big(\log\Big(Cr(l_n(T), l_n(M))\Big)\Big)\\&= \log\left(\log\left(\frac{|W_1 + W_2| \cdot |W_2 + W_3|}{|W_1 + W_2 + W_3| \cdot |W_2|}\right)\right)\end{aligned} \tag{2.19}$$

is uniformly bounded. To get a contradiction, let us consider $T' = K_2 \cup K_3 \cup K_4$ and $M' = K_3$, and $T'' = K_5 \cup K_1 \cup K_2$ and $M'' = K_1$. Then

$$\begin{aligned}&\log\Big(\log\Big(Cr(l_n(T'), l_n(M'))\Big)\Big)\\&= \log\left(\log\left(\frac{|W_2 + W_3| \cdot |W_3 + W_4|}{|W_2 + W_3 + W_4| \cdot |W_3|}\right)\right)\end{aligned} \tag{2.20}$$

and

$$\begin{aligned}&\log\Big(\log\Big(Cr(l_n(T''), l_n(M''))\Big)\Big)\\&= \log\left(\log\left(\frac{|W_5 + W_1| \cdot |W_1 + W_2|}{|W_5 + W_1 + W_2| \cdot |W_1|}\right)\right)\end{aligned} \tag{2.21}$$

are uniformly bounded. Since the interval W_4 is contained in $[a_R, l_n(0)]$, the length $|W_4|$ tends to zero as n goes to infinity. From Eq. (2.20), the length $|W_3|$ must tend to zero as n goes to infinity; from Eq. (2.19), the length $|W_2|$ tends to zero as n goes to infinity; from Eq. (2.21), the length $|W_1|$ tends to zero as n goes to infinity. This contradicts that $|W_1 + W_2 + W_3| \geq |a_R - a_L|$. ∎

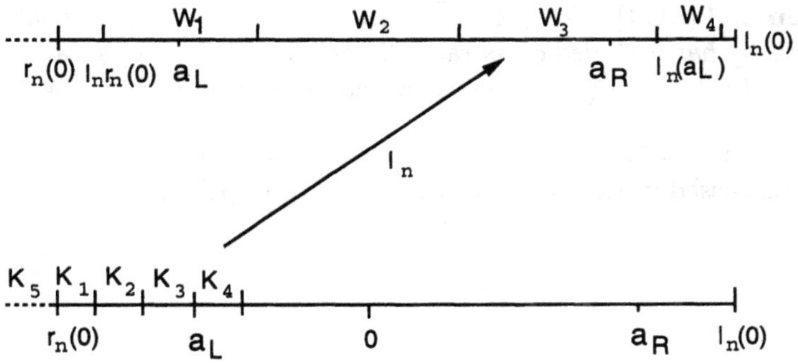

Fig. 2.8

A version of Denjoy's theorem is that

Theorem 2.2. *Every irrational circle mapping f in $\mathcal{S}$ whose cross-ratio distortion is bounded, and whose rotation number is ρ_f, is topologically conjugate to the rigid rotation $R(x) = x + \rho_f$ by the rotation number ρ_f.*

Proof. The proof is the same as that of Theorem 1.3. ■

Let f be a C^{1+Z} irrational circle mapping and let $f_t(x) = f(x+t) - t$ for $0 \le t < 1$. Let

$$\Big\{\{(l_{n,t}, L_{n,t}); (r_{n,t}, R_{n,t})\}\Big\}_{n=0}^{\infty}$$

be the corresponding sequence of renormalizations to f_t. From the fact that $l_{n,t}(x) = f_t^{\circ q_n}(x) - p_n$ for x in $L_{n,t}$, the cross-ratio distortion of $l_{n,t}$ on $M \subset T \subseteq (1+\mu)L_{n,t}$ (or $\subseteq (1+\mu)R_{n,t}$) satisfies the equation

$$\log\left(\frac{\log\Big(Cr\big(l_{n,t}(T), l_{n,t}(M)\big)\Big)}{\log\big(Cr(T,M)\big)}\right) = \sum_{i=0}^{q_n-1} \log\left(\frac{\log\big(Cr(T_{i+1}, M_{i+1})\big)}{\log\big(Cr(T_i, M_i)\big)}\right)$$

where $T_i = f_t^{\circ i}(T)$ and $M_i = f_t^{\circ i}(M)$ are identified with sets in $[f_t(0)-1, f_t(0)]$ by modulo the integers (see §1.8). Lemma 2.17 implies that,

$$|D(T, M, l_{n,t})| = \left|\log\left(\frac{\log\Big(Cr\big(l_{n,t}(T), l_{n,t}(M)\big)\Big)}{\log\big(Cr(T,M)\big)}\right)\right| \le C \sum_{i=0}^{q_n-1} |T_i|.$$

Sullivan [SU4] (see also [MV2]) proved from this calculation a version of Denjoy's theorem as follows.

Theorem 2.3 [SU4]. *Every C^{1+Z} irrational circle mapping f whose rotation number is ρ_f has bounded cross-ratio distortion and is topologically conjugate to the rigid rotation $R(x) = x + \rho_f$ by the rotation number ρ_f.* ∎

A condition which combines the Zygmund and bounded variation conditions is discussed in the paper of Hu and Sullivan [HUS].

2.6. Koebe's Distortion Theorem in One Complex Variable

Let $\mathbf{C}$ be the complex plane and let $\mathbf{D} = \{z \in \mathbf{C} \mid |z| < 1\}$ be the open unit disk of $\mathbf{C}$. A function defined on $\mathbf{D}$ is analytic if it can be written as a convergent power series on $\mathbf{D}$, that is,

$$f(z) = \sum_{n=0}^{\infty} a_n z^n$$

for z in $\mathbf{D}$. An analytic function f from $\mathbf{D}$ into $\mathbf{C}$ is called a schlicht function (or a univalent function or a conformal mapping) if it is one-to-one. This is equivalent to saying $f'(z) \neq 0$ for all z in $\mathbf{D}$. We first prove a theorem obtained by Koebe in 1907 (see [BIE]).

Theorem 2.4 (Koebe's $\frac{1}{4}$-Theorem). *Suppose $f : \mathbf{D} \to \mathbf{C}$ is a schlicht function. Then $f(\mathbf{D})$ contains an open disk centered at $f(0)$ with radius $|f'(0)|/4$.*

First we prove a lemma.

Lemma 2.19. *Suppose $f(z) = z + a_2 z^2 + \sum_{n=3}^{\infty} a_n z^n : \mathbf{D} \to \mathbf{C}$ is a schlicht function. Then*

$$|a_2| \leq 2.$$

Proof. Suppose $\mathbf{D}_r = \{z \in \mathbf{C} \mid |z| < r\}$ for $0 < r < 1$. Then $G_r = f(\mathbf{D}_r)$ is a bounded subset of the complex plane $\mathbf{C}$. Let G'_r be the complement of G_r. For $w = Re^{\Theta i}$ in G'_r,

$$\int\int_{G'_r} w^t \overline{w}^t R \, dR \, d\Theta > 0$$

provided the integral converges, whence

$$\lim_{r \to 1} \int\int_{G'_r} w^t \overline{w}^t R \, dR \, d\Theta \geq 0.$$

The integral certainly converges if $t < -1$. Let us set $z = re^{\theta i}$ and $w = f(z)$. On the boundary of G'_r, which is the image of $|z| = r$ under f, we set $R = R(\theta)$ and $\Theta = \Theta(\theta)$. Then

$$\int\int_{G'_r} w^t \overline{w}^t R \, dR \, d\Theta = \int\int_{G'_r} R^{2t+1} \, dR \, d\Theta = -\int_0^{2\pi} \frac{R^{2t+2}}{2t+2} \frac{d\Theta}{d\theta} \, d\theta > 0.$$

We have

$$\Theta = \arg w = \frac{\log w - \log \overline{w}}{2i}.$$

Therefore

$$\frac{d\Theta}{d\theta} = \frac{w'\overline{w}z + \overline{w}'w\overline{z}}{2w\overline{w}},$$

and

$$\int_0^{2\pi} \frac{R^{2t+2}}{2t+2}\frac{d\Theta}{d\theta}\,d\theta = \int_0^{2\pi} \frac{R^{2t}(w'\overline{w}z + \overline{w}'w\overline{z})}{4(t+1)}\,d\theta < 0.$$

This is easily transformed into

$$\int_0^{2\pi} \left(\overline{w}^{t+1} z \frac{dw^{t+1}}{dz} + w^{t+1}\overline{z}\frac{d\overline{w}^{t+1}}{d\overline{z}}\right) d\theta < 0.$$

But

$$w^{t+1} = z^{t+1}\Big(1 + (t+1)a_2 z + \Big((t+1)a_3 + \frac{t(t+1)}{2}a_2^2\Big)z^2 + \cdots\Big).$$

We substitute this series into the last integral and integrate term by term, observing that for any non-zero integer k, $\int_0^{2\pi} e^{ki\theta}d\theta = 0$. Thus we obtain

$$1 + (t+1)(t+2)|a_2|^2 r^2 + (t+1)(t+3)\left|a_3 + \frac{t}{2}a_2^2\right|^2 r^4 + \cdots > 0.$$

For $r \to 1$, this yields

$$1 + (t+1)(t+2)|a_2|^2 + (t+1)(t+3)\left|a_3 + \frac{t}{2}a_2^2\right|^2 + \cdots$$
$$+ (t+1)(t+k)\Big|a_k + g_k(a_2, \ldots, a_{k-1})\Big|^2 + \cdots \geq 0.$$

In particular, we take $t = -3/2$; then

$$1 - \frac{|a_2|^2}{4} \geq 0,$$

whence

$$|a_2| \leq 2.$$

■

Remark 2.13. In general, $|a_n| \leq n$ for all $n \geq 2$. This was the famous Bieberbach conjecture formulated in 1916 and proved by De Branges [BRA] in 1984.

Proof of Theorem 2.4. Let $f(z) \neq c$ for all $|z| < 1$. The function

$$f_1(z) = \frac{f(z) - f(0)}{f'(0)} = z + a_2 z^2 + \cdots \neq \frac{c - f(0)}{f'(0)}$$

for all $|z| < 1$. Set

$$f_2(z) = \frac{f(0) - c}{f'(0)} \frac{f(z) - f(0)}{f(z) - c} = z + \Big(a_2 - \frac{f'(0)}{f(0) - c}\Big) z^2 + \cdots .$$

Both of f_1 and f_2 are schlicht functions from $\mathbf{D}$ into $\mathbf{C}$. Thus

$$|a_2| \leq 2 \quad \text{and} \quad \left| a_2 - \frac{f'(0)}{f(0) - c} \right| \leq 2.$$

Therefore,

$$|c - f(0)| \geq \frac{|f'(0)|}{4}.$$

We have thus proved that no boundary point, of the image of $|z| < 1$ under the mapping f, has distance from $f(0)$ less than $|f'(0)|/4$. ∎

Lemma 2.20. *Suppose* $f(z) = z + \sum_{n=2}^{\infty} a_n z^n : \mathbf{D} \to \mathbf{C}$ *is a schlicht function. Then*

$$\frac{1 - |z|}{(1 + |z|)^3} \leq |f'(z)| \leq \frac{1 + |z|}{(1 - |z|)^3}.$$

Proof. For a point z in $\mathbf{D}$, let

$$g(\xi) = \frac{f(\frac{\xi + z}{1 + \overline{z}\xi}) - f(z)}{f'(z)(1 - z\overline{z})}.$$

This is a schlicht function from $\mathbf{D}$ into $\mathbf{C}$ and has a power series expansion valid in $|\xi| < 1$:

$$g(\xi) = \xi + b_2 \xi^2 + \cdots ,$$

where

$$b_2 = \frac{1}{2}\Big(\frac{f''(z)(1 - z\overline{z})}{f'(z)} - 2\overline{z}\Big).$$

Therefore,

$$\left| \frac{f''(z)(1 - z\overline{z})}{f'(z)} - 2\overline{z} \right| \leq 4.$$

Hence

$$\left|\frac{zf''(z)}{f'(z)} - \frac{2|z|^2}{1-|z|^2}\right| \le \frac{4|z|}{1-|z|^2}.$$

Thus

$$\frac{2|z|^2-4|z|}{1-|z|^2} \le \Re\Big(\frac{zf''(z)}{f'(z)}\Big) \le \frac{2|z|^2+4|z|}{1-|z|^2}.$$

Since

$$\Re\Big(\frac{zf''(z)}{f'(z)}\Big) = |z|\frac{\partial}{\partial|z|}\Big(\Re(\log f'(z))\Big) = |z|\frac{\partial}{\partial|z|}\Big(\log|f'(z)|\Big),$$

we obtain

$$\frac{2|z|-4}{1-|z|^2} \le \frac{\partial}{\partial|z|}\Big(\log\big(|f'(z)|\big)\Big) \le \frac{2|z|+4}{1-|z|^2}.$$

Integration now yields

$$\frac{1-|z|}{(1+|z|)^3} \le |f'(z)| \le \frac{1+|z|}{(1-|z|)^3}.$$

■

Theorem 2.5 (Koebe's Distortion Theorem). *Suppose $f:\mathbf{D}\to\mathbf{C}$ is a schlicht function. Then*

$$\Big(\frac{1-r}{1+r}\Big)^4 \le \frac{|f'(z_1)|}{|f'(z_2)|} \le \Big(\frac{1+r}{1-r}\Big)^4$$

for all $|z_1|\le r$ and all $|z_2|\le r$.

Proof. Let $g(z)=(f(z)-f(0))/f'(0)$. Then $g'(z)=f'(z)/f'(0)$. Thus, this theorem follows from Lemma 2.20. ■

Remark 2.14. From Lemma 2.20, if $f(z)=z+a_2z^2+\cdots+a_nz^n+\cdots:\mathbf{D}\to\mathbf{C}$ is a schlicht function, then

$$\frac{|z|}{(1+|z|)^2} \le |f(z)| \le \frac{|z|}{(1-|z|)^2}.$$

Remark 2.15. Theorems 2.4 and 2.5 are the sharpest possible. One sees this with the example

$$f(z)=\frac{z}{(1-z)^2}.$$

Lemma 2.21 (Schwarz Lemma). *For all $|z| < 1$, suppose that $f(z) = a_1 z + a_2 z^2 + \cdots$ converges, and that $|f(z)| \le 1$. Then for all $0 < |z| < 1$,*

$$|f(z)| \le |z| \quad \text{and} \quad |a_1| \le 1$$

with equalities unobtainable except if $f(z) = e^{i\alpha} z$ for some real number α.

Proof. Define the function

$$g(z) = \frac{f(z)}{z} = a_1 + a_2 z + \cdots .$$

Let $\mathbf{D}_r = \{z \in \mathbf{C} \mid |z| < r\}$ for $0 < r < 1$. By the maximum modulus principle, $g(z)$ cannot achieve a maximum of $|g(z)|$ in $\mathbf{D}_r$. Thus

$$|g(z)| \le \frac{1}{r}$$

for z in $\mathbf{D}_r$. This holds for any $z \in \mathbf{D}$ and any $|z| < r < 1$. Therefore,

$$|g(z)| \le 1,$$

that is,

$$|f(z)| \le |z| \quad \text{and} \quad |a_1| = |g(0)| \le 1$$

for $|z| < 1$.

If the equality sign holds for some $0 < |z_0| < 1$ (or if $|a_1| = 1$), then $|f(z_0)| = |z_0|$ (or $|g(0)| = 1$). Unless g is a constant function, g must map a neighborhood about z_0 (or 0) onto a neighborhood about $g(z_0)$ (or $g(0)$) since an analytic function maps an open set to an open set. Since $|g(z_0)| = 1$ (or $|g(0)| = 1$), this would imply that there are points z arbitrarily close to z_0 (or 0) for which $|g(z)| > 1$, which would contradict the first part of this argument. Hence, if the equality sign holds anywhere in $0 < |z| < 1$ (or if $|a_1| = 1$), then g must be a constant function. In this case, $|g(z)| = 1$ everywhere, whence $f(z) = e^{i\alpha} z$ on $\mathbf{D}$ for some real number α. ∎

Lemma 2.22. *Every schlicht function f from $\mathbf{D}$ onto $\mathbf{D}$ is a Möbius transformation*

$$f(z) = a\frac{z + b}{1 + \bar{b}z}$$

where a and b are complex numbers with $|a| = 1$ and $|b| < 1$.

Proof. Suppose $f(0) = 0$; by applying Lemma 2.21 to f and to f^{-1}, we get

$$|f(z)| \le |z| \text{ and } |z| \le |f(z)|.$$

Thus,

$$f(z) = az$$

with $|a| = 1$. In general, consider

$$g = \gamma \circ f,$$

where

$$\gamma(z) = \frac{z - f(0)}{1 - \overline{f(0)}z}.$$

Then g is a schlicht function from $\mathbf{D}$ onto $\mathbf{D}$ such that $g(0) = 0$. Thus $g(z) = az$ with $|a| = 1$, whence

$$f(z) = \gamma^{-1}(az) = \frac{az + f(0)}{1 + a\overline{f(0)}z} = a\frac{z + b}{1 + \bar{b}z}$$

where $b = \bar{a}f(0)$. ∎

Definition 2.9. *The hyperbolic disk $\mathcal{D}$ is the open unit disk $\mathbf{D}$ equipped with the metric*

$$d_H s = \rho(z)|dz| = \frac{|dz|}{1 - |z|^2}.$$

The corresponding distance d_H is

$$d_H(z_1, z_2) = \inf_l \int_l d_H s = \inf_l \int_l \frac{|dz|}{1 - |z|^2}$$

where l are over all C^1 curves in $\mathbf{D}$ connecting z_1 and z_2.

Remark 2.16. The metric $d_H s = \rho(z)|dz|$ is called the Poincaré metric (or hyperbolic metric). The distance d_H is called the (induced) hyperbolic distance.

From Lemma 2.21, we have that

Theorem 2.6. *Suppose f is a schlicht function from $\mathbf{D}$ into $\mathbf{D}$. Then for all z_1 and z_2 in $\mathbf{D}$,*

$$d_H\big(f(z_1), f(z_2)\big) \leq d_H(z_1, z_2) \tag{2.22}$$

with equality unobtainable except if

$$f(z) = a\frac{z + b}{1 + \bar{b}z} \tag{2.23}$$

for complex numbers a and b with $|a| = 1$ and $|b| < 1$.

Proof. For a fixed z in $\mathbf{D}$, let

$$g(\xi) = \frac{f(\frac{\xi+z}{1+\bar{z}\xi}) - f(z)}{1 - \overline{f(z)}f(\frac{\xi+z}{1+\bar{z}\xi})}.$$

Then $g(0) = 0$ and $|g(\xi)| \leq 1$. From Lemma 2.21, $|g'(0)| \leq 1$, whence

$$\frac{|f'(z)|}{1 - |f(z)|^2} \leq \frac{1}{1 - |z|^2}. \tag{2.24}$$

The equality sign holds if and only if $g(\xi) = c\xi$ with $|c| = 1$, which is equivalent to saying f takes a form in Eq. (2.23). Take two points z_1 and z_2 in $\mathbf{D}$. For any C^1 curve l in $\mathbf{D}$ connecting z_1 and z_2,

$$d_H(f(z_1), f(z_2)) \leq \int_{f(l)} \frac{|dz|}{1 - |z|^2} = \int_l \frac{|f'(z)|}{1 - |f(z)|^2} |dz| \leq \int_l \frac{|dz|}{1 - |z|^2}.$$

Therefore,

$$d_H\big(f(z_1), f(z_2)\big) \leq d_H(z_1, z_2).$$

The equality sign holds for some z_1 and z_2 if and only if f takes a form in (2.23). ∎

A domain Ω is an open and (path) connected set in the extended complex plane $\overline{\mathbf{C}} = \mathbf{C} \cup \{\infty\}$. A domain Ω is simple connected if every closed curve γ in Ω can be deformed in Ω to a point.

Theorem 2.7 (Riemann's Mapping Theorem). *There is a schlicht function H from $\mathbf{D}$ onto Ω for every simply connected domain Ω having at least two boundary points.*

The proof of this theorem can be found in many text books (e.g. [AH3,BIE]). The inverse of the mapping H in this theorem is called a Riemann mapping. Let $g = H^{-1} : \Omega \to \mathbf{D}$ be the inverse of H. The hyperbolic metric $d_{H,\Omega}s$ on Ω is

$$d_{H,\Omega}s = \rho(g(z))|g'(z)||dz|.$$

All results (with appropriate constants) in this section applies to a schlicht function defined on a simply connected domain Ω whose boundary $\partial\Omega$ contains at least two points. In particular, if

$$\Omega = \mathbf{UH} = \{z = x + yi \in \mathbf{C} \mid y > 0\}$$

is the upper-half plane, then the hyperbolic metric is

$$d_{H,\mathbf{UH}}s = \frac{|dz|}{|z - \overline{z}|}.$$

Any schlicht function $f : \mathbf{UH} \to \mathbf{UH}$ contracts this hyperbolic metric.

2.7. The Geometric Distortion Theorem

In this section we develop a general Koebe type distortion theorem in the plane $\mathbf{R}^2$ (and in any higher dimensional space $\mathbf{R}^m$ as well). Suppose W and U are two bounded domains in the complex plane $\mathbf{C}$ with $\overline{W} \subset U$. Let g be a C^1 map from U into $\mathbf{C}$. We use $T_z g$ to denote the derivative (or tangent map) of g at z. The map $g|\overline{W}$ is said to be $C^{1+\alpha}$ for some $0 < \alpha \leq 1$ if

$$g(w) = g(z) + T_z g(w - z) + R(w, z)$$

satisfies

$$|R(w, z)| \leq L_0 |w - z|^{1+\alpha}$$

for all $z \in \overline{W}$ and all $w \in U$, where $L_0 > 0$ is a constant. The map g is said to be contracting if there is a constant $0 < \lambda < 1$ such that

$$|T_z g(v)| \leq \lambda |v|$$

for all z in $\overline{W}$ and all v in $\mathbf{C}$. For a $C^{1+\alpha}$ contracting diffeomorphism $g : \overline{W} \to \overline{V}$, let $f : \overline{V} \to \overline{W}$ be its inverse. Then f is a $C^{1+\alpha}$ expanding diffeomorphism.

Remark 2.17. Let $\mathbf{R}^m$ be any m-dimensional Euclidean space for $m > 2$. Let W and V be two connected open sets in $\mathbf{R}^m$. A $C^{1+\alpha}$ contracting map $g : \overline{W} \to \overline{V}$ can be defined similarly (see [JI8ims]).

Suppose W_i and U_i, $i = 0, 1, \ldots, n-1$, are pairs of bounded domains in the complex plane $\mathbf{C}$ with $\overline{W}_i \subset U_i$. Suppose g_i are maps from U_i into $\mathbf{C}$ such that the restriction $g_i|\overline{W_i}$ from $\overline{W_i}$ onto $\overline{V_i}$ are $C^{1+\alpha}$ contracting diffeomorphisms for some $0 < \alpha \leq 1$ and for all i. To simplify the notations, we assume that $W = W_i$ and $U = U_i$ for all i and $\cup_{i=0}^{n-1} V_i \subset W$. We use

$$\mathcal{G} = \langle g_0, g_1, \ldots, g_{n-1} \rangle$$

to denote the semigroup generated by g_i for $0 \leq i \leq n-1$, and call it a (two-dimensional) finitely generated $C^{1+\alpha}$ contracting semigroup. Let $\Lambda = \cap_{g \in \mathcal{G}} g(\overline{W})$ be the limit set of $\mathcal{G}$. An m-dimensional finitely generated $C^{1+\alpha}$ contracting semigroup can be defined similarly for any $m > 2$.

Suppose $z = x + yi$ is a point in $\mathbf{C}$; let $\overline{z} = x - yi$ be the conjugate of z. By the complex analysis (see [AH1]), we know that for $z \in \overline{W}$ and $w \in \mathbf{C}$ with $|w| = 1$,

$$\big| |(g_i)_z| - |(g_i)_{\overline{z}}| \big| \leq |T_z g_i(w)| \leq |(g_i)_z| + |(g_i)_{\overline{z}}|.$$

Let

$$l_i(z) = |(g_i)_z| + |(g_i)_{\overline{z}}|, \quad s_i(z) = \big||(g_i)_z| - |(g_i)_{\overline{z}}|\big|$$

and $K_i(z) = l_i(z)/s_i(z)$, the conformal dilatation of g_i at z. Let $l = \max\{l_i(z)\}$, let $s = \min\{s_i(z)\}$, and let $K = \max\{K_i(z)\} < +\infty$ where max and min are over all z in $\overline{W}$ and all $0 \leq i < n$. Then $0 < s \leq l < 1$.

Definition 2.10. *We say a finitely generated $C^{1+\alpha}$ contracting semigroup $\mathcal{G}$ is regular if*

$$K < \frac{1}{l^{\alpha}}.$$

Remark 2.18. The statement that an m-dimensional finitely generated $C^{1+\alpha}$ contracting semigroup $\mathcal{G}$ is regular can be defined similarly (see [JI8ims]).

Let $D(z,r) = \{w \in \mathbf{C} \mid |w-z| \leq r\}$ be the closed disk of radius r centered at z.

Theorem 2.8 (Geometric Distortion Theorem [JI8]). *Suppose $\mathcal{G} = \langle g_0, g_1, \ldots, g_{n-1}\rangle$ is a finitely generated regular $C^{1+\alpha}$ contracting semigroup. There are two functions $\delta = \delta(\varepsilon) > 0$ and $C = C(\varepsilon) \geq 1$, with $\delta(\varepsilon) \to 0$ and $C(\varepsilon) \to 1$ as $\varepsilon \to 0+$, such that*

$$\begin{aligned} g(z) + C^{-1} \cdot \Big(T_z g\big(D(0,r)\big)\Big) &\subseteq g\big(D(z,r)\big) \\ &\subseteq g(z) + C \cdot \Big(T_z g\big(D(0,r)\big)\Big) \qquad (2.25) \end{aligned}$$

for any $0 < r \leq \delta(\varepsilon)$, any $g \in \mathcal{G}$ and any $z \in \overline{W}$ (see Fig. 2.9).

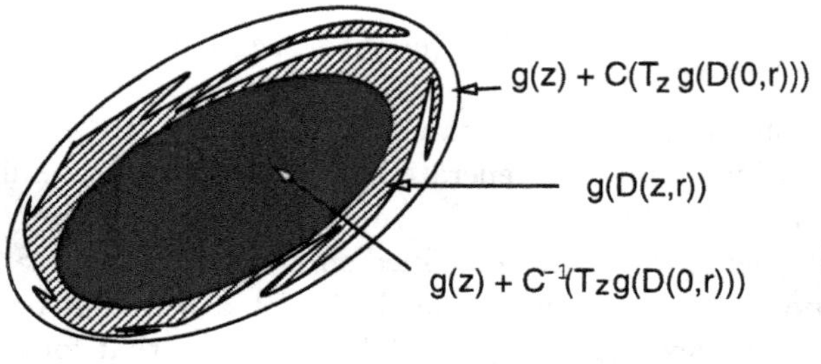

Fig. 2.9

Proof. Since $\overline{W}$ is compact, there is a strictly positive function $\delta = \delta(\varepsilon)$ with $\delta(\varepsilon) \to 0$ as $\varepsilon \to 0+$, such that every g_i is defined on $D(z, \delta)$ for z in $\overline{W}$ and such that

$$g_i(w) = g_i(z) + T_z g_i(w - z) + R_i(w, z)$$

satisfies

$$|R_i(w, z)| \leq \frac{\varepsilon}{2}\Big(\inf_{w \in \overline{W}} ||T_z g_i||\Big)|w - z|$$

for z in $\overline{W}$, $|w - z| \leq \delta$, and $0 \leq i < n$. This implies (see Fig. 2.10) that for z in $\overline{W}$ and $0 < r \leq \delta$,

$$\begin{aligned} & g_i(z) + (1+\varepsilon)^{-1} \cdot \Big(T_z g_i\big(D(0, r)\big)\Big) \\ & \subseteq g_i\big(D(z, r)\big) \\ & \subseteq g_i(z) + (1+\varepsilon) \cdot \Big(T_z g_i\big(D(0, r)\big)\Big). \end{aligned} \qquad (2.26)$$

Suppose $L_0 > 0$ and $0 < \beta < \alpha$ are constants such that

$$|R_i(w, z)| \leq L_0 |w - z|^{1+\alpha}$$

and

$$K_i(z) \leq \left(\frac{1}{l_i(z)}\right)^{\beta}$$

for all $0 \leq i < n$, all z in $\overline{W}$, and all w with $|w - z| < \delta$. Let $\kappa_m = \sum_{i=0}^{m} l^{(\alpha-\beta)i}$. We take $\delta = \delta(\varepsilon) \leq 1$ so small that

$$\Theta_\varepsilon = \frac{L_0}{s}\big(1 + \varepsilon + \kappa_\infty\big)^{1+\alpha} \delta^{(\alpha-\beta)} \leq 1$$

and then take

$$C_m(\varepsilon) = 1 + \varepsilon + \delta^{\beta} \kappa_m.$$

It is clear that $C_m(\varepsilon) \to 1$ as $\varepsilon \to 0+$.

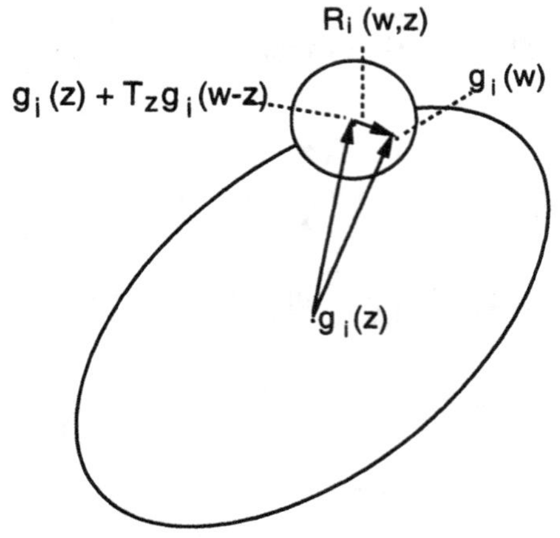

Fig. 2.10

Claim. *For* $g = g_{i_0} \circ g_{i_1} \circ \cdots \circ g_{i_m}$ *in* $\mathcal{G}$,

$$g(z) + C_m^{-1} \cdot \Big(T_z g\big(D(0,r)\big)\Big) \subseteq g\big(D(z,r)\big) \subseteq g(z) + C_m \cdot \Big(T_z g\big(D(0,r)\big)\Big).$$

Proof of claim. For $m = 0$, it is from Formula (2.26). Suppose the claim holds for $m = 0, 1, \ldots, M-1$ ($M \geq 1$). Then for $g = g_{i_0} \circ g_{i_1} \circ \cdots \circ g_{i_M} = g_{i_0} \circ G$,

$$g_{i_0}\Big(G(z) + C_{M-1}^{-1} \cdot \big(T_z G\big(D(0,r)\big)\big)\Big) \subseteq g\big(D(z,r)\big)$$
$$\subseteq g_{i_0}\Big(G(z) + C_{M-1} \cdot \big(T_z G\big(D(0,r)\big)\big)\Big).$$

For any $w - z$ in $D(0,r)$, we know that

$$g_{i_0}\Big(G(z) + C_{M-1}^j \cdot \big(T_z G(w-z)\big)\Big) = g(z) + C_{M-1}^j \cdot \big(T_z g(w-z)\big) + R$$

where $R = R_{i_0}\Big(G(z) + C_{M-1}^j \cdot \big(T_z G(w-z)\big),\ G(z)\Big)$ and $j = 1$ or -1. Then

$$|R| \leq L_0 C_{M-1}^{1+\alpha} ||T_z G||^{1+\alpha} |w-z|^{1+\alpha}.$$

But for $z_0 = z$ and $z_i = g_{M-i} \circ \cdots \circ g_{i_M}(z)$, $i = 1, 2, \ldots, M$,

$$||T_z G|| = \prod_{1 \leq k \leq M} ||T_{z_{M-k}} g_{i_k}|| \leq \prod_{1 \leq k \leq M} l_{i_k}(z_{M-k}).$$

Since $K_i(z) \leq \big(1/l_i(z)\big)^{\beta}$ for all i, we have

$$||T_zG||^{1+\alpha} \leq \Big(\prod_{1\leq k\leq M} s_{i_k}(z_{M-k})\Big) l^{(\alpha-\beta)M}.$$

Let $B_M = (L_0/s)C_{M-1}^{1+\alpha}\delta^{\alpha} l^{(\alpha-\beta)M}$; then

$$|R| \leq B_M\Big(\prod_{0\leq k\leq M} s_{i_k}(z_{M-k})\Big)|w-z|.$$

Since $B_M \leq \Theta_\epsilon \delta^\beta l^{(\alpha-\beta)M} \leq \delta^\beta l^{(\alpha-\beta)M}$, we have $C_{M-1}+B_M \leq C_M$. We can conclude now from the estimates that $g(w)-g(z)$ is in $C_M \cdot \Big(T_z g\big(D(0,r)\big)\Big)$. Similarly, if $|w-z| = r$, $g(w)-g(z)$ is outside of $C_M^{-1} \cdot \Big(T_z g\big(D(0,r)\big)\Big)$ (refer to Fig. 2.10). The proof of the claim is completed.

Take $C = C_\infty(\varepsilon)$. Then δ and C are the functions we want. ∎

Remark 2.19. Theorem 2.8 can be proved for an m-dimensional finitely generated $C^{1+\alpha}$ contracting semigroup for any $m > 2$ (see [JI8ims]).

Without considering the best constant, Theorem 2.8 is a generalization of Koebe's distortion theorem (Theorem 2.5) as follows. Let $\mathbf{D} = \{z \in \mathbf{C} \mid |z| < 1\}$ be the open unit disk. Let $g_i : \mathbf{D} \to \mathbf{D}$ be a conformal mapping for $i = 0$, 1, ..., $n-1$. Since g_i is analytic, $(g_i)_{\overline{z}} = 0$ on $\mathbf{D}$ for all $0 \leq i \leq n-1$. Then

$$\mathcal{G}_0 = \langle g_0, g_1, \cdots, g_{n-1}\rangle$$

is a conformal semigroup, this means that the maximal conformal dilatation $K = 1$. Without loss of generality, let us assume that $g_i(0) = 0$. Suppose g_i is not a Möbius transformation (see Lemma 2.22) for every $0 \leq i \leq n-1$.

For any $0 < r_0 < 1$, let $\mathbf{D}_{r_0} = \{z \in \mathbf{C} \mid |z| < r_0\}$. From Theorem 2.6, there are constants $C > 0$ and $0 < \lambda < 1$ such that

$$\big|\big(g_{i_0} \circ g_{i_1} \circ \cdots \circ g_{i_{m-1}}\big)'(z)\big| \leq C\lambda^m$$

for all $z \in \overline{\mathbf{D}}_{r_0}$ and all sequences of $i_0 i_1 \ldots i_{m-1}$ of 0's, 1's, ..., $(n-1)$'s. Without loss of generality, we assume that

$$|g_i'(z)| \leq \lambda < 1$$

for all $z \in \overline{\mathbf{D}}_{r_0}$ and all $0 \leq i \leq n-1$.

For the conformal semigroup $\mathcal{G}_0$, we have that $K = 1$ and that

$$l = \max_{z \in \overline{\mathbf{D}}_{r_0}, 0 \leq i \leq n-1} \left\{ |g_i'(z)| \right\} < 1.$$

Therefore, $\mathcal{G}_0$ is a finitely generated regular C^{1+1} contracting semigroup defined on $\overline{\mathbf{D}}_{r_0}$. For any $z \in \overline{\mathbf{D}}_{r_0}$, let $D(z,r) = \{w \in \mathbf{C} \mid |w - z| \leq r\}$ be the closed disk of radius r centered at z. From Theorem 2.8, there are constants $C = C(r_0) > 0$ and $\delta = \delta(r_0) > 0$ such that

$$\begin{aligned} g(z) + C^{-1} \cdot g'(z) \cdot D(0,r) &\subseteq g(D(z,r)) \\ &\subseteq g(z) + C \cdot g'(z) \cdot D(0,r) \end{aligned} \tag{2.27}$$

for any $z \in \overline{\mathbf{D}}_{r_0}$, any $0 < r < \delta$, and any $g \in \mathcal{G}_0$. Formula (2.27) is the geometric form of Koebe's distortion theorem (Theorem 2.5).

To obtain a non-conformal example of a finitely generated regular $C^{1+\alpha}$ contracting semigroup for any $0 < \alpha \leq 1$, we consider a C^1 perturbation of the conformal semigroup $\mathcal{G}_0$ in the $C^{1+\alpha}$ function space as follows: let $\tilde{g}_i$ be a $C^{1+\alpha}$ diffeomorphism defined on $\mathbf{D}$ and let

$$||\tilde{g}_i - g_i||_1 = \max_{z \in \overline{\mathbf{D}}_{r_0}} ||T_z \tilde{g}_i - T_z g_i||$$

where $T_z \tilde{g}_i$ and $T_z g_i$ $(= g_i'(z))$ are the derivatives of $\tilde{g}_i$ and g_i at z. There is a number $\tau = \tau(\alpha, \mathcal{G}_0) > 0$ such that if $||\tilde{g}_i - g_i||_1 \leq \tau$ for all $0 \leq i \leq n-1$, then $\tilde{\mathcal{G}}_0 = \langle \tilde{g}_0, \tilde{g}_1, \ldots, \tilde{g}_{n-1} \rangle$ is a finitely generated regular $C^{1+\alpha}$ contracting semigroup. The study of geometric and thermodynamical properties as well as rigidity property of $\tilde{\mathcal{G}}_0$ is an interesting research problem to me. We discuss some examples in the next section.

2.8. A Regular and Markov $C^{1+\alpha}$ Contracting Semigroup

Let $P_0(z) = z^2$ and let $\mathbf{S}^1 = \{z \in \mathbf{C} \mid |z| = 1\}$ be the unit circle. It is easy to see that $P_0(\mathbf{S}^1) = \mathbf{S}^1$. Suppose $U = \{z \in \mathbf{C} \mid 3/4 < |z| < 4/3\}$ is a neighborhood of $\mathbf{S}^1$. Then the maximal invariant set of P_0 in U is exactly $\mathbf{S}^1$. Now we consider a perturbation of P_0,

$$f_\epsilon(z) = z^{\frac{p+2}{2}}(\overline{z})^{\frac{p-2}{2}} + b\overline{z} + c,$$

where $p > 1$ is a real number, where b and c are complex numbers, and where $\epsilon = (p-2, b, c)$. Suppose

$$J_\epsilon = \{z \in U \mid f_\epsilon^{\circ n}(z) \in U \text{ for all } 0 \leq n < \infty\}$$

is the maximal invariant set of f_ϵ in U. From the structural stability theorem (see [SHU] and [PRZ]) in smooth dynamical systems, for every $p > 1$, there exists a small number $\delta > 0$ such that, for $|b| < \delta$ and for $|c| < \delta$, there is a homeomorphism h_ϵ from U onto its image such that $h_\epsilon(\mathbf{S}^1) = J_\epsilon$ and

$$f_\epsilon = h_\epsilon \circ P_0 \circ h_\epsilon^{-1}$$

on U. Therefore, J_ϵ is topologically a Jordan curve (see Fig. 2.11 and Fig. 2.12). We know that $\mathbf{S}^1$ is a round circle with dimension one. Now questions are when is J_ϵ a fractal set ? what is the dimension of J_ϵ ? and how can we calculate it ? These questions are the main motivations for us working out Theorem 2.8 in [JI8].

Suppose $\mathcal{G} = \langle g_0, g_1, \ldots, g_{n-1}\rangle$ is a finitely generated $C^{1+\alpha}$ contracting semigroup. Let Λ be the limit set of $\mathcal{G}$. It is said to be Markov for a real number $\delta_0 > 0$ if there are connected and simple connected, pairwise disjoint, open sets $\Omega_0, \Omega_1, \ldots, \Omega_{q-1}$ such that

a. $\max_{0 \leq j \leq q-1} \mathrm{diam}(\Omega_j) \leq \delta_0$,

b. $\cup_{j=0}^{q-1} \overline{\Omega_j} \supset \Lambda$, and

c. $f_i(\overline{\Omega_j \cap \Lambda}) = \left(\cup_{t=1}^{k_j} \overline{\Omega_{i_t}}\right) \cap \Lambda$ for every $0 \leq j < q$ and every $\Omega_j \subset V_i$, where $f_i = g_i^{-1}$.

Suppose $\mathcal{G} = \langle g_0, \ldots, g_{n-1}\rangle$ is a finitely generated regular and Markov $C^{1+\alpha}$ contracting semigroup. Without loss of generality, we assume $q = n$ and $g_i = (f_i|\Omega_i)^{-1}$ for $0 \leq i \leq n-1$. Let $A = (a_{ij})$ be the $n \times n$ matrix of 0's and 1's such that $a_{ij} = 1$ if $f_i(\Omega_i \cap \Lambda) \supset \Omega_j \cap \Lambda$ and $a_{ij} = 0$ otherwise. A sequence $w_k = i_0 i_1 \ldots i_{k-1}$ of symbols $\{0, 1, \ldots, n-1\}$ is said to be admissible if $a_{i_j\, i_{j+1}} = 1$ for $j = 0, 1, \ldots, k-1$ (k may be ∞). Let Σ_k be the space of all admissible sequences w_k of length k, let $\sigma(i_0 i_1 \ldots) = i_1 \ldots$ be the shift map on Σ_∞, and let $\pi(i_0 i_1 \ldots) = \cap_{k=0}^{\infty} g_{i_k}(\overline{W})$ be the projection from Σ_∞ to Λ (see §1.2). (We note that if we give Σ_∞ the product topology, then $\pi : \Sigma_\infty \to \Lambda$ is continuous and onto, but may not be one-to-one, and $\pi \circ \sigma(i_0 i_1 \ldots) = f_{i_0} \circ \pi(i_0 i_1 \ldots)$. The function π is called a semi-conjugacy.) We call functions

$$\phi_{up}(w) = \log\big(l_i \circ \pi(w)\big) \qquad \text{and} \qquad \phi_{lo}(w) = \log\big(s_i \circ \pi(w)\big),$$

for $w = i i_1 \ldots \in \Sigma_\infty$, the upper and lower potential functions of $\mathcal{G}$ where $l_i(z) = |(g_i)_z| + |(g_i)_{\overline{z}}|$ and $s_i(z) = \big||(g_i)_z| - |(g_i)_{\overline{z}}|\big|$. They are Hölder continuous (refer to §1.3).

Let $P(\phi)$ be the pressure function defined on C^H, the space of Hölder continuous functions on Σ_∞. Then $P(\phi)$ can be defined (refer to Chapter Six) as

$$P(\phi) = \lim_{k \to \infty} \frac{1}{k} \log\Big(\sum_{w \in Fix(\sigma^{\circ k})} \exp\Big(\sum_{j=0}^{k-1} \phi(\sigma^{\circ j}(w))\Big)\Big)$$

where $Fix(\sigma^{\circ k})$ is the set of all fixed points of $\sigma^{\circ k}$. For $\phi = \phi_{up}$ or ϕ_{lo}, $P(t\phi)$ is a continuous, strictly monotone, and convex function on the real line $\mathbf{R}$ and tends to $-\infty$ and $+\infty$ as t goes to $+\infty$ and $-\infty$. (In general, $P(\phi)$ is also a continuous function of ϕ if we consider C^H as a topological space.) There is a unique $t_{up} > 0$ (respectively, $t_{lo} > 0$) such that $P(t_{up}\phi_{up}) = 0$ (respectively, $P(t_{lo}\phi_{lo}) = 0$). Let $HD(\Lambda)$ be the (Hausdorff) dimension of the limit set Λ of $\mathcal{G}$. Applying Theorem 2.8 and Gibbs theory, we have

Theorem 2.9 [JI8]. *Suppose $\mathcal{G} = \langle g_0, \ldots, g_{n-1}\rangle$ is a finitely generated regular and Markov $C^{1+\alpha}$ contracting semigroup. Then $t_{lo} \le HD(\Lambda) \le t_{up}$.*

Suppose $\mathcal{G}_\epsilon = \langle g_{0,\epsilon}, \ldots, g_{n-1,\epsilon}\rangle$ is a family of finitely generated regular and Markov $C^{1+\alpha}$ contracting semigroups such that every $g_{i,\epsilon}(z)$ is C^1 on both variables ϵ and z. Let $HD(\epsilon) = HD(\Lambda_\epsilon)$. From Theorem 2.9, we have

Corollary 2.1 [JI8]. *If all g_{i,ϵ_0} are conformal (that is, $K_{\epsilon_0} = 1$), then $HD(\epsilon)$ is continuous at ϵ_0.*

For a quadratic polynomial $P_c(z) = z^2 + c$ with small $|c|$, the maximal invariant set J_c of P_c in $U = \{z \in \mathbf{C} \mid 3/4 < |z| < 4/3\}$ is also called a Julia set. We discuss Julia sets in more details in Chapter Five. The next theorem is proved by Ruelle [RU4]. He proved the theorem by applying Theorem 2.5 and Gibbs Theory, and obtained the formula in the theorem by solving the equation $P(t\phi_c) = 0$ for $\phi_c = \log|P_c'|$.

Theorem 2.10 [RU4]. *Suppose $P_c(z) = z^2 + c$ is a quadratic polynomial. Then there is a small number $c_0 > 0$ such that for $|c| < c_0$, the Hausdorff dimension $HD(J_c)$ of the Julia set J_c is an analytic function of c. In fact, for $|c| < c_0$.*

$$HD(J_c) = 1 + \frac{|c|^2}{4 \log 2} + 0(|c|^2).$$

Suppose $f_\epsilon(z) = z^2 + b\overline{z} + c$ (or $f_\epsilon(z) = z^{\frac{p+2}{2}}(\overline{z})^{\frac{p-2}{2}} + c$), where b and c are complex with small $|b|$ and $|c|$ (or $p > 1$ is real, c is complex with small $|c|$) and where $\epsilon = (b, c)$ (or $\epsilon = (p-2, c)$). Let J_ϵ be the maximal invariant set of f_ϵ in $U = \{z \in \mathbf{C} \mid 3/4 < |z| < 4/3\}$ (see Fig. 2.11 and Fig. 2.12). Let V_0, V_1, V_2, and V_3 be four simply connected domains such that $J_\epsilon \subset \cup_{i=0}^{3} V_i \subseteq U$ and $f_\epsilon|V_i$ is injective for $i = 0$, 1, 2, and 3. Let $g_{i,\epsilon} = (f_\epsilon|V_i)^{-1} : W_i \to V_i$ be the inverse for every $0 \le i \le 3$. Let $\mathcal{G}_\epsilon = \langle g_{0,\epsilon}, g_{1,\epsilon}, g_{2,\epsilon}, g_{3,\epsilon}\rangle$. Then $\mathcal{G}_\epsilon$ is a family of finitely generated $C^{1+\alpha}$ contracting semigroups such that every $g_{i,\epsilon}(z)$ is C^1 on both variables ϵ and z. The limit set of $\mathcal{G}_\epsilon$ is the maximal invariant set J_ϵ of f_ϵ. One can check that $\mathcal{G}_\epsilon$ is Markov.

For $\epsilon_0 = \epsilon_0(c) = (0, c)$, g_{i,ϵ_0} are conformal for all $0 \le i \le 3$, i.e., $K_{\epsilon_0} = 1$. For any small perturbations $g_{i,\epsilon}$ of g_{i,ϵ_0} for $0 \le i \le 3$, $\mathcal{G}_\epsilon$ is regular. Thus, following Corollary 2.1 and Theorem 2.10, we have that

Corollary 2.2 [JI8]. *There exists a constant $a_0 > 0$ such that for each c with $0 < |c| \le a_0$, there is a $\tau(c) > 0$ such that for every b with $|b| \le \tau(c)$ (or $|p-2| \le \tau(c)$),*

$$HD(J_\epsilon) > 1.$$

Remark 2.20. For $f_\epsilon(z) = z^{\frac{p+2}{2}}(\overline{z})^{\frac{p-2}{2}} + c$, Biefeleld, Sutherland, Tangerman and Veerman [BSTV] showed that there exists a constant $a_0 > 0$ such that for every $0 < p - 2 \le a_0$, there is an $\eta(p) > 0$ such that J_ϵ is a smooth curve for $|c| < \eta(p)$ (see the first picture in Fig. 2.12). Many computer pictures and proposed research problems in this direction can be found in [BSTV].

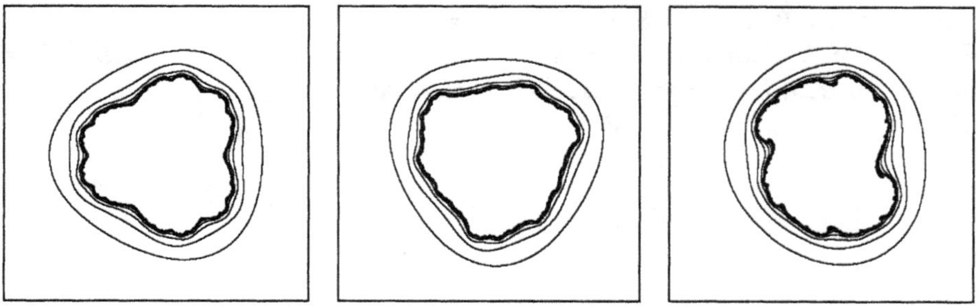

$\epsilon = (0.2, 0)$ $\epsilon = (-0.2i, 0)$ $\epsilon = (0.1, 0.2 + 0.1i)$

Preimages of a circle with large radius under iterates of
$f_\epsilon(z) = z^2 + b\overline{z} + c$ where $\epsilon = (b, c)$.

Fig. 2.11

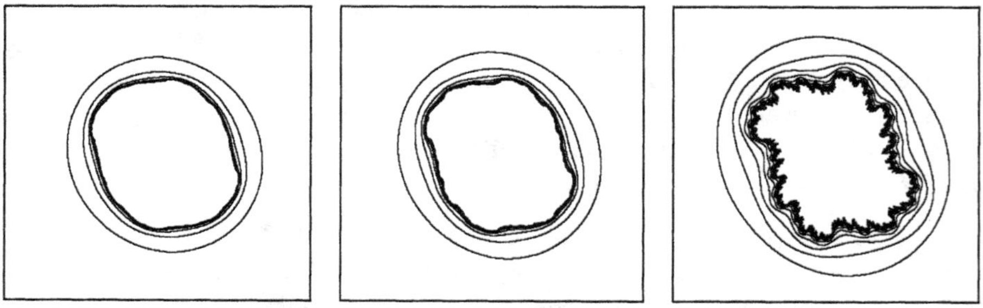

$\epsilon = (0.5, 0.1 + 0.2i)$ $\epsilon = (0.1, 0.1 + 0.2i)$ $\epsilon = (-0.5, 0.1 + 0.2i)$

Preimages of a circle with large radius under iterates of
$f_\epsilon(z) = z^{\frac{p+2}{2}} |z|^{\frac{p-2}{2}} + c$ where $\epsilon = (p - 2, c)$.

Fig. 2.12

Chapter Three

The Geometry of One-Dimensional Maps

This chapter expands the author's thesis [JI3]. One well-recognized program is to "fill in" the dictionary between the theory of real and complex one-dimensional dynamical systems and the theory of Kleinian groups (see the papers of, among others, Bowen [BO2], Shub and Sullivan [SHS], Mostow [MOS], Tukia [TUK], and Thurston [THU], McMullen [MC1,MC2], Gardiner and Sullivan [GS1,GS2]). Seeking to advance this program, we define in §3.3, for one real dimension, a geometrically finite map, which is a certain real one-dimensional $C^{1+\alpha}$ map with finitely many critical points.

The $C^{1+\alpha}$-Denjoy-Koebe distortion lemma for critical geometrically finite one-dimensional maps is proved in §3.4. We discuss, in §3.5, bounded geometry, and bounded nearby geometry, for a geometrically finite one-dimensional map. This property enables us to examining further the quasisymmetric property of a conjugacy. We define scaling functions for Markov one-dimensional maps in §3.2. In §3.6 and §3.7, we prove that the scaling function of a non-critical geometrically finite one-dimensional map exists and is Hölder continuous, while the scaling function of a critical such map exists but is discontinuous. In §3.8, we use the scaling functions together with exponents and asymmetries to smoothly classify geometrically finite one-dimensional maps within a topological conjugacy class. In this section, we first prove a rigidity result about geometrically finite one-dimensional maps which says that if the conjugacy between two geometrically finite one-dimensional maps is differentiable at one point with non-zero derivative, then the conjugacy must be smooth. In §3.9, we study Ulam-von Neumann transformations, which is a special conjugacy class within geometrically finite one-dimensional maps. Introduced next is an interesting technique connected with the orbifold metric of a Ulam-von Neumann transformation. In §3.10, we use the scaling functions of Ulam-von Neumann transformations to study the asymptotic behavior of scaling functions of hyperbolic Cantor sets.

In §3.11, the geometric factor μ of a C^{1+bv} irrational circle mapping f is defined via renormalization. It is a smooth invariant and takes values in the open interval $(0,1)$. In the same section, the combinatorics of f, which is a topological invariant, is defined as a sequence $(k_1, k_2, \dots,)$ of integers related to the sequence of fast renormalizations for f. The compatibility of the geometry and the combinatorics of f is also defined in this section. Herman's theorem (see the papers of Herman [HER], Yoccoz [YO3], Khanin and Sinai [KHS], Stark [STA], Katznelson and Ornstein [KAO], and Veerman and Tangerman [VET]) says that every $C^{2+\alpha}$ $(0 < \alpha \leq 1)$ irrational circle diffeomorphism f with rotation number ρ is $C^{1+\beta}$ $(0 < \beta \leq 1)$ conjugate to the rigid rotation R_ρ with rotation number ρ, provided that the rotation number ρ satisfies a certain number-theoretic property. In §3.12, we prove that Herman's theorem holds for a smooth irrational circle diffeomorphism whose geometry and combinatorics are compatible.

3.1. Markov One-Dimensional Maps and Symbolic Dynamical Systems

Let M be an oriented compact one-dimensional C^2 Riemannian manifold and let $f : M \to M$ be a continuous self-map.

Definition 3.1. *The map f is said to be Markov if there is a set $\eta_1 = \{I_1, \ldots, I_k\}$ of closed intervals of M such that*

a. *$I_1, \ldots, I_k$ have pairwise disjoint interiors,*

b. *the union $\cup_{i=1}^{k} I_i$ of all intervals in η_1 is M,*

c. *the restriction $f|I$ to every interval I in η_1 is injective and continuous, and*

d. *the image $f(I)$ of every interval I in η_1 under f is the union of some intervals in η_1.*

Remark 3.1. A set of closed intervals satisfying **a**, **b**, **c**, and **d** is called a Markov partition of M with respect to (w.r.t.) f. For a Markov map f, there are many Markov partitions of M w.r.t. f. But there is a natural one for a map which is discussed from §3.3 to §3.10.

Let f be a Markov map and $\eta_1 = \{I_1, \ldots, I_k\}$ be a fixed Markov partition of M w.r.t. f. Let g_i be the inverse of the restriction $f|I_i$ for $i = 1, \ldots, k$. A sequence $w_n = i_0 \ldots i_{n-1}$ of $1's, \ldots, k's$ is admissible if $I_{i_m} \subset f(I_{i_{m-1}})$ for every $m = 1, \ldots, n-1$. An infinite admissible sequence $w_n = i_0 i_1 \cdots$ is defined similarly. For an admissible sequence $w_n = i_0 \ldots i_{n-1}$, define $g_{w_n} = g_{i_0} \circ \cdots \circ g_{i_{n-1}}$ and define $I_{w_n} = g_{w_n}(f(I_{i_{n-1}}))$. We use η_n to denote the set of intervals I_{w_n} for all admissible sequences w_n of length n. We call η_n the nth-partition of M. (It is also a Markov partition.) Let λ_n denote the maximum of the lengths of intervals in η_n. We always assume $(*)$ λ_n tends to zero as n goes to infinity. Let $\eta = \{\eta_n\}_{n=1}^{\infty}$ be the sequence of nested partitions induced from (f, η_1). Then (Σ, σ) is the phase space of f with η_1 where $\Sigma = \{a = i_0 i_1 \ldots\}$ is the set of all infinite admissible sequences with the product topology and $\sigma(i_0 i_1 \ldots) = i_1 \ldots$ is the shift map of Σ.

Proposition 3.1. *Let f be a Markov map and let η_1 be a fixed Markov partition. Then there is a continuous map h from Σ onto M such that $f \circ h = h \circ \sigma$ on Σ.*

Proof. For every $a = i_0 i_1 \ldots$ in Σ, let $w_n = i_0 i_1 \ldots i_{n-1}$. Every w_n is admissible and $I_{w_{n+1}} \subseteq I_{w_n}$ for $n > 0$. Hence $\cap_{n=0}^{\infty} I_{w_n}$ is a non-empty set.

But from our assumption that λ_n tends to zero as n goes to infinity, this set contains only one number x_a. Set $h(a) = x_a$. Then

$$f(h(a)) = f(x_a) = f(\cap_{n=0}^{\infty} I_{w_n}) = \cap_{n=0}^{\infty} f(I_{w_n}) = \cap_{n=0}^{\infty} I_{\sigma(w_n)} = h(\sigma(a)).$$

Now let us show that h is continuous. Two points a and b in Σ are n-close if the first n digits of them are the same. Suppose $a = i_0 \dots i_{n-1} i_n \dots$ and $b = i_0 \dots i_{n-1} i'_n \dots$ are n-close. Then $h(a)$ and $h(b)$ are in the same interval I_{w_n} where $w_n = i_0 \dots i_{n-1}$. This implies that $|h(a) - h(b)| \leq \lambda_n$. So $h(b)$ tends to $h(a)$ in M as b tends to a in Σ because λ_n tends to zero as n goes to infinity. This means that h is continuous at every point a in Σ.

The map h is onto because $\cup_{I \in \eta_n} I = M$ for every $n \geq 0$. Actually, h is one-to-one except for countably many points which are the preimages, under h, of endpoints of intervals of η_n, for $n > 0$. ■

The dynamical system σ of Σ is a topological model of a topological conjugacy class as follows:

Proposition 3.2. *Let f be a Markov map and let η_1 be a fixed Markov partition of M to f. Suppose (Σ, σ) is the phase space of f with η_1. A Markov map g is topologically conjugate to f if and only if there is a Markov partition η'_1 of M to g such that the phase space of g with η'_1 is (Σ, σ).*

Proof. Suppose g is topologically conjugate to f. There is a homeomorphism H of M such that $H \circ f = g \circ H$. Let $\eta'_1 = H(\eta_1)$. Then η'_1 is a Markov partition of M to g and the phase space of g with η'_1 is (Σ, σ).

Now suppose there is a Markov partition η'_1 of M to g such that the phase space of g with η'_1 is also (Σ, σ). From Proposition 3.1, there are two continuous maps h_1 and h_2 from Σ onto M such that $f \circ h_1 = h_1 \circ \sigma$ and $g \circ h_2 = h_2 \circ \sigma$. Let $H = h_2 \circ h_1^{-1}$. It is defined on M except for countably many points which are endpoints of intervals in the partitions $\{\eta_n\}$. The map H is also uniformly continuous. So it can be extended to a continuous map from M to M. Using the same argument, $H^{-1} = h_1 \circ h_2^{-1}$ can be also extended to a continuous map from M to M. Hence H is a homeomorphism of M. ■

Let f be a Markov map and let $\eta_1 = \{I_1, \dots, I_k\}$ be a fixed Markov partition of M w.r.t. f. Let $\eta = \{\eta_n\}_{n=1}^{\infty}$ be the sequence of nested partitions induced from (f, η_1).

Definition 3.2. *The sequence $\eta = \{\eta_n\}_{n=1}^{\infty}$ of nested partitions tends to zero exponentially if there are constants $C > 0$ and $0 < \mu < 1$ such that $\lambda_n \leq C\mu^n$ for all positive integers n.*

Definition 3.3. *The sequence $\eta = \{\eta_n\}_{n=1}^{\infty}$ of partitions is said to be of bounded geometry if there is a constant $C > 0$ such that the ratio*

$$\frac{|J|}{|I|} \geq C$$

for every integer $n > 0$ and every pair $J \subset I$ with $J \in \eta_{n+1}$ and $I \in \eta_n$. It is said to be of bounded nearby geometry if there is a constant $C > 0$ such that the ratio

$$\frac{|J_1|}{|J_2|} \geq C$$

for every $n > 0$ and every pair J_1 and J_2 in η_n with a common endpoint.

Remark 3.2. We use BG_f to denote the largest possible value of C in this definition.

3.2. Markov Maps, Dual Symbolic Spaces, and Scaling Functions

Suppose f is a Markov map and $\eta_1 = \{I_1, \ldots, I_k\}$ is a fixed Markov partition of M w.r.t. f. In this section, we consider another symbolic space induced from f with η_1.

Let Γ_n be the set of all admissible sequences w_n of length n. A (n,m)-right cylinder about $w_n^0 = i_{n-1}^0 \ldots i_0^0$ for $0 \le m \le n-1$ is

$$\{w_n = i_{n-1} \ldots i_0 \in \Gamma_n \mid i_l = i_l^0, l = 0, \ldots, m\}.$$

All the (n,m)-right cylinders form a topological basis of Γ_n. Let Γ_n^* be the set Γ_n with this topological basis and $\left(\Sigma^*, \sigma^*\right)$ be the inverse limit of the sequence $\{\left(\Gamma_n^*, I_n^*\right)\}_{n=1}^{\infty}$ where $I_n^* : \Gamma_n^* \to \Gamma_{n-1}^*$ is the inclusion and $\sigma^* : \Sigma^* \to \Sigma^*$ is the shift. We call $\Sigma^* = \{a^* = \ldots i_1 i_0\}$ the dual phase space of f with η_1. The scaling function of f with η_1 is a function defined on the dual phase space Σ^*. For $a^* = \ldots i_1 i_0$ and $w_n = i_{n-1} \ldots i_1 i_0$, we have $\sigma^*(a^*) = \ldots i_1$; we also denote $i_{n-1} \ldots i_1$ by $\sigma^*(w_n)$. Let $a^* = \ldots w_n$ in Σ^*. Define

$$s(w_n) = \frac{|I_{w_n}|}{|I_{\sigma^*(w_n)}|}.$$

Definition 3.4. *If* $\lim_{n \to +\infty} s(w_n)$ *exists for every* a^* *in* Σ^*, *then we define the scaling function*

$$s(a^*) = \lim_{n \to +\infty} s(w_n)$$

on Σ^*.

Let a^* be a periodic point of σ^* of period m. Then

$$a^* = w_m^{\infty} = \ldots w_m \ldots w_m$$

where $w_m = i_{m-1} \ldots i_0$ is an admissible sequence of length m. We may define $h_*(a^*) = \cap_{i=1}^{\infty} I_{w_m^i}$ since $I_{w_m^{i+1}} \subseteq I_{w_m^i}$; $h_*(a^*)$ is a periodic point of f of period m. The eigenvalue E_p of a periodic point p of a map f of period m is, by definition, $E_p = (f^{\circ m})'(p)$.

Proposition 3.3. *Let* f *be a Markov map and let* η_1 *be a fixed Markov partition of* M *w.r.t.* f. *Assume that the scaling function* s *on* Σ^* *of* f *with* η_1 *exists. Then for every periodic point* a^* *of* σ^* *of period* m,

$$\frac{1}{|E_p|} = \prod_{l=0}^{m-1} s((\sigma^*)^{\circ l}(a^*)),$$

where $p = h_*(a^*)$ is a periodic point of f of period m.

Proof. Suppose $a^* = w_m^\infty$. Then $f^{\circ m}(I_{w_m^{i+1}}) = I_{w_m^i}$. So

$$\left|(f^{\circ m})'(b_i)\right|^{-1} = \frac{|I_{w_m^{i+1}}|}{|I_{w_m^i}|} = \prod_{l=0}^{m-1} s((\sigma^*)^{\circ l}(w_m^{i+1}))$$

for some b_i in $I_{w_m^{i+1}}$. As i goes to infinity, b_i tends to p and $s((\sigma^*)^{\circ l}(w_m^{i+1}))$ tends to $s((\sigma^*)^{\circ l}(a^*))$. Hence

$$\left|(f^{\circ m})'(p)\right|^{-1} = \prod_{l=0}^{m-1} s((\sigma^*)^{\circ l}(a^*)).$$

∎

Proposition 3.4. *Let f be a Markov map and let η_1 be a fixed Markov partition of M w.r.t. f. The scaling function s, on Σ^*, of f with η_1 (if it exists) is a C^1-invariant.*

Proof. Let $g = h \circ f \circ h^{-1}$ where h is a C^1-diffeomorphism of M. Then g is a Markov map and $\eta_1' = h(\eta_1)$ is a Markov partition of M to g. Let s_f be the scaling function on Σ^* of f with η_1 and let s_g be the scaling function s on Σ^* of g with η_1'.

For every $a^* = \ldots w_n$, w_n is admissible and

$$s_g(w_n) = \frac{|h(I_{w_n})|}{|h(I_{\sigma^*(w_n)})|} = \frac{|h'(b_n)|}{|h'(b_n')|}\frac{|I_{w_n}|}{|I_{\sigma^*(w_n)}|} = \frac{|h'(b_n)|}{|h'(b_n')|} s_f(w_n)$$

where b_n and b_n' are in I_{w_n}. As n goes to infinity, $s_g(w_n) \to s_g(a^*)$, $s_f(w_n) \to s_f(a^*)$, and $|h'(b_n)|/|h'(b_n')| \to 1$. Hence $s_g(a^*) = s_f(a^*)$. So $s_f = s_g$ on Σ^*.

∎

3.3. Geometrically Finite One-Dimensional Maps

Let M be a one-dimensional compact C^2-Riemannian manifold. Suppose that $f : M \to M$ is continuous and piecewise C^1. A singular point a of f is either a non-differentiable point or a differentiable point with zero derivative. A singular point a of f is said to be power law if there is a $\gamma \geq 1$ such that

$$\lim_{x \to a+} \frac{f'(x)}{|x-a|^{\gamma-1}} \quad \text{and} \quad \lim_{x \to a-} \frac{f'(x)}{|x-a|^{\gamma-1}}$$

exist with non-zero limits A_+ and A_-. The numbers γ and $A = A_+/A_-$ are called the exponent and the asymmetry of f at a.

Remark 3.3. Both γ and A are orientation-preserving C^1-invariants, i.e., they are the same for f and $h \circ f \circ h^{-1}$ whenever h is an orientation-preserving C^1-diffeomorphism of M.

Henceforth, we assume that f has only power law singular points and, without loss of generality, that f maps the boundary of M (if it is not empty) into itself and that the one-sided derivatives of f at all boundary points of M are non-zero. We note that in the general case, a boundary point of M should count as a singular point anyway. We call a singular point with exponent $\gamma > 1$ a critical point. Remember that a singular point with exponent $\gamma = 1$ is a non-differentiable point of f and that a critical point is a differentiable point of f with zero derivative. Let

$$NP = \{a_1, \ldots, a_{d'}\}$$

be the set of non-differentiable points of f and let

$$CP = \{c_1, \ldots, c_d\}$$

be the set of critical points of f. Let

$$EX = \{\gamma_1, \ldots, \gamma_d\}$$

be the set of corresponding exponents at critical points of f. Then $SP = NP \cup CP$ is the set of singular points of f. Let $SO = \cup_{n=0}^{\infty} f^{\circ n}(SP)$ be the set of singular orbits of f. Let $PCO = \cup_{n=1}^{\infty} f^{\circ n}(CP)$ be the set of post-critical orbits of f.

Definition 3.5. *The map f is said to be C^{1+} if there is a number $0 < \alpha \leq 1$ such that*

i. *for every interval I in the complement of SP in M, $f|I$ is C^1 and $(f|I)'$ is α-Hölder continuous, and*

ii. *for every critical point c_i of f, there is an open neighborhood U_i of c_i such that*

$$\frac{f'(x)}{|x - c_i|^{\gamma_i - 1}}$$

is α-Hölder continuous when restricted to $\{x < c\} \cap U_i$ and to $\{x > c\} \cap U_i$.

If the set SO of singular orbits is non-empty and finite, then f is a Markov map and there is a natural Markov partition η_1 which consists of the closures of intervals in the complement of SO in M. Let η_n be the nth-partition of M induced from f with η_1 and let λ_n be the maximum of the lengths of intervals in η_n.

Definition 3.6. *A one-dimensional map $f : M \to M$ is said to be geometrically finite if*

(1) *f is C^{1+},*

(2) *the set of singular orbits SO is non-empty and finite,*

(3) *no critical point is periodic, and*

(4) *there are constants $C > 0$ and $0 < \mu < 1$ such that $\lambda_n \leq C\mu^n$ for all integers $n > 0$.*

Remark 3.4. A critical point c is called preperiodic if it is not periodic and there is an integer $n > 0$ such that $f^{\circ n}(c)$ is a periodic point of f. Conditions (**2**) and (**3**) imply that every critical point is preperiodic.

For a geometrically finite one-dimensional map f, we always take the natural Markov partition η_1 which is the set of the closures of intervals in the complement of SO in M. Let η_n be the nth-partition induced from f with η_1 and let Σ^* be the dual phase space of f with η_1. Within this context, we can talk about the dual phase space and the scaling function of f (if it exists). We can also fix an integer $n_0 > 0$ such that the closure $\overline{U}_i$ of every open interval U_i in **ii.** of Definition 3.5 is the union of two intervals in η_{n_0} and such that $U = \cup_{i=1}^{d} U_i$ is disjoint with $PCO \setminus CP$ (refer to §2.3 and Fig. 2.6). Let V be the closure of the complement of U in M (refer to §2.3). These notations are fixed for the rest of this chapter.

Definition 3.7. *A geometrically finite one-dimensional map f is said to be non-critical if it has no critical point; and f is said to be critical if it has critical points.*

Let us give an example of a geometrically finite one-dimensional map. A periodic point p of a map f is called expanding if the absolute value $|E_p|$ of the eigenvalue of f at p is bigger than one.

Example 3.1. *A C^3-map f having negative Schwarzian derivative and only preperiodic power law critical points so that no interval in η_1 is periodic and the boundary of M (if it exists) consists of expanding fixed points of f.*

Proof. To prove that f is geometrically finite, we only need to check **(4)** in Definition 3.6. We prove bounded geometry directly under the assumption that the boundary of M is empty and $PCO \cap CP = \emptyset$. If the boundary of M is not empty, the proof is similar (and may refer to the proof of Theorem 2.1). If $PCO \cap CP \neq \emptyset$, one can refer to the proof of Lemma 3.3′. In this example, f has only critical singular points. that is, $SP = CP$.

An interval I in η_n is called critical if one of its endpoint is a critical point of f. Since no interval in η_1 is periodic, there is an integer $n_1 \geq 1$ such that every critical interval in η_{n_1} is disjoint with the set of post-critical orbits PCO. Let U_1 be the union of all critical intervals in η_{n_1} and V_1 be the closure of $M \setminus U_1$. Suppose $W_1, \ldots, W_d$ are the connected components of V_1 and $\tilde{W}_1, \ldots, \tilde{W}_d$ are the connected components of $M \setminus CP$. Then $W_i \subset \tilde{W}_i$.

Let $C_0 = \min\{|J|/|I| \mid J \subseteq I \text{ with } J \in \eta_{n_1+1}, I \in \eta_{n_1}\}$. Now for any $m > n_1$ and any $J \subset I$ with $J \in \eta_{m+1}$ and $I \in \eta_m$, let $n = m - n_1$, $J_i = f^{\circ(n-i)}(J)$ and $I_i = f^{\circ(n-i)}(I)$ for $0 \leq i \leq n$, we discuss it in two cases: (1) I_i is in V_1 for every $0 < i \leq n$ and (2) there is a $0 < k \leq n$ such that I_k is in U_1 and I_i is in V_1 for every $0 < i < k$.

Let us use C to denote a constant (even though it may be different in different formulas). Applying Lemma 2.6 (refer to Remark 2.4), there is a constant $\lambda > 1$ such that

$$P_{f(\tilde{W}_j)}(f(J)) \geq \lambda P_{\tilde{W}_j}(J)$$

for any $1 \leq j \leq d$ and J in W_j. Using Möbius transformations to normalize $f(\tilde{W}_j)$ and $\tilde{W}_j$ to the interval $[0,1]$ and using Lemma 2.6, we can prove that there are constants $C > 0$ and $0 < \mu < 1$ such that in case (1), $|I_i| \leq C\mu^i$. Applying Lemma 1.2 in case (1), there is a constant $C > 0$ such that the distortion

$$\left|\log\left(\frac{|(f^{\circ n})'(\xi_1)|}{|(f^{\circ n})'(\xi_2)|}\right)\right| \leq C$$

for any ξ_1 and ξ_2 in I. Moreover,

$$\frac{|J|}{|I|} \geq e^{-C} C_0.$$

This also proved that in case (2),

$$\frac{|J_{k-1}|}{|I_{k-1}|} \geq C.$$

In case (2), because $f|I_k$ is comparable to the power function

$$p(x) = |x - c_{i_k}|^{\gamma_{i_k}} + f(c_{i_k}),$$

there is a constant $C > 0$, such that $|J_k|/|I_k| \geq C$. The inverse g of $f^{\circ(n-k)}|I$ is a diffeomorphism from I_k to $I_n = I$ having positive Schwarzian derivative. Applying Lemma 2.4,

$$\left|N(g)(x)\right| \leq \frac{2}{d(U_1, PCO)}$$

for x in I_k. Applying Lemma 2.1, there is a constant $C > 0$ such that

$$\frac{|J|}{|I|} \geq C\frac{|J_k|}{|I_k|} \geq C.$$

Hence there is a positive constant $C > 0$ such that

$$\frac{|J|}{|I|} \geq C$$

in both of the cases (1) and (2). This proved that $\eta = \{\eta_n\}_{n=1}^{\infty}$ is of bounded geometry.

Since no interval in η_{n_1} is periodic, there is an integer $n_2 > n_1$ such that every I in η_1 contains at least two intervals in η_{n_2}. Therefore, we can find a constant $0 < \mu < 1$ such that

$$|J| \leq \mu|I|$$

for $J \subset I$ with $J \in \eta_{n+n_2-n_1}$ and $I \in \eta_n$. This implies that η_n tends to zero exponentially as n goes to infinity. ∎

An example of a geometrically finite one-dimensional map about a C^{1+1} map $f : M \to M$ having only preperiodic power law critical points and only expanding periodic points can be found in [JI3]. For any $0 \leq \alpha < 1$, there is an example of a $C^{1+\alpha}$ map $f : M \to M$ having only preperiodic power law critical points and only expanding periodic points but which is not geometrically finite (see [JI3]).

3.4. The Distortion of a Critical Geometrically Finite One-Dimensional Map

The distortion of a non-critical geometrically finite one-dimensional map can be easily controlled by using Lemma 1.2.

Lemma 3.1. *Suppose f is a non-critical geometrically finite one-dimensional map. Then there are constants $C > 0$ and $0 < \mu < 1$ such that for any integers n, $m > 0$, any interval I in η_{n+m}, and any x, $y \in I$*

$$\left|\log\left(\frac{|(f^{\circ m})'(x)|}{|(f^{\circ m})'(y)|}\right)\right| \leq C\mu^n.$$

Proof. For any x and y in I, let $x_i = f^{\circ i}(x)$ and $y_i = f^{\circ i}(y)$. From **(4)** of Definition 3.6, there are constants $C_1 > 0$ and $0 < \mu_1 < 1$ such that

$$|x_i - y_i| \leq C_1 \mu_1^{n+m-i}$$

for $0 \leq i < m$. Let $a_0 = \min_{x \in J \in \eta_1} |f'(x)|$ and let

$$b_0 = \sup_{J \in \eta_1, x \neq y \in J} \frac{|f'(x) - f'(y)|}{|x-y|^\alpha}.$$

Since f is non-critical and $C^{1+\alpha}$, we have $0 < a_0, b_0 < \infty$. Applying Lemma 1.2 and the chain rule, there are constants $C > 0$ and $0 < \mu < 1$ such that

$$\left|\log\left(\frac{|(f^{\circ m})'(x)|}{|(f^{\circ m})'(y)|}\right)\right| \leq C\mu^n.$$

∎

Since $\min_{I \in \eta_1, x \in I} |f'(x)| = 0$ for a critical geometrically finite one-dimensional map f, we can not using Lemma 1.2 directly to control the distortion of this map. To overcome this difficulty, we prove two distortion lemmas in this section.

Let f be a critical geometrically finite one-dimensional map. The integer n_0 and the sets U and V are the same as in §3.3. For a point x in M, let $x_i = f^{\circ i}(x)$.

Lemma 3.2. *There are constants $C > 0$ and $0 < \mu < 1$ such that for any integers $n \geq n_0$ and $m > 0$, and for any points x and y in an interval of η_{n+m}, if x_i and y_i are in V for all $0 \leq i < m$, then*

$$\left|\log\left(\frac{|(f^{\circ m})'(x)|}{|(f^{\circ m})'(y)|}\right)\right| \leq C\mu^n.$$

Proof. Suppose f is $C^{1+\alpha}$ for some $0 < \alpha \leq 1$. Let $a_0 = \min_{x\in V} |f'(x)|$ and

$$b_0 = \sup_{x\neq y\in V\cap I, I\in\eta_1} \frac{|f'(x) - f'(y)|}{|x-y|^\alpha}.$$

Then $0 < a_0, b_0 < \infty$. From Lemma 1.2,

$$\left|\log\left(\frac{|(f^{\circ m})'(x)|}{|(f^{\circ m})'(y)|}\right)\right| \leq \frac{b_0}{a_0}\sum_{i=0}^{m-1} |x_i - y_i|^\alpha.$$

Since x and y are in an interval in η_{n+m}, x_i and y_i are in an interval in η_{n+m-i}. Hence $|x_i - y_i| \leq C\mu^{n+m-i}$ because of **(4)** of Definition 3.6. Hence there are constants $C > 0$ and $0 < \mu < 1$ such that

$$\left|\log\left(\frac{|(f^{\circ m})'(x)|}{|(f^{\circ m})'(y)|}\right)\right| \leq C\mu^n.$$

∎

The next lemma is one of the key lemmas. To present a clear idea of the proof, we first prove the following lemma under the assumption that $PCO \cap CP = \emptyset$. For a point x in M, let $x_i = f^{\circ i}(x)$.

Lemma 3.3 ($C^{1+\alpha}$-Denjoy-Koebe distortion lemma [JI4]). *There are constants $C > 0$ and $0 < \mu < 1$ such that for any integers $n \geq n_0$ and $m > 0$ and any points x and y in an interval of η_{n+m}, if x_m and y_m are in U, then*

$$\left|\log\left(\frac{|(f^{\circ m})'(x)|}{|(f^{\circ m})'(y)|}\right)\right| \leq C\mu^n.$$

Proof. The ratio $|(f^{\circ m})'(x)|/|(f^{\circ m})'(y)|$ equals the product

$$\prod_{i=0}^{m-1} \frac{|f'(x_i)|}{|f'(y_i)|}.$$

We divide this product into two sub-products,

$$\prod_{x_i,y_i\in U}\frac{|f'(x_i)|}{|f'(y_i)|} \quad \text{and} \quad \prod_{x_i,y_i\in V}\frac{|f'(x_i)|}{|f'(y_i)|}.$$

Following the proof of Lemma 3.2, there are constants C_1, $C_1' > 0$ and $0 < \mu_1 < 1$ such that

$$\Big|\log\Big(\prod_{x_i,y_i\in V}\frac{|f'(x_i)|}{|f'(y_i)|}\Big)\Big| \leq C_1 \sum_{x_i,y_i\in V}|x_i-y_i|^{\alpha} \leq C_1'\mu_1^n.$$

To estimate the first product $\prod_{x_i,y_i\in U}|f'(x_i)|/|f'(y_i)|$, we write it as the product of three factors:

$$\mathcal{I} = \prod_{x_i,y_i\in U}\Big(\frac{|x_i-c_{k_i}|^{\gamma_{k_i}}}{|f(x_i)-f(c_{k_i})|}\frac{|f(y_i)-f(c_{k_i})|}{|y_i-c_{k_i}|^{\gamma_{k_i}}}\Big)^{t_{k_i}},$$

$$\mathcal{II} = \prod_{x_i,y_i\in U}\frac{|y_i-c_{k_i}|^{\gamma_{k_i}-1}}{|f'(y_i)|}\frac{|f'(x_i)|}{|x_i-c_{k_i}|^{\gamma_{k_i}-1}},$$

and

$$\mathcal{III} = \prod_{x_i,y_i\in U}\Big(\frac{|f(x_i)-f(c_{k_i})|}{|f(y_i)-f(c_{k_i})|}\Big)^{t_{k_i}},$$

where x_i and y_i are in U_{k_i} and $t_{k_i} = (\gamma_{k_i}-1)/\gamma_{k_i}$. Applying Lemma 1.2 and **(1)** of Definition 3.6, and following the proof of Lemma 3.2, there are constants $C_2, C_2' > 0$ and $0 < \mu_2 < 1$ such that

$$\Big|\log\mathcal{I}\Big|, \Big|\log\mathcal{II}\Big| \leq C_2 \sum_{x_i,y_i\in U}|x_i-y_i|^{\alpha} \leq C_2'\mu_2^n.$$

Now the proof of Lemma 3.3 concentrates on the estimate of $\mathcal{III}$. Let

$$\frac{f(x_i)-f(c_{k_i})}{f(y_i)-f(c_{k_i})} = 1 + \frac{f(x_i)-f(y_i)}{f(y_i)-f(c_{k_i})}.$$

Then

$$\mathcal{III} = \exp\Big(\sum_{s=1}^{r-1}\frac{1}{t_{k_{i_s}}}\log\Big|1+\frac{f(x_{i_s})-f(y_{i_s})}{f(y_{i_s})-f(c_{k_{i_s}})}\Big|\Big)$$

where $i_1 < i_2 < \cdots < i_{r-1} < m$. Let $i_r = m$. For each i_s, $1 \leq s < r$, consider the interval L_s bounded by y_{i_s} and $c_{k_{i_s}}$ and the map $h_s = f^{\circ(i_{s+1}-i_s)}$. Let

$R_s \subseteq L_s$ be the maximal interval containing y_{i_s} such that h_s on R_s is injective. One of the endpoints of R_s is y_{i_s} and the other is a preimage e of a singular point c_{j_s} under $f^{\circ k_s}$ for some $0 \leq k_s < i_{s+1} - i_s$. Let us assume that c_{j_s} is a critical point. (If it is not a critical point, we can further reduce to this situation because $c_{k_{i_s}}$ is a critical point.) Let $l_s = i_{s+1} - i_s - k_s$. Then h_s on the minimal interval J_s containing x_{i_s} and R_s is injective and maps J_s onto an interval containing the points y_{i_s+1}, x_{i_s+1}, and $f^{\circ l_s}(c_{j_s})$. We enlarge every interval J of V into a closed interval $J' \supset J$ such that $J' \cap CP = \emptyset$ and such that the length of $J' \cap U$ is greater than a constant $a > 0$. Let $V' = \cup_{J \in V} J'$ be the union of all these enlarged intervals and let $U' = M \setminus V'$. If $f^{\circ i}(J_s) \subseteq V'$ for all $1 \leq i < i_{s+1} - i_s$, by following the proof of Lemma 3.2, there is a constant $C_3 > 0$ such that

$$\frac{|f(x_{i_s}) - f(y_{i_s})|}{|f(y_{i_s}) - f(c_{k_{i_s}})|} \leq C_3 \frac{|x_{i_s+1} - y_{i_s+1}|}{|y_{i_s+1} - f^{\circ l_s}(c_{j_s})|}.$$

Since y_{i_s+1} is in U and $f^{\circ l_s}(c_{j_s})$ is in PCO,

$$\frac{|f(x_{i_s}) - f(y_{i_s})|}{|f(y_{i_s}) - f(c_{k_{i_s}})|} \leq C_3 \frac{|x_{i_s+1} - y_{i_s+1}|}{D}$$

where $D > 0$ is the distance between U and the post-critical orbits PCO. Otherwise, let $0 < k < i_{s+1} - i_s$ be the smallest integer such that $f^{\circ k}(J_s) \cap U' \neq \emptyset$. Since $f^{\circ i}(J_s) \subseteq V'$ for all $1 \leq i < k$, following the proof of Lemma 3.2, there is a constant $C_4 > 0$ such that

$$\frac{|f(x_{i_s}) - f(y_{i_s})|}{|f(y_{i_s}) - f(c_{k_{i_s}})|} \leq C_4 \frac{|x_{i_s+k} - y_{i_s+k}|}{|y_{i_s+k} - f^{\circ k}(e)|}.$$

Since y_{i_s+k} is in V and $f^{\circ k}(e)$ is in U',

$$\frac{|f(x_{i_s}) - f(y_{i_s})|}{|f(y_{i_s}) - f(c_{k_{i_s}})|} \leq C_4 \frac{|x_{i_s+k} - y_{i_s+k}|}{D'}$$

where $D' > 0$ is the distance between V and U'. Hence, there are constants $C_5 > 0$, $C_5' > 0$ and $0 < \mu_4 < 1$ such that

$$\left|\log III\right| \leq C_5 \sum_{i=0}^{m-1} |x_i - y_i| \leq C_5' \mu_4^n.$$

Combining all the estimates, we have constants $C > 0$ and $0 < \mu < 1$ satisfying the lemma. ∎

If $PCO \cap CP \neq \emptyset$, we need to consider a critical chain: a subset $\{c_{i_1}, \ldots, c_{i_l}\}$ of CP is a critical chain if for every $1 \leq k < l$, there is an integer $j_k > 0$ such that $f^{\circ j_k}(c_{i_k}) = c_{i_{k+1}}$ and $\{f^{\circ j}(c_{i_k})\}_{j=1}^{j_k-1} \cap CP = \emptyset$.

Let x and y be two points in an interval in η_{n+m} where $n \geq n_0$ and $m > 0$. Let $x_i = f^{\circ i}(x)$ and $y_i = f^{\circ i}(y)$ for $0 \leq i \leq m$. Suppose x_m and y_m are in U. Using the same notation as in the proof of Lemma 3.3, we consider

$$\mathcal{III} = \exp\Big(\sum_{s=1}^{r-1} \frac{1}{t_{k_{i_s}}} \log\Big|1 + \frac{f(x_{i_s}) - f(y_{i_s})}{f(y_{i_s}) - f(c_{k_{i_s}})}\Big|\Big)$$

where $i_1 < i_2 < \cdots < i_{r-1} < m$ and x_{i_s} and y_{i_s} are in $U_{k_{i_s}}$. (Note that $c_{k_{i_s}} \in U_{k_{i_s}}$). Let $i_r = m$ and denote $c(s) = c_{k_{i_s}}$ for $1 \leq s \leq r$. We divide $\{c(s)\}_{s=1}^{m}$ into maximal critical chains $\mathcal{I}_1 = \{c(1), \ldots, c(s_1)\}$, $\mathcal{I}_2 = \{c(s_1+1), \ldots, c(s_2)\}$, ..., $\mathcal{I}_q = \{c(s_{q-1}+1), \ldots, c(s_q)\}$. Remember that $c(s_q) = c_{k_m}$. The number of points in every maximal critical chain is less than or equal to the number of points in CP. Now we can generalize Lemma 3.3.

Lemma 3.3′. *There are constants $C > 0$ and $0 < \mu < 1$ such that for any integers $n \geq n_0$ and $m > 0$ and for any points x and y in an interval of η_{n+m}, if x_m and y_m are in U and if the last critical chain $\mathcal{I}_q$ contains only one critical point c_{k_m}, then*

$$\Big|\log\Big(\frac{|(f^{\circ m})'(x)|}{|(f^{\circ m})'(y)|}\Big)\Big| \leq C\mu^n.$$

Proof. We use the same notation as in the proof of Lemma 3.3. The estimates of $\mathcal{I}$ and $\mathcal{II}$ are the same as in that proof. Here we add maximal critical chains in the estimates of $\mathcal{III}$. Consider a critical point $c(l)$ in a maximal critical chain $\mathcal{I}_i = \{c(s_{j-1}+1), \ldots, c(s_j)\}$ for $1 \leq j < q$. By arguments similar to those in the proof of Lemma 3.3 and by the fact that f is comparable to $B_{\pm}|x - c(l)|^{\gamma(l)} + f(c(l))$ near $c(l)$, there is a positive constant C_6 such that

$$\frac{|f(x_{i_l}) - f(y_{i_l})|}{|f(y_{i_l}) - f(c_{k_{i_l}})|} \leq C_6 \frac{|x_{i_{s_j}+1} - y_{i_{s_j}+1}|^{\frac{1}{\tau_j}}}{D''}$$

where $\tau_j = \prod_{l=s_{j-1}+1}^{s_j} \gamma(l) \leq \gamma = \prod_{i=1}^{d} \gamma_i$ and where D'' is a constant. Therefore, there are constants $C_5 > 0$ and $0 < \mu_5 < 1$ such that

$$\Big|\log \mathcal{III}\Big| \leq C_5 \mu_5^n.$$

Combining the estimates for $\mathcal{I}$ and $\mathcal{II}$ together, there are constants $C > 0$ and $0 < \mu < 1$ satisfying the lemma. ∎

Remark 3.5. From the proof of Lemma 3.3 and Lemma 3.3′, one can see that the distortion of f along an orbit is controlled by $|x_m - y_m|/d(y_m, PCO)$ even if the orbit may visit the neighborhood U of the set CP of critical points of f many times. This is a property like Koebe's Distortion Theorem (Theorem 2.5).

3.5. Bounded Geometry, and Bounded Nearby Geometry

We prove bounded geometry and bounded nearby geometry for a geometrically finite one-dimensional map.

Theorem 3.1 [JI7]. *Suppose f from M into itself is geometrically finite and $\eta = \{\eta_n\}_{n=1}^{\infty}$ is the sequence of nested partitions induced from f. Then η is of bounded geometry and bounded nearby geometry.*

Proof. Let n_0 be the integer and U and V be the closed sets in §3.3 after Remark 3.4. Let $C_1 > 0$ be the minimum of ratios $|J|/|I|$ for all $J \subset I$ with $J \in \eta_{j+1}$ and $I \in \eta_j$ and all $1 \leq j \leq n_0$.

For a pair $J \subset I$ with $J \in \eta_{k+1}$ and $I \in \eta_k$, let $n = k - n_0 > 0$ and $J_i = f^{\circ i}(J)$ and $I_i = f^{\circ i}(I)$ for $i = 0, \ldots, n$. Then $J_n \in \eta_{n_0+1}$ and $I_n \in \eta_{n_0}$. We consider the intervals $\{I_0, \ldots, I_n\}$ in two cases: (i) no one of them is in U and (ii) at least one of them is in U.

In (i), applying Lemma 3.2 (or Lemma 3.1), there is a constant $C_2 > 0$, such that

$$\frac{|(f^{\circ n})'(y)|}{|(f^{\circ n})'(x)|} \geq C_2$$

for x and y in I. This implies that

$$\frac{|J|}{|I|} \geq C_3 = C_2 C_1.$$

In (ii), let $l \leq n$ be the greatest integer so that $I_l \subset U$. We note that $I_i \subset V$ for $i = l+1, \ldots, n$. Applying Lemma 3.2 (or Lemma 3.1) again as in (i), we have that

$$\frac{|J_{l+1}|}{|I_{l+1}|} \geq C_3.$$

Suppose I_{l_j} is contained in U_{i_j} for $l_0 \leq j \leq l$ and suppose $\{c_{i_{l_0}}, \ldots, c_{i_l}\}$ is a maximal critical chain. Because $f|U_{i_j}$ is comparable to the map $q_{i_j}(x) = |x - c_{i_j}|^{\gamma_{i_j}} + f(c_{i_j})$ for $l_0 \leq j \leq l$, there is a constant $C_4 > 0$ (only depends on C_3) such that

$$\frac{|J_{l_0}|}{|I_{l_0}|} \geq C_4.$$

Now applying Lemma 3.3′, we have a constant $C_5 > 0$ such that

$$\frac{|(f^{\circ(n-l_0)})'(y)|}{|(f^{\circ(n-l_0)})'(x)|} \geq C_5$$

for x and y in I because $I_{l_0} = f^{\circ(n-l_0)}(I) \subseteq U$. This implies that

$$\frac{|J|}{|I|} \geq C_6 = C_5 C_4.$$

Hence η is of bounded geometry.

To prove η is of bounded nearby geometry, let $n_1 > n_0$ be an integer such that if a pair J_1 and J_2 in η_{n_1} with a common endpoint then either both of them are in U or no endpoints of J_1 and J_2 are critical points of f. Suppose $C_7 > 0$ is the minimum of ratios $|J_1|/|J_2|$ for J_1 and J_2 in η_j with a common endpoint for $1 \leq j \leq n_1$.

Now for $k \geq n_1$ and J_1 and J_2 in η_k with a common endpoint, let $J_{1,i} = f^{\circ i}(J_1)$ and $J_{2,i} = f^{\circ i}(J_2)$ for $i = 0, \ldots, n = k - n_1$. We consider $\{J_{1,i}\}_{i=0}^{n}$ and $\{J_{2,i}\}_{i=0}^{n}$ in two cases: (a) for some $0 < l \leq n$, $J_{1,l} = J_{2,l}$ and (b) $J_{1,i}$ and $J_{2,i}$ are all different for $i = 0, \ldots, n$.

In (a), let l be the smallest such integer, then the common endpoint p of $J_{1,l-1}$ and $J_{2,l-1}$ is a singular point of f. It is easy to see that there is a constant $C_8 > 0$ such that

$$\frac{|J_{1,l-1}|}{|J_{2,l-1}|} \geq C_8.$$

Let us assume that the common endpoint p is a critical point. (If p is not a critical point, the proof can be reduced to (b)). Since $J_{1,l-1} \cup J_{2,l-1}$ is a subinterval of U, by applying Lemma 3.3$'$, we have a constant $C_9 > 0$ such that

$$\frac{|(f^{\circ(l-1)})'(y)|}{|(f^{\circ(l-1)})'(x)|} \geq C_9$$

for x and y in $J_1 \cup J_2$. This implies that

$$\frac{|J_1|}{|J_2|} \geq C_{10} = C_9 C_8.$$

In (b), by using almost the same arguments as those in the proof of bounded geometry, we have a constant $C_{11} > 0$ such that

$$\frac{|J_1|}{|J_2|} \geq C_{11}.$$

Hence η is of bounded nearby geometry. ∎

Two maps f and g from M into itself are topologically conjugate if there is a homeomorphism h of f such that $f \circ h = h \circ g$. Furthermore, they are quasisymmetrically conjugate if the conjugacy h is quasisymmetric (see Definition 2.6).

Theorem 3.2 [JI7]. *Suppose f and g from M into itself are geometrically finite and topologically conjugate. They are then quasisymmetrically conjugate.*

Proof. Suppose h is the conjugacy between f and g, namely $h \circ f = g \circ h$. Let $\eta_f = \{\eta_{n,f}\}_{n=0}^{+\infty}$ and $\eta_g = \{\eta_{n,g}\}_{n=0}^{+\infty}$ be the sequence of nested partitions induced from f and g respectively. Let B_f and B_g are the constants in Remark 3.2 for η_f and η_g.

For any $x < y$ in M, let $z = (x+y)/2$ be the midpoint of x and y and $N > 0$ be the smallest integer such that there is an interval I in $\eta_{N,f}$ contained in $[x, y]$. Let $\tilde{I}$ be the interval in $\eta_{N-1,f}$ containing I. Then the union of $\tilde{I}$ and one of its adjacent intervals in $\eta_{N-1,f}$ contains $[x, y]$ (see Fig. 3.1, Fig. 3.2 and Fig. 3.3). Because of bounded geometry and bounded nearby geometry of η_g (refer to Fig. 3.1, 3.2 and 3.3), there is a constant $C_1 = C_1(B_g) > 0$ such that

$$\frac{|h(I)|}{|h([x,z])|} \geq C_1 \qquad \text{and} \qquad \frac{|h(I)|}{|h([z,y])|} \geq C_1.$$

Because $\eta_{n,f}$ tends to zero exponentially and η_f is of bounded geometry and bounded nearby geometry, we can find a constant integer $N_1 = N_1(B_f) > 0$ such that there are intervals J_1 and J_2 in η_{N+N_1} contained in $[x, z]$ and $[z, y]$, respectively. This implies that $h(J_1)$ and $h(J_2)$ are contained in $h([x, z])$ and $h([z, y])$ respectively. Because of bounded geometry and bounded nearby geometry of η_g again, there is a constant $C = C(N_1, B_g) > 0$ (see Fig. 3.1, 3.2 and 3.3) such that

$$C^{-1} \leq \frac{|h(x) - h(z)|}{|h(z) - h(y)|} \leq C,$$

which shows that h is quasisymmetric.

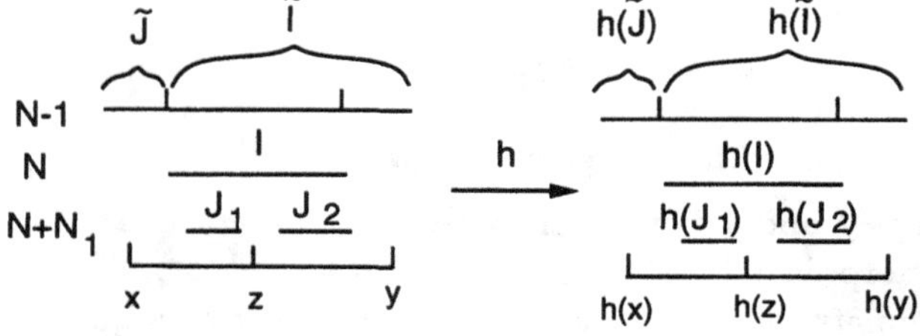

Fig. 3.1

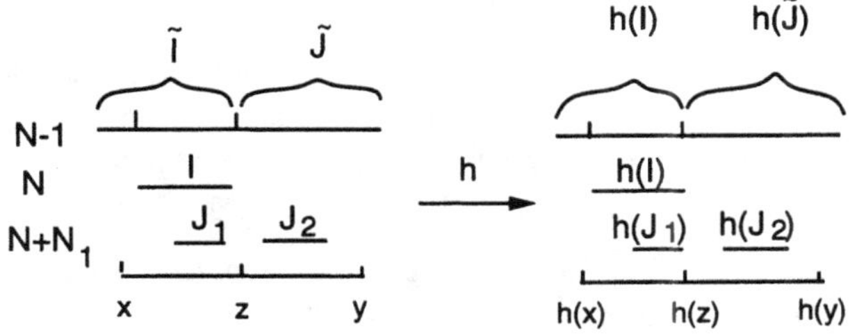

Fig. 3.2

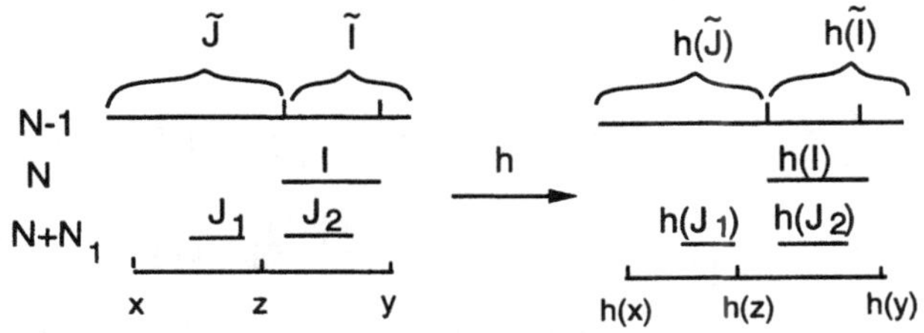

Fig. 3.3

■

Remark 3.6. Theorem 3.2 is first proved for Ulam-von Neumann transformations (see Remark 2.10 in §2.3 and §3.9) and folding mappings with a unique preperiodic critical point in [JI3]. Jakobson [JAK] proved independently a similar result for non-recurrent quadratic polynomials by inducing expansion.

3.6. Non-Critical Geometrically Finite One-Dimensional Maps

Let f be a geometrically finite one-dimensional map and let Σ^* be the dual phase space of f. A function s on Σ^* is said to be Hölder if there are constants $C > 0$ and $0 < \nu < 1$ such that $|s(a_1^*) - s(a_2^*)| \leq C\nu^n$ whenever the first n digits of a_1^* and a_2^* in Σ^* are the same.

Theorem 3.3 [JI11]. *Let f be a non-critical geometrically finite one-dimensional map. Then the scaling function s on Σ^* of f exists and is Hölder.*

Proof. Suppose $a^* = \ldots w_n$ is a point in Σ^* where w_n is the first n digits of a^* starting from the right. For any $n, m > 0$, define $I_{w_n} = f^{\circ m}(I_{w_{n+m}})$ and $I_{\sigma^*(w_n)} = f^{\circ m}(I_{\sigma^*(w_{n+m})})$. Hence

$$|s(w_{n+m}) - s(w_n)| = \left|\frac{(f^{\circ m})'(x)}{(f^{\circ m})'(y)} - 1\right| s(w_n)$$

$$= \left|\frac{(f^{\circ m})'(x)}{(f^{\circ m})'(y)} - 1\right| \left|\frac{(f^{\circ n})'(x')}{(f^{\circ n})'(y')}\right| s(w_1)$$

where x and y are in $I_{w_{n+m}}$ and x' and y' are in I_{w_n}. From Lemma 3.1, there are constants $C > 0$ and $0 < \mu < 1$ such that

$$\left|\frac{(f^{\circ m})'(x)}{(f^{\circ m})'(y)} - 1\right| \leq C\mu^n \qquad \text{and} \qquad \left|\frac{(f^{\circ n})'(x')}{(f^{\circ n})'(y')}\right| s(w_1) \leq C.$$

Thus $|s(w_{n+m}) - s(w_n)| \leq C^2\mu^n$ for all $m > 0$, whence $\{s(w_n)\}_{n=0}^{\infty}$ is a Cauchy sequence. This implies that the scaling function $s(a^*) = \lim_{n\to\infty} s(w_n)$ exists.

To prove that the scaling function s on Σ^* is Hölder, we consider two points $a^* = \ldots w_{n+m}$ and $b^* = \ldots w'_{n+m}$ whose first n digits are the same; that is, $w_n = w'_n$. Following the previous argument,

$$|s(w_{n+m}) - s(w'_{n+m})| = \left|\frac{(f^{\circ m})'(x)}{(f^{\circ m})'(y)} - \frac{(f^{\circ m})'(x')}{(f^{\circ m})'(y')}\right| s(w_n)$$

$$= \left|\frac{(f^{\circ m})'(x)}{(f^{\circ m})'(y)} - \frac{(f^{\circ m})'(x')}{(f^{\circ m})'(y')}\right| \left|\frac{(f^{\circ n})'(x'')}{(f^{\circ n})'(y'')}\right| s(w_1)$$

where x and y are in $I_{w_{n+m}}$, x' and y' are in $I_{w'_{n+m}}$, and x'' and y'' are in I_{w_n}. Hence

$$|s(w_{n+m}) - s(w'_{n+m})| \leq 2C^2\mu^n.$$

As m goes to infinity, we have

$$|s(a^*) - s(b^*)| \leq 2C^2\mu^n;$$

that is, s is Hölder continuous. ∎

3.7. Critical Geometrically Finite One-Dimensional Maps

Let f be a geometrically finite one-dimensional map and let Σ^* be the dual phase space of f. For a critical geometrically finite one-dimensional map, we have

Theorem 3.4 [JI11]. *Let f be a critical geometrically finite one-dimensional map. Then the scaling function s, on Σ^*, of f exists.*

Furthermore, because of the existence of critical points for a critical geometrically finite one-dimensional map, we have

Corollary 3.1 [JI11]. *Let f be a critical geometrically finite one-dimensional map. Then its scaling function s is discontinuous on Σ^* (see Fig. 3.5).*

Proof of Theorem 3.4. Let $CP = \{c_1, \ldots, c_d\}$ be the set of critical points of f and let $EX = \{\gamma_1, \ldots, \gamma_d\}$ be the set of corresponding exponents. Let U and V be the sets in §3.3 after Remark 3.4.

Suppose $a^* = \ldots w_n$ is a point in Σ^* and $v_n = \sigma^*(w_n)$. Then $w_n = v_n i_0$. We discuss the sequence $\{I_{v_n}\}_{n=1}^{\infty}$ (refer to §2.3) in two cases: (1) there is an integer $n_1 \geq n_0$ such that $n \geq n_1$ implies $I_{v_n} \subset V$, (2) there is an increasing subsequence $\{n_k\}_{k=1}^{\infty}$ of integers such that $I_{v_{n_k}} \subseteq U$ for all n_k.

In (1), by arguments similar to those in the proof of Theorem 3.3, we can prove that $\{s(w_n)\}_{n=1}^{\infty}$ is a Cauchy sequence. Hence the limit $s(a^*) = \lim_{n\to\infty} s(w_n)$ exists.

In (2), consider all $I_{v_{m_k}} \subseteq U_{m_k} \subset U$ and $\{c_{m_k}\}_{k=1}^{\infty}$. We can find an increasing sequence $\{m_{k_i}\}_{i=1}^{\infty}$ such that the last critical chain of $\{c_{m_k}\}_{k=k_i}^{\infty}$ contains only one critical point for every $1 \leq i < \infty$. Let $n_i = m_{k_i}$. For any $n \geq n_i$, the intervals $I_{w_{n_i}}$ and $I_{v_{n_i}}$ are the images of I_{w_n} and I_{v_n} under $f^{\circ(n-n_i)}$ and the intervals $I_{w_{n_1}}$ and $I_{v_{n_1}}$ are the images of $I_{w_{n_i}}$ and $I_{v_{n_i}}$ under $f^{\circ(n_i-n_1)}$. Thus

$$|s(w_n) - s(w_{n_i})| = \left|\frac{(f^{\circ(n-n_i)})'(x)}{(f^{\circ(n-n_i)})'(y)} - 1\right| \frac{|I_{w_{n_i}}|}{|I_{v_{n_i}}|}$$

$$= \left|\frac{(f^{\circ(n-n_i)})'(x)}{(f^{\circ(n-n_i)})'(y)} - 1\right| \left|\frac{(f^{\circ(n_i-n_1)})'(x')}{(f^{\circ(n_i-n_1)})'(y')}\right| s(w_1)$$

where x and y are in I_{v_n} and x' and y' are in $I_{v_{n_i}}$. By Lemma 3.3′, there are constants $C_1 > 0$ and $0 < \mu_1 < 1$ such that

$$\left|\frac{(f^{\circ(n-n_i)})'(x)}{(f^{\circ(n-n_i)})'(y)} - 1\right| \leq C\mu^{n_i} \qquad \text{and} \qquad \left|\frac{(f^{\circ(n_i-n_1)})'(x')}{(f^{\circ(n_i-n_1)})'(y')}\right| s(w_1) \leq C$$

for $n \geq n_i$. Thus

$$|s(w_n) - s(w_{n_i})| \leq C^2 \mu^{n_i}$$

for $n \geq n_i$. The sequence $\{s(w_n)\}_{n=1}^{\infty}$ is thus a Cauchy sequence and

$$\lim_{n\to\infty} s(w_n) = s(a^*)$$

exists. ∎

Proof of Corollary 3.1. Suppose c is a critical point of f. It is not periodic and its orbit $\{f^{\circ n}(c)\}_{n=0}^{\infty}$ is finite. There is a periodic point $p \neq c$ and an integer $l > 0$ such that $f^{\circ i}(c) \neq p$ for $0 \leq i < l$ and $f^{\circ l}(c) = p$. Let $a^* = w_m^{\infty}$ be a point in Σ^* such that $\{p\} = \cap_{i=1}^{\infty} I_{w_m^i}$ where m is the period of p. From **(4)** of Definition 3.6 and from Proposition 3.3, we have $|E_p| = |(f^{\circ m})'(p)| > 1$ and the existence of an integer $0 \leq m_0 < m$ such that for $c^* = (\sigma^*)^{\circ m_0}(a^*)$, $0 < s(c^*) < 1$. We prove that s is discontinuous at c^*.

Suppose $c^* = u_m^{\infty}$. Without loss of generality, we assume that $\{f^{\circ n}(c)\}_{n=1}^{\infty}$ contains no critical points and that c is not the image of any critical point under any iterate of f. For any u_m^i, there is a $b^* = \dots v_n u_m^i$ in Σ^* such that $I = I_{v_n u_m^i}$ contains c. For large $i > 0$, I is contained in U and $f^{\circ j}(I)$ is contained in V for every $0 < j \leq n$. Since $f|I$ is asymptotically $B_{\pm}|x - c|^{\gamma} + f(c)$ as n tends to infinity where $\gamma > 1$, Lemma 3.2 implies that,

$$\lim_{i\to\infty} s(v_n u_m^i) = \lim_{i\to\infty} \left(s(u_m^i)\right)^{\gamma} = \left(s(c^*)\right)^{\gamma}.$$

From Lemma 3.3′ and from the proof of Theorem 3.4, $|s(b^*) - s(v_n w_m^i)| \leq C\mu^{n+m^i}$. We have

$$\lim_{b^*\to c^*} s(b^*) = \left(s(c^*)\right)^{\gamma}.$$

Hence c^* is a jump discontinuity of s. ∎

Remark 3.7. From the proof of Corollary 3.1, we can actually find out all continuous points and discontinuous points of the scaling function of a critical geometrically finite one-dimensional map. A point $a^* = \dots w_n$ in Σ_f^* is said to be recurrent if there is an increasing subsequence $\{n_i\}_{i=1}^{\infty}$ of integers such that $I_{w_{n_i}} \subseteq U$ for every n_i. It is said to be non-recurrent if there is an integer $n > 0$ such that all preimages of I_{w_n} under iterates of f is contained in V. The point a^* is said to be wandering if there is an integer $m > 0$ such that $I_{w_n} \subseteq V$ for $n \geq m$, and if for every integer $n \geq m$, there is another integer $\overline{n}$ such that there is an interval in the preimage of I_{w_n} under $f^{\circ \overline{n}}$ contained in U. We can prove that s on Σ^* is continuous at all recurrent a^*, all non-recurrent a^*, and all wandering points a^* with $s(a^*) = 1$, and that s is discontinuous at wandering points a^* with $0 < s(a^*) < 1$. All discontinuities are jump discontinuities (see Fig. 3.5).

Remark 3.8. From the proof of Corollary 3.1, if $PCO \cap CP = \emptyset$, then the exponent $\gamma > 1$ of f at a critical point c can be calculated from the scaling function s. From this and from Proposition 3.4, one can also observed that the exponent γ is a C^1-invariant.

3.8. Complete Smooth Invariants

A geometrically finite one-dimensional map f is simple if $PCO \cap CP = \emptyset$. To avoid notational complication, we consider only simple geometrically finite one-dimensional maps in this section. Let $\mathcal{F}$ be an (orientation-preserving) topological conjugacy class in the space of simple geometrically finite one-dimensional maps, i.e., $\mathcal{F}$ is a subset of simple geometrically finite one-dimensional maps such that every pair of maps f and g in $\mathcal{F}$ are topologically conjugate by an orientation-preserving homeomorphism h. Since all maps f in $\mathcal{F}$ have the same dual phase space Σ^* we may speak of the dual phase space Σ^* of a topological conjugacy class $\mathcal{F}$.

Remark 3.9. Take a fixed orientation-reversing diffeomorphism h_0 of M. If f and g are topologically conjugate by an orientation-reversing homeomorphism h, then f and $\tilde{g} = h_0 \circ g \circ h_0^{-1}$ are topologically conjugate by an orientation-preserving homeomorphism.

Suppose f and g are in $\mathcal{F}$ and h is the conjugacy from f to g, i.e., $h \circ f = g \circ h$. In §3.5, we saw that h must be a quasisymmetric homeomorphism. This implies that h is α-Hölder continuous for some $0 < \alpha \leq 1$ (see [AH1]). Usually, h is not Lipschitz because f has a lot Lipschitz invariants, for example, all eigenvalues of f at periodic points. However, we have

Theorem 3.5 [JI11]. *Let f and g be maps in $\mathcal{F}$ and let h be the conjugacy from f to g. Then h is bi-Lipschitz continuous if the scaling functions s_f and s_g on Σ^* of f and g are the same.*

Proof. Let η_1 be the natural Markov partition of M w.r.t. f. Let η_n be the nth-partition induced by f with η_1. Let m_0 be the number of the intervals in η_1 and define $k_0 = m_0 + n_0$ where n_0 is the fixed integer in §3.3. For any integer $n > 0$ and any interval $I_{w_{k_0} w_n}$ in η_{n+k_0},

$$\frac{|h(I_{w_{k_0} w_n})|}{|I_{w_{k_0} w_n}|} = \frac{|s_g(w_{k_0} w_n)|}{|s_f(w_{k_0} w_n)|} \frac{|h(I_{w_{k_0} v_n})|}{|I_{w_{k_0} v_n}|}$$

where $w_{k_0} v_n = \sigma^*(w_{k_0} w_n)$.

Let $a^* = \dots u_m w_{k_0} w_n$ be a point in Σ^*. Our discussion of the sequence $\{I_{u_m w_{k_0} w_n}\}_{m=0}^{\infty}$ considers three cases: **(1)** $I_{w_{k_0} w_n}$ is contained in U; **(2)** $I_{u_m w_{k_0} w_n}$ is contained in V for every $m \geq 0$; **(3)** there is an integer $m \geq 1$ such that $I_{u_i w_{k_0} w_n}$ is contained in V for every $0 \leq i \leq m$ and $I_{u_{m+1} w_{k_0} w_n}$ is contained in U.

In case **(1)**, we use Lemma 3.3 for both f and g to find constants $C_1 > 0$ and $0 < \mu_1 < 1$ such that

$$\left|\log\left(\frac{s_f(a^*)}{s_f(w_{k_0}w_n)}\right)\right| \leq C_1\mu_1^n$$

and

$$\left|\log\left(\frac{s_g(a^*)}{s_g(w_{k_0}w_n)}\right)\right| \leq C_1\mu_1^n.$$

Because $s_g = s_f$, there are constants $C_2 > 0$ and $0 < \mu_2 < 1$ such that

$$\left|\log\left(\frac{s_f(w_{k_0}w_n)}{s_g(w_{k_0}w_n)}\right)\right| \leq C_2\mu_2^n.$$

In case **(3)**, suppose I is the interval in η_{k_0} having a critical point c of f as an endpoint and containing $I_{u_{m+1}w_{k_0}w_n}$. Then there is an integer $0 < l < m_c$ such that $f^{\circ l}(c)$ is a periodic point p of f and $f^{\circ l}(I_{u_{m+1}w_{k_0}w_n}) \subset f^{\circ l}(I)$. We note that $f^{\circ l}(I)$ is an interval in η_{k_0-l} and that $f^{\circ l}(I_{u_{m+1}w_{k_0}w_n})$ is an interval in η_{m+1+k_0+n-l}. Remember that p is an endpoint of $f^{\circ l}(I)$. We can now find a point $b^* = \ldots w'_j$ in Σ^* such that the first $m+1+k_0+n-l$ digits of a^* and b^* (starting from the right) are the same and such that the interval $I_{w'_j}$ in η_j is contained in V for every $j > 0$ and tends to the periodic orbit $\cup_{i=0}^{\infty} f^{\circ i}(p)$ as j goes to infinity. Thus we get another sequence $\{I_{w'_j}\}$ which is in case **(2)**. If $m+1-l > 0$, then $I_{w_{k_0}w_n}$ is in $\{I_{w'_j}\}$. If $m+1-l < 0$, then we have that $I_{w_{k_0-l}w_n}$ is in $\{I_{w'_j}\}$; from Lemma 3.2, there are now constants $C_3 > 0$ and $0 < \mu_3 < 1$ such that

$$\left|\log\left(\frac{s_f(w_{k_0}w_n)}{s_f(w_{k_0-l}w_n)}\right)\right| \leq C_3\mu_3^n.$$

Only case **(2)** remains. By applying Lemma 3.2, there are constants $C_4 > 0$ and $0 < \mu_4 < 1$ such that

$$\left|\log\left(\frac{s_f(b^*)}{s_f(w_{k_0}w_n)}\right)\right| \leq C_4\mu_4^n$$

and

$$\left|\log\left(\frac{s_g(b^*)}{s_g(w_{k_0}w_n)}\right)\right| \leq C_4\mu_4^n.$$

Because $s_g = s_f$, there are constants $C_5 > 0$ and $0 < \mu_5 < 1$ such that

$$\left|\log\left(\frac{s_f(w_{k_0}w_n)}{s_g(w_{k_0}w_n)}\right)\right| \leq C_5\mu_5^n.$$

Let C_0 be the minimum of all ratios $\log(|h(I_w)|/|I_w|)$ for I_w in η_{k_0}. From the arguments above, there are constants $C > 0$ and $0 < \mu < 1$ such that

$$\left|\log \frac{|h(I_{w_{k_0}w_n})|}{|I_{w_{k_0}w_n}|}\right| \leq C_0 + C\sum_{i=1}^{n} \mu^i \leq C_0 + \frac{C\mu}{1-\mu}$$

for all $I_{w_{k_0}w_n}$ in η_{k_0+n} and all $n \geq 0$. Hence there is a constant, still denoted it as $C > 0$, such that

$$C^{-1} \leq \frac{|h(I_{w_{k_0}w_n})|}{|I_{w_{k_0}w_n}|} \leq C$$

for all $I_{w_{k_0}w_n}$ in η_{n+k_0} and all $n \geq 0$. Since the union of boundary points of all intervals in η_{n+k_0} for all $n \geq 0$ is dense in M,

$$C^{-1} \leq \frac{|h(x) - h(y)|}{|x - y|} \leq C$$

for every pair x and y in M. In other words, h is bi-Lipschitz continuous. ∎

Let f be a geometrically finite one-dimensional map. Let SP be the set of all singular points of f and let $PSO = \cup_{i=1}^{\infty} f^{\circ i}(SP)$ be the post-singular orbits of f. A singular point $p \in SP$ is fold if $f'(x)f'(2p - x) < 0$ for $x \neq p$ near p. Let FSP denote the set of all fold singular points of f and let $FSO = \cup_{i=1}^{\infty} f^{\circ i}(FSP)$ denote the post fold singular orbits. A subset $\{b_1, \ldots, b_k\} \subseteq SP$ is called a cycle if $f(b_i) = b_{i+1\,(\mathrm{mod}\,k)}$. Let PSP denote the union of all cycles in SP. Note that $FSO \cup PSP \subseteq PSO$. Let ξ be the set of the closures of intervals of $M \setminus (FSO \cup PSP)$. Let η_1 be the natural Markov partition of M w.r.t. f. Let η_n be the nth-partitions of M induced by f with η_1. We say f is mixing if for any interval I in η_1, there is an integer $n > 0$ such that $f^{\circ n}(I) = M$. Since the mixing condition is topologically invariant, we may speak of a topological conjugacy class $\mathcal{F}$ being mixing. To avoid notational complication, we prove the following rigidity result under mixing condition.

Theorem 3.6 [JI11]. *Let f and g be maps in a mixing topological conjugacy class $\mathcal{F}$. Let h be the conjugacy from f to g, i.e., $h \circ f = g \circ h$. Then $h|I$ for any I in ξ is a $C^{1+\beta}$-diffeomorphism for some $0 < \beta \leq 1$ if and only if (i) h is differentiable at one point p in M with non-zero derivative and (ii) the exponents and the asymmetries of f and g at all corresponding critical points are the same.*

Remark 3.10. This theorem is generalized for a quasi-hyperbolic one-dimensional map in [JI12].

We prove Theorem 3.6 by means of several lemmas. Let $\mathcal{F}$ be a mixing topological conjugacy class in the space of simple geometrically finite one-dimensional maps and let f and g be maps in $\mathcal{F}$. Suppose both f and g are $C^{1+\alpha}$ for some $0 < \alpha \leq 1$. Let h be the conjugacy from f to g: $h \circ f = g \circ h$. Let η_1 be the natural Markov partition of M to f. Let η_n be the nth-partition of M induced by f with η_1 for $1 \leq n < \infty$. Let Σ^* be the dual phase space of $\mathcal{F}$. Let $CP = \{c_1, \ldots, c_d\}$ be the set of critical points of f. Let $EX = \{\gamma_1, \ldots, \gamma_d\}$ be the set of corresponding exponents of f at critical points and let $AS = \{A_1, A_2, \ldots, A_d\}$ be the set of corresponding asymmetries of f. Let n_0 and U and V be the fixed integer and the sets in §3.3 for f (after Remark 3.4).

Suppose $I = I_{w_{n_0}} \in \eta_{n_0}$ and $a^* = \ldots w_n w_{n_0}$ is a point in Σ^*. Then $I_{w_n w_{n_0}}$ is an interval in η_{n+n_0}. For any x and y in I, let x_n and y_n in $I_{w_n w_{n_0}}$ be the preimage of x and y under $f^{\circ n}$ for $n \geq 0$.

Lemma 3.4. *There is a constant $C > 0$ such that if $I_{w_n w_{n_0}}$ are in V for all $1 \leq n \leq N$, then*

$$\Big|\log\Big(\prod_{n=1}^{N} \frac{|f'(y_n)|}{|f'(x_n)|}\Big)\Big| \leq C|x-y|^\alpha$$

and

$$\Big|\log\Big(\prod_{n=1}^{N} \frac{|g'(h(x_n)|)}{|g'(h(y_n))|}\Big)\Big| \leq C|h(x)-h(y)|^\alpha.$$

Proof. From the proof of Lemma 3.2, there is a constant $C_1 > 0$ such that

$$\Big|\log\Big(\prod_{i=1}^{N} \frac{|f'(y_i)|}{|f'(x_i)|}\Big)\Big| \leq C_1 \sum_{i=1}^{N} |x_i - y_i|^\alpha.$$

By applying Lemma 3.2, we have a constant $C_2 > 0$ such that

$$\frac{|x_n - y_n|}{|I_{w_{n+n_0}}|} \leq C_2 \frac{|x-y|}{|I_{w_{n_0}}|}$$

for all $1 \leq n \leq N$. So there are constants $C_3 > 0$ and $0 < \mu_1 < 1$ such that

$$|x_n - y_n| \leq C_3 \mu_1^n |x-y|$$

for all $0 \leq n \leq N$. This implies that

$$\Big|\log\Big(\prod_{n=1}^{N} \frac{|f'(y_n)|}{|f'(x_n)|}\Big)\Big| \leq C|x-y|^\alpha$$

for some constant $C > 0$. Similarly,

$$\left|\log\Big(\prod_{n=1}^{N}\frac{|g'(h(y_n))|}{|g'(h(x_n))|}\Big)\right| \leq C|h(x)-h(y)|^{\alpha}.$$

∎

Lemma 3.5. *There is a constant $C > 0$ such that if $I_{w_{n_0}}$ is in U, then for any $N \geq 1$,*

$$\left|\log\Big(\prod_{n=1}^{N}\frac{|f'(y_n)|}{|f'(x_n)|}\Big)\right| \leq C|x-y|^{\alpha}$$

and

$$\left|\log\Big(\prod_{n=1}^{N}\frac{|g'(h(x_n))|}{|g'(h(y_n))|}\Big)\right| \leq C|h(x)-h(y)|^{\alpha}.$$

Proof. From the proof of Lemma 3.3, there is a constant $C_1 > 0$ such that

$$\left|\log\Big(\prod_{n=1}^{N}\frac{|f'(y_n)|}{|f'(x_n)|}\Big)\right| \leq C_1\sum_{n=1}^{N}|x_n-y_n|^{\alpha}.$$

By applying Lemma 3.3, we have a constant $C_2 > 0$ such that

$$\frac{|x_n-y_n|}{|I_{w_{n+n_0}}|} \leq C_2\frac{|x-y|}{|I_{w_{n_0}}|}$$

for all $1 \leq n \leq N$. So there are constants $C_3 > 0$ and $0 < \mu_1 < 1$ such that

$$|x_n-y_n| \leq C_3\mu_1^n|x-y|$$

for all $0 \leq n \leq N$. This implies that

$$\left|\log\Big(\prod_{n=1}^{N}\frac{|f'(y_n)|}{|f'(x_n)|}\Big)\right| \leq C|x-y|^{\alpha}$$

for a constant $C > 0$. Similarly,

$$\left|\log\Big(\prod_{n=1}^{N}\frac{|g'(h(y_n))|}{|g'(h(x_n))|}\Big)\right| \leq C|h(x)-h(y)|^{\alpha}.$$

∎

Let q be a point in M. If $h'(f(q)) \neq 0$ exists and if q is not a singular point, from the equation $h \circ f = g \circ h$, we have that

$$h'(q) = \frac{f'(q)}{g'(h(q))} h'(f(q)) \neq 0$$

exists too. Suppose q is a singular point. Let $\gamma_{f,q}$ and $\gamma_{g,h(q)}$ be the exponents of f and g at q and at $h(q)$. Let $A_{f,q} = A_{+,f}/A_{-,f}$ and $A_{g,h(q)} = A_{+,g}/A_{-,g}$ be the asymmetries of f and g at q and at $h(q)$. If $\gamma_{f,q} = \gamma_{g,h(q)}$ and if $h'(f(q)) \neq 0$ exists, then

$$\left(h'(q-)\right)^{\gamma_{f,q}} = \frac{A_{-,f}}{A_{-,g}} h'(f(q)) \text{ and } \left(h'(q+)\right)^{\gamma_{f,q}} = \frac{A_{+,f}}{A_{+,g}} h'(f(q))$$

exist. Furthermore, if $A_{f,q} = A_{g,h(q)}$, then $h'(q-) = h'(q+)$. Therefore, if $h'(f(q)) \neq 0$ exists, if $\gamma_{f,q} = \gamma_{g,h(q)}$, and if $A_{f,q} = A_{g,h(q)}$, then $h'(q) \neq 0$ exists.

For a point p in M, let $GO(p) = \cup_{i=0}^{\infty} f^{-i}(p)$ be the grand backward orbit of p under f. Then $GO(p)$ is a dense subset of M since f is mixing. From the argument in the previous paragraph, if h is differentiable at one point in M with non-zero derivative and and if the exponents and the asymmetries of f and g at the corresponding singular points are the same, then we can find a point p in $M \setminus PSO$ such that $h'(q) \neq 0$ exists for any $q \in GO(p)$. Note that if $p \in M \setminus PSO$, then $GO(p) \subseteq M \setminus PSO$.

Lemma 3.6. *If h is differentiable at a point p in $M \setminus PSO$ with non-zero derivative, then $h'|GO(p)$ is continuous at p.*

Proof. Since $h'(p) \neq 0$ exists, there is an open interval $p \in W$ such that $h|W$ is bi-Lipschitz. Because the eigenvalue of an expanding periodic point of f is a bi-Lipschitz invariant, we have

$$(f^{\circ n})'(q) = (g^{\circ n})'(h(q))$$

if $f^{\circ n}(q) = q \in W$.

Suppose $p \in I_k = I_{w_k w_{n_0}} \in \eta_{k+n_0}$ and suppose $I_k \subseteq W$. For any $x \in I_k \cap GO(p)$ with $f^{\circ n}(x) = p$, let $x_i = f^{\circ(n-i)}(x)$. Let $x_i \in I_{i+k} \in \eta_{i+k+n_0}$. Then $I_{n+k} \subseteq f^{\circ n}(I_{n+k}) = I_k$. There is a point $q \in I_{n+k}$ such that $f^{\circ n}(q) = q$. From $h \circ f = g \circ h$, we have

$$\frac{h'(p)}{h'(x)} = \frac{(g^{\circ n})'(h(x))}{(f^{\circ n})'(x)} = \frac{(g^{\circ n})'(h(x))}{(g^{\circ n})'(h(q))} \frac{(f^{\circ n})'(q)}{(f^{\circ n})'(x)}.$$

Let $q_i = f^{\circ(n-i)}(q)$ and consider

$$\frac{(f^{\circ n})'(q)}{(f^{\circ n})'(x)} = \prod_{i=1}^{n} \frac{f'(q_i)}{f'(x_i)}.$$

If $I_k \subseteq U$, then from Lemma 3.5,

$$\left|\log\left(\frac{|(f^{\circ n})'(q)|}{|(f^{\circ n})'(x)|}\right)\right| \leq C_1|p-q|^\alpha \leq C\mu_1^k$$

where $C_1 > 0$ and $0 < \mu_1 < 1$ are constants. Otherwise, consider the smallest integer $m > 0$ such that $I_{m+k} \subseteq U_j \subset U$. From Lemma 3.4, there are constants $C_2 > 0$ and $0 < \mu_2 < 1$ such that

$$\left|\log\left(\prod_{i=1}^{m-1} \frac{|f'(q_i)|}{|f'(x_i)|}\right)\right| \leq C_2|p-q|^\alpha \leq C_2\mu_2^k.$$

From Lemma 3.5, there are constants $C_3 > 0$ and $0 < \mu_3 < 1$ such that

$$\left|\log\left(\prod_{i=m+1}^{n} \frac{|f'(q_i)|}{|f'(x_i)|}\right)\right| \leq C_3|x_m - q_m|^\alpha \leq C_3\mu_3^{k+m}.$$

Because $f|U_j$ and $f'|U_j$ are comparable to $|z-c_j|^{\gamma_j} + f(c_j)$ and to $|z-c_j|^{\gamma_j - 1}$, there are constants $C_4 > 0$ and $0 < \mu_4 < 1$ such that

$$\begin{aligned}\left|\log\left(\frac{|f'(q_m)|}{|f'(x_m)|}\right)\right| &\leq C_4|x_m - q_m|^\alpha + \frac{1}{\gamma_j - 1}\left|\log\left(\frac{|x_m - c_j|}{|q_m - c_j|}\right)\right| \\ &\leq C_4\mu_4^{m+k} + \frac{1}{\gamma_j - 1}\left|\log\left(\frac{|x_m - c_j|}{|q_m - c_j|}\right)\right|\end{aligned}$$

If the distance $d_k = \mathrm{dist}(I_k, PCO)$ between I_k and PCO is greater than the length $|I_k|$ of I_k, then

$$|x_m - q_m| \leq C_6 \min\{|x_m - c_j|, q_m - c_j|\}$$

and

$$\begin{aligned}\left|\log\left(\frac{|x_m - c_j|}{|q_m - c_j|}\right)\right| &\leq C_7\frac{|x_m - q_m|}{|q_m - c_j|} \leq C_8\left(\frac{|x_{m+1} - q_{m+1}|}{|q_{m+1} - f(c_j)|}\right)^{\frac{1}{\gamma_j}} \\ &\leq C_9\left(\frac{|p-q|}{d_k}\right)^{\frac{1}{\gamma_j}} \leq C_{10}\left(\frac{\mu_{10}^k}{d_k}\right)^{\frac{1}{\gamma_j}},\end{aligned}$$

where $C_i > 0$ for $i = 7, 8, 9, 10$ and $0 < \mu_{10} < 1$ are constants. So we have constants $C > 0$ and $0 < \mu < 1$ such that

$$\left|\log\left(\frac{|(f^{\circ n})'(q)|}{|(f^{\circ n})'(x)|}\right)\right| \leq C(\mu^k + \mu^{k+m}).$$

Similarly, we can get

$$\left|\log\left(\frac{|(g^{\circ n})'(h(x))|}{|(g^{\circ n})'(h(q))|}\right)\right| \leq C(\mu^k + \mu^{k+m}).$$

Thus

$$\left|\log\left(\frac{h'(p)}{h'(x)}\right)\right| \leq 2C(\mu^k + \mu^{k+m}).$$

This implies that $h'|GO(p)$ is continuous at p. ∎

Corollary 3.2. *If h is differentiable at a point p in $M \setminus PSO$ with non-zero derivative, then $h'|GO(p)$ is continuous at every point x in $GO(p)$.*

Proof. We use the same notation as in Lemma 3.6. For any $x \in GO(p)$, let $f^{\circ n}(x) = p$, let $p \in I_k = I_{w_k w_{n_0}} \in \eta_{k+n_0}$, and let $x \in I_{n+k} \in \eta_{n+k+n_0}$ such that $f^{\circ n}(I_{n+k}) = I_k$. For any $y \in I_{n+k}$, we have

$$\frac{h'(x)}{h'(y)} = \frac{(f^{\circ n})'(x)}{(f^{\circ n})'(y)} \frac{(g^{\circ n})'(h(y))}{(g^{\circ n})'(h(x))} \frac{h'(p)}{h'(f^{\circ n}(y))}.$$

Similar to the proof of Lemma 3.6, we have

$$\left|\log\left(\frac{h'(x)}{h'(y)}\right)\right| \leq C\left(\mu^k + \mu^{k+m}\right) + \left|\log\left(\frac{h'(p)}{h'(f^{\circ n}(y))}\right)\right|.$$

Now Lemma 3.6 implies that $h'|GO(p)$ is continuous at x. ∎

An interval $I_{w_{n_0}}$ in η_{n_0} is called critical if one of its end-points is a critical point of f. Otherwise, it is called non-critical.

Lemma 3.7. *If h is differentiable at a point p in $M \setminus PSO$ with non-zero derivative, then the restriction of h to every critical interval in η_{n_0} is $C^{1+\beta}$ for some $0 < \beta \leq 1$.*

Proof. Suppose I is a critical interval in η_{n_0}. Then $I \subset U$. Since f is mixing, there is a preimage I_n of I under $f^{\circ n}$ such that I_n tends to p.

For any x and y in I, let x_n and y_n in I_n be the preimages of x and y under $f^{\circ n}$. From the equation $h \circ f = g \circ h$, we have

$$\frac{h'(x)}{h'(y)} = \frac{(g^{\circ n})'(h(x_n))}{(f^{\circ n})'(x_n)} \frac{(f^{\circ n})'(y_n)}{(g^{\circ n})'(h(y_n))} \frac{h'(x_n)}{h'(y_n)}.$$

Thus,

$$\left|\log\left(\frac{h'(x)}{h'(y)}\right)\right| \leq \left|\log\left(\frac{(g^{\circ n})'(h(x_n))}{(g^{\circ n})'(h(y_n))}\right)\right| + \left|\log\left(\frac{(f^{\circ n})'(y_n)}{(f^{\circ n})'(x_n)}\right)\right| + \left|\log\left(\frac{h'(x_n)}{h'(y_n)}\right)\right|$$

$$\leq C\Big(|x-y|^{\alpha} + |h(x)-h(y)|^{\alpha}\Big) + \left|\log\left(\frac{h'(x_n)}{h'(y_n)}\right)\right|.$$

Since $h'(x_n) \to h'(p)$ and $h'(y_n) \to h'(p)$ as n tends to infinity, we have

$$\left|\log\left(\frac{h'(x)}{h'(y)}\right)\right| \leq C\Big(|x-y|^{\alpha} + |h(x)-h(y)|^{\alpha}\Big).$$

This implies that $h'|(I \cap GO(p))$ is uniformly continuous. Therefore, it can be extended to a continuous function h' on I. Furthermore, the last inequality implies again that h is $C^{1+\beta}$ for $\beta = \alpha$. ∎

Lemma 3.8. *If h is differentiable at a point p in $M \setminus PSO$ with non-zero derivative and if the exponents of f and g at corresponding critical points are the same, then the restriction of h to every interval $I \subseteq V$ in η_{n_0} is $C^{1+\beta}$ for some $0 < \beta \leq 1$.*

Proof. Suppose I_n in η_{n_0+n} is a preimage of I under $f^{\circ n}$. Since f is mixing, we can choose I_n such that I_n tends to p. Let $I_{n,i} = f^{\circ(n-i)}(I_n)$ for $0 \leq i \leq n$.

For any x and y in I and $n > 0$, let x_n and y_n in I_n be the preimages of x and y under $f^{\circ n}$. From the equation $h \circ f = g \circ h$, we have

$$\frac{h'(x)}{h'(y)} = \frac{(g^{\circ n})'(h(x_n))}{(f^{\circ n})'(x_n)} \frac{(f^{\circ n})'(y_n)}{(g^{\circ n})'(h(y_n))} \frac{h'(x_n)}{h'(y_n)}.$$

Thus,

$$\left|\log\left(\frac{h'(x)}{h'(y)}\right)\right| \leq \left|\log\left(\frac{|(g^{\circ n})'(h(x_n))|}{|(g^{\circ n})'(h(y_n))|}\right)\right| + \left|\log\left(\frac{|(f^{\circ n})'(y_n)|}{|(f^{\circ n})'(x_n)|}\right)\right| + \left|\log\left(\frac{h'(x_n)}{h'(y_n)}\right)\right|.$$

Let $m = m(n) > 0$ be the smallest integer such that $I_{m,n} \subseteq U_j \subset U$. Then

$$\begin{aligned}\left|\log\left(\frac{h'(x)}{h'(y)}\right)\right| \le & \left|\sum_{i=1}^{m-1}\Big(\log|f'(y_i)| - \log|f'(x_i)|\Big)\right| \\ & + \left|\sum_{i=1}^{m-1}\Big(\log|g'(h(x_i))| - \log|g'(h(y_i))|\Big)\right| \\ & + \left|\log\left(\frac{|g'(h(x_m))|}{|f'(x_m)|}\frac{|f'(y_m)|}{|g'(h(y_m))|}\right)\right| \\ & + \left|\sum_{i=m+1}^{n}\Big(\log|f'(y_i)| - \log|f'(x_i)|\Big)\right| \\ & + \left|\sum_{i=m+1}^{n}\Big(\log|g'(h(x_i))| - \log|g'(h(y_i))|\Big)\right|.\end{aligned}$$

From Lemma 3.4, there is a constant $C_1 > 0$ such that

$$\left|\sum_{i=1}^{m-1}\Big(\log|f'(y_i)| - \log|f'(x_i)|\Big)\right| \le C_1|x-y|^\alpha$$

and

$$\left|\sum_{i=1}^{m-1}\Big(\log|g'(h(x_i))| - \log|g'(h(y_i))|\Big)\right| \le C_1|h(x)-h(y)|^\alpha.$$

From Lemma 3.5, there are constants $C_2 > 0$, $C_3 > 0$, and $C_4 > 0$ such that

$$\begin{aligned}\left|\sum_{i=m+1}^{n}\Big(\log|f'(y_i)| - \log|f'(x_i)|\Big)\right| &\le C_2|x_m - y_m|^\alpha \\ &\le C_3|x_{m-1} - y_{m-1}|^{\frac{\alpha}{\gamma_j}} \le C_4|x-y|^{\frac{\alpha}{\gamma_j}}.\end{aligned}$$

Similarly,

$$\begin{aligned}\left|\sum_{i=m+1}^{n}\Big(\log|g'(h(x_i))| - \log|g'(h(y_i))|\Big)\right| &\le C_2|h(x_m) - h(y_m)|^{\frac{\alpha}{\gamma_j}} \\ &\le C_4|h(x)-h(y)|^{\frac{\alpha}{\gamma_j}}.\end{aligned}$$

Now we consider

$$S = \frac{|g'(h(x_m))|}{|f'(x_m)|}\frac{|f'(y_m)|}{|g'(h(y_m))|}.$$

Define

$$S = S_1 \cdot S_2 \cdot S_3$$

where

$$S_1 = \frac{|g'(h(x_m))|}{|h(x_m) - h(c_j)|^{\gamma_j - 1}} \frac{|h(y_m) - h(c_j)|^{\gamma_j - 1}}{|g'(h(y_m))},$$

$$S_2 = \frac{|x_m - c_j|^{\gamma_j - 1}}{|f'(x_m)|} \frac{|f'(y_m)|}{|y_m - c_j|^{\gamma_j - 1}},$$

and

$$S_3 = \Big(\frac{|h(x_m) - h(c_j)|}{|x_m - c_j|}\Big)^{\gamma_j - 1} \Big(\frac{|y_m - c_j|}{|h(y_m) - h(c_j)|}\Big)^{\gamma_j - 1}.$$

Lemma 3.7 implies that

$$\Big|\log S_3\Big| \leq C_5 |x_m - y_m|^{\alpha} \leq C_6 |x_{m-1} - y_{m-1}|^{\frac{\alpha}{\gamma_j}} \leq C_7 |x - y|^{\frac{\alpha}{\gamma_j}}$$

where C_i for $i = 5$, 6, 7 are constants. From **(1)** of Definition 3.6,

$$\Big|\log S_2\Big| \leq C_8 |x_m - y_m|^{\alpha} \leq C_9 |x_{m-1} - y_{m-1}|^{\frac{\alpha}{\gamma_j}} \leq C_{10} |x - y|^{\frac{\alpha}{\gamma_j}}$$

and

$$\begin{aligned}\Big|\log S_1\Big| &\leq C_8 |h(x_m) - h(y_m)|^{\alpha} \\ &\leq C_9 |h(x_{m-1}) - h(y_{m-1})|^{\frac{\alpha}{\gamma_j}} \leq C_{10} |h(x) - h(y)|^{\frac{\alpha}{\gamma_j}}\end{aligned}$$

where C_i for $i = 8$, 9, and 10 are constants. Thus,

$$\Big|\log\Big(\frac{h'(x)}{h'(y)}\Big)\Big| \leq C\Big(|x - y|^{\frac{\alpha}{\gamma}} + |h(x) - h(y)|^{\frac{\alpha}{\gamma}}\Big)$$

where $\gamma = \max\{\gamma_i\}_{i=1}^{d}$. So $h'|(I \cap GO(p))$ is uniformly continuous. It can be extended to a continuous function on I. Furthermore, the last inequality implies that $h|I$ is $C^{1+\beta}$ for $\beta = \alpha/\gamma$. ∎

Proof of Theorem 3.6. The "only if" part follows from direct calculation. Let us prove the "if" part. From Lemmas 3.7 and 3.8, h is $C^{1+\beta}$ for some $0 < \beta \leq 1$ when restricted to any interval I in η_{n_0}. Using the equation $h \circ f = g \circ h$, we have that h is $C^{1+\beta}$ when it is restricted to any interval I in the natural Markov partition η_1 of M to f. The rest is to check that h is continuous on I for any I in ξ. If $b \notin PSO$ is a singular point, then we can find an open interval W contained in an interval I_0 of η_1 such that $b \in f(W)$ and $f : W \to f(W)$ is a diffeomorphism. This implies that h is continuous at

b. Further, if b is not a fold singular point, from the equation $h \circ f = g \circ h$, we have that for $x \neq b$ near b,

$$h'(f(x)) = \frac{g'(h(x))}{f'(x)} h'(x).$$

This implies that

$$h'(f(b)+) = \frac{A_{+,g}}{A_{+,f}} \left(h'(b)\right)^{\gamma} \qquad \text{and} \qquad h'(f(b)-) = \frac{A_{-,g}}{A_{-,f}} \left(h'(b)\right)^{\gamma}$$

where $A_{\pm,f}$ and $A_{\pm,g}$ are the numbers in the beginning of §3.3 for f and g and where γ is the exponent of f and g at b and $h(b)$. Because f and g have the same asymmetry at b, the last equation implies that

$$h'(f(b)-) = h'(f(b)+).$$

So h is continuous at $f(b)$. Similar argument implies that h is continuous at all points in $PSO \setminus (FSO \cup PSP)$. Therefore, the restriction of h to any interval I in ξ is $C^{1+\beta}$. ■

One consequence of Theorems 3.5 and 3.6 is that scaling functions together with exponents and asymmetries are complete C^1-invariants as follows:

Theorem 3.7 [JI11]. *Let f and g be maps in a mixing topological conjugacy class $\mathcal{F}$ and let Σ^* be the dual phase space of $\mathcal{F}$. Let h be the topological conjugacy from f to g, i.e., $f \circ h = h \circ g$. Then $h|I$ for any I in ξ is a $C^{1+\beta}$-diffeomorphism for some $0 < \beta \leq 1$ if and only if the scaling functions s_f and s_g, on Σ^*, and the exponents and the asymmetries of f and g at corresponding critical points are the same.*

Proof. The "only if" part follows from Proposition 3.4, Remark 3.3 and direct calculation. Let us prove the "if" part. From Theorem 3.5, h is bi-Lipschitz. A bi-Lipschitz function is absolutely continuous; therefore, $h'(x) > 0$ exists for almost all points in M. Theorem 3.7 now follows from Theorem 3.6. ■

3.9. Generalized Ulam-von Neumann Transformations

In this section, we study a special conjugacy class, Ulam-von Neumann Transformations, in geometrically finite one-dimensional maps.

Suppose $M = [-1, 1]$ and f is a piecewise C^1 self map of M with a unique singular point s in $(-1, 1)$ and $f(-1) = f(1) = -1$. Conjugating by a Möbius transformation (see Remark 2.3), we assume $s = 0$. We call f a Ulam-von Neumann transformation if f is geometrically finite and $f(0) = 1$ (see Fig. 3.4). An example of a Ulam-von Neumann transformation is $q(x) = 1 - 2x^2$.

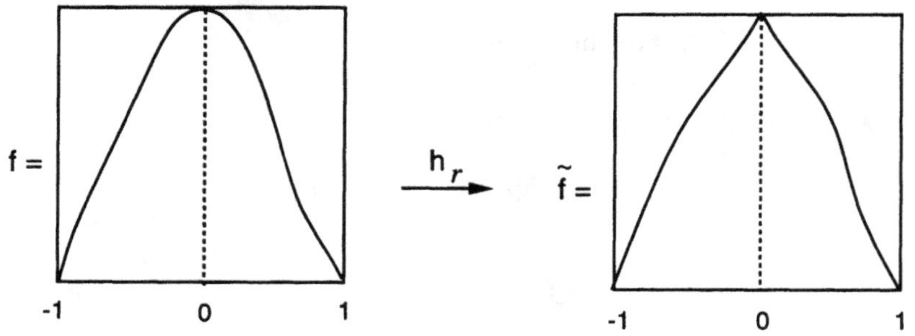

Fig. 3.4

For a Ulam-von Neumann transformation f, the fixed Markov partition is always $\eta_1 = \{I_0, I_1\}$ where $I_0 = [-1, 0]$ and $I_1 = [0, 1]$. The exponent and the asymmetry of f at 0 are denoted as $\gamma_f \geq 1$ and

$$A_f = \lim_{x \to 0+} \frac{f'(x)}{f'(-x)} \neq 0.$$

The phase space of f is $\Sigma_2 = \prod_{n=0}^{\infty}\{0, 1\} = \{w = a_0 a_1 \ldots a_k \ldots \mid a_k = 0 \text{ or } 1\}$. Therefore from Proposition 3.2, it is easy to see that

Theorem 3.8. *Any two Ulam-von Neumann transformations are topologically conjugate.*

For Ulam-von Neumann transformations, we would like to introduce an interesting technique in [JI3]. Suppose f is a Ulam-von Neumann transformation and the exponent of f at 0 is $\gamma \geq 1$. We define the orbifold metric associated to f to be

$$d_\gamma y = \frac{dx}{(1 - x^2)^{\frac{\gamma - 1}{\gamma}}}$$

on M. The corresponding change of coordinate on M is $y = h_\gamma(x)$, where

$$h_\gamma(x) = -1 + b \int_{-1}^{x} \frac{dx}{(1-x^2)^{\frac{\gamma-1}{\gamma}}}$$

and b is the constant such that $h_\gamma(1) = 1$. The representation (see Fig. 3.4) of f under the orbifold metric associated to f is

$$\tilde{f} = h_\gamma \circ f \circ h_\gamma^{-1}.$$

Lemma 3.9. *The map $\tilde{f}$ is also a Ulam-von Neumann transformation and $\gamma_{\tilde{f}} = 1$.*

Proof. Conditions **(2)** and **(3)** in Definition 3.6 are easy to check. Now let us check **(1)**. If y is not one of 0, 1 and -1, then $\tilde{f}$ is differentiable at y. Suppose x is the preimage of y under h_γ. By the chain rule,

$$\tilde{f}'(y) = \frac{f'(x)(1-x^2)^{\frac{\gamma-1}{\gamma}}}{\left(1-(f(x))^2\right)^{\frac{\gamma-1}{\gamma}}}.$$

Using this equation, we can get that $\tilde{f}'(0-)$ and $\tilde{f}'(0+)$ exist and equal nonzero numbers and that $\tilde{f}'(-1) = (f'(-1))^{\frac{1}{\gamma}}$ and $\tilde{f}'(1) = -|f'(1)|^{\frac{1}{\gamma}}$. This implies that $\gamma_{\tilde{f}} = 1$. Since f is geometrically finite, it is $C^{1+\alpha'}$ for some $0 < \alpha' \leq 1$. Thus $\tilde{f}$ is $C^{1+\alpha}$ for some $0 < \alpha \leq 1$ when it is restricted on I_0 and I_1. Since h_γ is a $(1/\gamma)$-Hölder continuous function, so $\tilde{f}$ satisfies **(4)** of Definition 3.6 too. ∎

Remark 3.11. The inverse of h_γ is C^1. Suppose $\tilde{f}$ is a Ulam-von Neumann transformation and $\gamma_{\tilde{f}} = 1$. Then for any $\gamma > 1$,

$$f = h_\gamma^{-1} \circ \tilde{f} \circ h_\gamma$$

is a Ulam-von Neumann transformation and $\gamma_f = \gamma$.

Remark 3.12. Let f be a Ulam-von Neumann transformation and let $\gamma \geq 1$ be the exponent of f. Let B mean a Borel subset of M and let $m(B)$ mean the Lebesgue measure of B. A measure ν on M is an invariant measure of f if $\nu(f^{-1}(B)) = \nu(B)$ for all B. A measure ν is equivalent to the Lebesgue measure m (denote as $\nu \sim m$) if $\nu(B) = 0$ if and only if $m(B) = 0$. If $\nu \sim m$, we can consider the Radon-Nikodým derivative $\rho = d\nu/dm$. Then

$\nu(B) = \int_B \rho(x)dx$. If, furthermore, ρ is a continuous function on $(-1, 1)$, then ν is an invariant measure of f if and only if

$$\rho(x) = \sum_{u \in f^{-1}(x)} \frac{\rho(u)}{f'(u)}$$

for all x in $(-1, 1)$. Let $\tilde{f}$ be the representation of f under the orbifold metric $d_\gamma y$ and let $y = h_\gamma(x)$ be the corresponding change of coordinate. Then $\tilde{f}$ is a Ulam-von Neumann transformation with $\gamma_{\tilde{f}} = 1$. One can prove, by using Ruelle's Perron Frobenius operators in §6.1 and by similar arguments in the proof of the existence of the Gibbs measures in §6.1, that $\tilde{f}$ has a unique invariant probability measure

$$\tilde{\nu}(B) = \int_B \tilde{\rho}(y)dy$$

where $\tilde{\rho}$ is a continuous function on M and where $\tilde{\rho}(y) > 0$ for all y in M. Let $\rho(x) = b\tilde{\rho}(h_\gamma(x))$ where

$$b = \left(\int_{-1}^{1} \frac{\tilde{\rho}(h_\gamma(x))}{(1-x^2)^{\frac{\gamma-1}{\gamma}}} dx \right)^{-1}.$$

Then the measure

$$\nu(B) = \int_B \frac{\rho(x)}{(1-x^2)^{\frac{\gamma-1}{\gamma}}} dx$$

is the unique invariant probability measure, which is equivalent to the Lebesgue measure, of f.

Theorem 3.9. *A Ulam-von Neumann transformation f is ergodic.*

Proof. Suppose the exponent of f at 0 is $\gamma \geq 1$. Let $\tilde{f}$ be the representation of f under the orbifold metric $d_\gamma y$. Let $\eta_{n,f}$ and $\tilde{\eta}_{n,f}$ be the nth-partitions induced from f and $\tilde{f}$ respectively. Let $m(\cdot)$ be the Lebesgue measure. (Or let $\nu(\cdot) \sim m(\cdot)$ be the invariant measure of f.)

Suppose X is a f-invariant subset of M and $m(X) > 0$. Then $\tilde{X} = h_\gamma(X)$ is a $\tilde{f}$-invariant subset of M and $m(\tilde{X}) > 0$. Suppose p is a Lebesgue density point of $\tilde{X}$. Then there is a nested sequence of intervals I_n in $\tilde{\eta}_{n,f}$ such that

$$\lim_{n\to\infty} \frac{m(I_n \cap \tilde{X})}{m(I_n)} = 1.$$

From **(4)** of Definition 3.6 and Lemma 3.1, there is a constant $C > 0$ such that

$$C^{-1} \leq \frac{|(\tilde{f}^{\circ n})'(x)|}{|(\tilde{f}^{\circ n})'(y)|} \leq C$$

for all x and y in I_n. Thus

$$\lim_{n \to \infty} \frac{m(\tilde{f}^{\circ n}(I_n \cap \tilde{X}))}{m(\tilde{f}^{\circ n}(I_n))} = 1.$$

But $\tilde{f}^{\circ n}(I_n) = M$ and $\tilde{f}^{\circ n}(I_n \cap \tilde{X}) \subset \tilde{X}$. So $m(\tilde{X}) = m(M)$. This implies that $m(X) = m(M)$, i.e., X has full measure. Hence f is ergodic. ∎

Let f and g be two Ulam-von Neumann transformations and let γ_f and γ_g be the exponents of f and g at 0. Let $\tilde{f} = h_{\gamma_f} \circ f \circ h_{\gamma_f}^{-1}$ and $\tilde{g} = h_{\gamma_g} \circ g \circ h_{\gamma_g}^{-1}$. Let H be the conjugacy from f to g, i.e., $f \circ H = H \circ g$. Let $\tilde{H} = h_{\gamma_f} \circ H \circ h_{\gamma_g}^{-1}$. Then $\tilde{H}$ is the conjugacy from $\tilde{f}$ to $\tilde{g}$.

Theorem 3.10 [JI9]. *Let f and g be two Ulam-von Neumann transformations and let H be the conjugacy from f to g. Suppose the eigenvalues of two Ulam-von Neumann transformations f and g at all corresponding periodic points are the same and $\gamma_f = \gamma_g$. Then $\tilde{H}$ is bi-Lipschitz.*

Proof. From the proof of Lemma 3.9 and the fact that $h_\gamma|(-1,1)$ is C^1 where $\gamma = \gamma_f = \gamma_g$, the eigenvalues of $\tilde{f}$ and $\tilde{g}$ at all corresponding periodic points are the same too. From Lemma 3.1, there is a constant $C > 0$ such that

$$C^{-1} \leq \frac{|(\tilde{f}^{\circ n})'(x)|}{|(\tilde{f}^{\circ n})'(y)|} \leq C$$

for any $n > 0$ and any x and y in $I \in \tilde{\eta}_{n,f}$, and

$$C^{-1} \leq \frac{|(\tilde{g}^{\circ n)})'(x)|}{|(\tilde{g}^{\circ n})'(y)|} \leq C$$

for any $n > 0$ and any x and y in $I \in \tilde{\eta}_{n,g}$. Let $|I|$ be the length of an interval I. For any interval I in $\tilde{\eta}_{n,g}$, the ratio

$$\frac{|\tilde{H}(I)|}{|I|} = \frac{|(\tilde{g}^{\circ n})'(\xi)|}{|(\tilde{f}^{\circ n})'(\tau)|}$$

for some ξ in I and τ in $\tilde{H}(I)$ because both images of I and $\tilde{H}(I)$ under the $\tilde{g}^{\circ n}$ and $\tilde{f}^{\circ n}$ are M. But $\tilde{g}^{\circ n}$ has a fixed point p in I and, from the condition in the lemma,

$$(\tilde{g}^{\circ n})'(p) = (\tilde{f}^{\circ n})'(\tilde{H}(p)).$$

Therefore,

$$C^{-2} \leq \frac{|\tilde{H}(I)|}{|I|} \leq C^2.$$

Now for any x and y in M, the interval bounded by x and y can be written as a union of some intervals $\{I_i\}$ of $\{\tilde{\eta}_{n,\tilde{g}}\}_{n=0}^{\infty}$ where $\{I_i\}$ have pairwise disjoint interiors. Thus we have

$$C^{-2} \leq \frac{|\tilde{H}(x) - \tilde{H}(y)|}{|x-y|} = \frac{\sum_i |\tilde{H}(I_i)|}{\sum_i |I_i|} \leq C^2$$

which means that $\tilde{H}$ is bi-Lipschitz. ∎

Since the post-singular orbit $PSO = \cup_{i=1}^{\infty} f^{\circ i}(0)$ of a Ulam-von Neumann transformation f is just the boundary $\{-1, 1\}$ of M, Theorem 3.6 in this case can be read as

Corollary 3.3 [JI9]. *Let f and g be two Ulam-von Neumann transformations and let H be the conjugacy from f to g. Then H is a $C^{1+\beta}$ diffeomorphism for some $0 < \beta \leq 1$ if and only if H is differentiable at one point in $[-1, 1]$ with non-zero derivative and the exponents and the asymmetries of f and g at 0 are the same.*

Now Theorem 3.10 and Corollary 3.3 imply that

Theorem 3.11 [JI9]. *Let g and f be two Ulam-von Neumann transformations and let H be the conjugacy from g to f. Then H is a $C^{1+\beta}$-diffeomorphism for some $0 < \beta \leq 1$ if and only if the eigenvalues at all corresponding periodic points and the exponents and the asymmetries of f and g at 0 are the same.*

Remark 3.13. Let f be a Ulam-von Neumann transformation. For any $x \in [-1, 1]$, there is a unique sequence $\{x_i\}_{i=0}^{\infty} \subset [-1, 0]$ such that $x_0 = x$ and $f(x_i) = x_{i-1}$ for $i \geq 1$. One can check that $x_i \to -1$ as $i \to \infty$. Let g be another Ulam-von Neumann transformation and let H be the conjugacy from f to g. If H is C^1, then

$$H'(x) = \Big(\prod_{i=1}^{\infty} \frac{g'(H(x_i))}{f'(x_i)}\Big) H'(-1)$$

for all $x \in [-1,1]$. Using this equality, one can further discuss higher smoothness of H if f and g are both $C^{k+\alpha}$ where $k \geq 2$ is an integer and $0 \leq \alpha \leq 1$ is a real number (see [JI12]).

Remark 3.14. Let $\mathbf{S}^1$ be the unit circle in the complex plane. An orientation-preserving circle endomorphism is a C^1 self-map f of $\mathbf{S}^1$ such that $|f'(z)| \neq 0$ for all z in $\mathbf{S}^1$. An example of an orientation-preserving circle endomorphism is $Q_d(z) = z^d$ restricted on $\mathbf{S}^1$ where $d \geq 2$ is an integer. An orientation-preserving circle endomorphism is expanding if there are two constants $C > 0$ and $\lambda > 1$ such that $|(f^{\circ n})'(z)| \geq C\lambda^n$ for all $z \in \mathbf{S}^1$ and all positive integers $n > 0$. It is known that any orientation-preserving C^1-expanding circle endomorphism is topologically conjugate to Q_d for some integer $d \geq 2$ (see [SHU]). The integer $d \geq 2$ is called the degree of f. Thus any two orientation-preserving $C^{1+\alpha}$-expanding circle endomorphisms f and g with the same degree are topologically conjugate. An orientation-preserving $C^{1+\alpha}$-expanding circle endomorphism f of degree $d \geq 2$ has a fixed point p in $\mathbf{S}^1$. Let η_0 be the set of closures of intervals in $\mathbf{S}^1 \setminus f^{-1}(p)$. Then η_0 is a Markov partition of $\mathbf{S}^1$ w.r.t. f. The results in §3.1 to §3.9 apply to this Markov map (f, η_0) (refer to [SHS]). Another direction in this research is to weak the smoothness condition. Cui [CUI] worked out a nice result by considering uniformly asymptotically affine expanding circle maps and by considering the space of nearby scaling functions, which can be defined as the limiting version of the bounded nearby geometry in Definition 3.3. Pinto and Sullivan [PNS] had a similar result.

Remark 3.15. De La Llave [LL1,LL2], Marco and Moriyon [MAM], and De La Llave, Marco, and Moriyon [LMM] studied the smooth classification of Anosov diffeomorphisms of the torus by all eigenvalues at periodic points. Zhang and Li [ZHL] studied some rigidity phenomenon in the family of diffeomorphisms of the real line $\mathbf{R}$ by embedding some of these diffeomorphisms into flows acting on $\mathbf{R}$.

3.10. The Asymptotic Scaling Function of a Family of Hyperbolic Cantor Sets

Suppose $M = [-1,1]$ and f is a C^1 folding mapping from M to the real line $\mathbf{R}$ with a unique power law critical point 0 and $\gamma_f > 0$ is the exponent of f at 0. We assume that f is increasing on $[-1,0]$ and decreasing on $[0,1]$, and $f(0) \geq 1$, $f(-1) = f(1) = -1$. Let g_0 from M to $[-1,0]$ and g_1 from M to $[0,1]$ be the left and right inverse branches of f (see Fig. 2.5). For a sequence $w_n = i_0 \dots i_{n-1}$ of 0's and 1's, let

$$g_{w_n} = g_{i_0} \circ \cdots \circ g_{i_{n-1}}$$

be the composition and let

$$I_{w_n} = g_{w_n}(M)$$

be the image of M under g_{w_n}. We use η_n to denote the set of intervals I_{w_n} for all w_n of length n and $\lambda_n = \max_{I \in \eta_n} |I|$ to denote the maximum of the lengths of the intervals in η_n.

Definition 3.8. *We say f is in $\mathcal{H}$, if*

a) $f(0) > 1$,
b) *f is $C^{1+\alpha}$ for some $0 < \alpha \leq 1$, and*
c) *there are constants $C > 0$ and $0 < \mu < 1$ such that $\lambda_n \leq C\mu^n$ for all integers $n > 0$.*

Suppose f is a map in $\mathcal{H}$. Then $\Lambda_f = \cap_{n=0}^{+\infty} f^{-n}(M)$, the non-escaping set of f, is a Cantor set in the real line $\mathbf{R}$ (see the proof of Theorem 1.1). The dual symbolic space $\Sigma^* = \prod_{-\infty}^{i=0}\{0,1\}$ and the scaling function s of (f, Λ_f) are defined similarly to those in §1.3. Similar to Theorem 1.2, we have

Theorem 3.12. *Suppose f is a map in $\mathcal{H}$. The scaling function s of (f, Λ_f) exists and is Hölder continuous on Σ^*.*

Although the non-escaping set of a Ulam-von Neumann transformation is the interval M, we still have the dual symbolic space $\Sigma^* = \prod_{-\infty}^{i=0}\{0,1\}$. Let s_f be the scaling function of f defined on Σ^*. In §3.6 and §3.7, we showed that this new scaling function exists and is Hölder continuous if $\gamma_f = 1$ and is discontinuous if $\gamma_f > 1$. Moreover, for the scaling function s_f of a Ulam-von Neumann transformation f with $\gamma_f > 1$, we can restate Theorem 3.4 and Corollary 3.1 more precisely (see Fig. 3.5).

Let

$$I^* = \{a^* = \dots i_n \dots i_0 \in \Sigma^* \mid i_{n_k} = 1 \text{ for infinitely many } k\}$$

and O^* be the complement of I^* in Σ^*. The set O^* is countable.

Theorem 3.13 [JI6]. *Suppose f is a Ulam-von Neumann transformation and $\gamma_f > 1$. Then the scaling function s_f exists, and is continuous at every point a^* in I^* and discontinuous at every point a^* in O^*. Moreover, every a^* in O^* is a jump discontinuity of s and the restriction $s|I^*$ is uniformly continuous (see Fig. 3.5).*

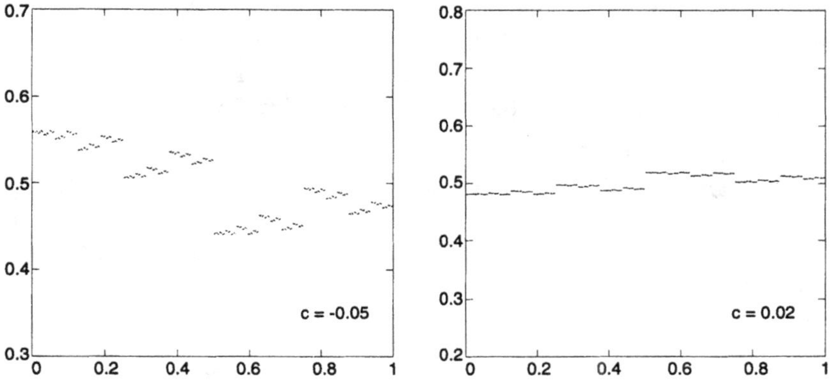

The graphs of the scaling functions for $f_c(x) = -x^2 + 2 + cx^2(4 - x^2)$

Fig. 3.5

Suppose f is a Ulam-von Neumann transformation and $\gamma = \gamma_f > 1$ and $d_\gamma y$ is the orbifold metric on M associated with f. Then $\tilde{f} = h_\gamma \circ f \circ h_\gamma^{-1}$ is a Ulam-von Neumann transformation and $\gamma_{\tilde{f}} = 1$. So the scaling function $s_{\tilde{f}}$ of $\tilde{f}$ is Hölder continuous. This scaling function is actually the scaling function of f under the orbifold metric $d_\gamma y$. We use $\tilde{s}_f$ to denote $s_{\tilde{f}}$.

Corollary 3.4 [JI6]. *Suppose f is a Ulam-von Neumann transformation and $\gamma = \gamma_f > 1$. The scaling function $\tilde{s}_f$ under the orbifold metric $d_\gamma y$ exists and is Hölder continuous, and moreover, it is the continuous extension of $s|I^*$ to Σ^* (see Fig. 3.5).*

Let $\mathcal{BH}$ be the space of Ulam-von Neumann transformations and let $\epsilon_0 > 0$ be a real value.

Definition 3.9. *A family $\{f_\epsilon\}_{0 \leq \epsilon \leq \epsilon_0}$ in $\mathcal{H} \cup \mathcal{BH}$ with $f_\epsilon(0) = 1 + \epsilon$ is said to be asymptotically non-hyperbolic if*

i *$F(x, \epsilon) = f_\epsilon(x)$ is a C^1 function from $[-1, 1] \times [0, \epsilon_0]$ to $\mathbf{R}$ and $f'_\epsilon(x)$ are uniformly α-Hölder continuous functions of x for $0 \leq \epsilon \leq \epsilon_0$,*

ii *f_ϵ has the same exponent γ at 0 for every ϵ in $[0, \epsilon_0]$ and $r_-(x, \epsilon) = f'_\epsilon(x)/|x|^{\gamma-1}$ on $[-1, 0] \times [0, \epsilon_0]$ and $r_+(x, \epsilon) = f'_\epsilon(x)/|x|^{\gamma-1}$ on $[0, 1] \times$*

$[0, \epsilon_0]$ are continuous and $r_+(x, \epsilon)$ and $r_-(x, \epsilon)$ are uniformly α-Hölder continuous functions of x for $0 \leq \epsilon \leq \epsilon_0$, and

iii *there are two constants $C > 0$ and $0 < \mu < 1$ such that $\lambda_{n,\epsilon} \leq C\mu^n$ for all $n > 0$ and all $0 \leq \epsilon \leq \epsilon_0$ (refer to* **c)** *of Definition 3.8).*

Let $\{f_\epsilon\}_{0 \leq \epsilon \leq \epsilon_0}$ be an asymptotically non-hyperbolic family in $\mathcal{H} \cup \mathcal{BH}$. Let $\Lambda_\epsilon = \cap_{n=0}^{\infty} f_\epsilon^{-n}([-1, 1])$ be the non-escaping set of f_ϵ for $0 \leq \epsilon \leq \epsilon_0$. Let s_ϵ be the scaling function of $(f_\epsilon, \Lambda_\epsilon)$ for $0 < \epsilon \leq \epsilon_0$ and let s_0 be the scaling function of f_0 which is a Ulam-von Neumann transformation. The following theorem gives the asymptotic behavior of the scaling function of $(f_\epsilon, \Lambda_\epsilon)$.

Theorem 3.14 [JI6]. *Let $\{f_\epsilon\}_{0 \leq \epsilon \leq \epsilon_0}$ in $\mathcal{H} \cup \mathcal{BH}$ be an asymptotically non-hyperbolic family of folding mappings and let $\{s_\epsilon\}_{0 \leq \epsilon \leq \epsilon_0}$ be the family of corresponding scaling functions. Then*

(I) *for $\delta_0 \in (0, \epsilon_0]$, s_ϵ converges to s_{δ_0} uniformly on Σ^* as ϵ tends to δ_0,*

(II) *for every $a^* \in \Sigma^*$, $s_\epsilon(a^*)$ tends to $s_0(a^*)$ as ϵ goes to zero, and*

(III) *$s_\epsilon | I^*$ converges uniformly to $s_0 | I^*$ as ϵ goes to zero.*

Suppose $\{f_\epsilon\}_{0 \leq \epsilon \leq \epsilon_0}$ in $\mathcal{H} \cup \mathcal{BH}$ is an asymptotically non-hyperbolic family of folding mappings. Let U_ϵ be the closed interval bounded by $f_\epsilon^{-1}(0)$ and V_ϵ be the closure of the complement of U_ϵ in M for every ϵ in $[0, 1]$ (refer to Fig. 2.6).

Lemma 3.10. *There is a constant $C_1 > 0$ such that for every ϵ in $[0, \epsilon_0]$, if $f_\epsilon^{\circ i}(x)$ and $f_\epsilon^{\circ i}(y)$ are in the same connected component of V_ϵ for $i = 0, 1, \ldots, n-1$, then*

$$\log\left(\frac{|(f_\epsilon^{\circ n})'(x)|}{|(f_\epsilon^{\circ n})'(y)|}\right) \leq C_1 |f_\epsilon^{\circ n}(x) - f_\epsilon^{\circ n}(y)|^\alpha.$$

Proof. It follows from Lemma 1.2 since $f_\epsilon'(x) \neq 0$ for x in V_ϵ (refer to the proof of Lemma 3.4). ■

Lemma 3.11. *There is a constant $C_2 > 0$ such that for every ϵ in $[0, \epsilon_0]$, every pair x and y in an interval of $\eta_{n,\epsilon}$ and every $0 < m \leq n$, if $f_\epsilon^{\circ m}(x)$ and $f_\epsilon^{\circ m}(y)$ are in U_ϵ, then*

$$\log\left(\frac{|(f_\epsilon^{\circ m})'(x)|}{|(f_\epsilon^{\circ m})'(y)|}\right) \leq C_2 |f_\epsilon^{\circ m}(x) - f_\epsilon^{\circ m}(y)|^\alpha.$$

Proof. It follows Lemma 3.3 since $d(U_\epsilon, \partial[-1, 1])$ is greater than a constant $C > 0$ for all $0 \leq \epsilon \leq 1$ (refer to the proof of Lemma 3.5). ■

Proof of Theorem 3.14. **(I)** is from Lemma 3.1 and the proof of Theorem 3.3.

A straightforward calculation from Lemmas 3.10 and 3.11 to the cases that $a^* = \ldots w_n$ is in O^* and that $a^* = \ldots w_n$ is in I^*, respectively, implies that there are constants $C = C(a^*) > 0$ and $0 < \alpha \leq 1$ and a subsequence $\{n_i\}$ of the integers such that for all $m \geq k \geq n_i$, and ϵ and ϵ' in $[0, \epsilon_0]$,

$$|s_\epsilon(w_k) - s_{\epsilon'}(w_m)| \leq C\Big(|I_{w_{n_i},\epsilon}|^\alpha + |I_{w_{n_i},\epsilon'}|^\alpha\Big) + |s_\epsilon(w_{n_i}) - s_{\epsilon'}(w_{n_i})|.$$

(II) follows from this inequality directly. **(III)** follows from this inequality and the fact,

$$0 < \sup_{a^* \in I^*} C(a^*) < +\infty.$$

■

3.11. Bounded Nearby Geometry for Circle Mappings

Suppose f is an irrational circle mapping in $\mathcal{S}$ (see §1.4) and suppose

$$\Big\{\{(l_n, L_n); (r_n, R_n)\}\Big\}_{n=0}^{\infty}$$

is the corresponding sequence of commuting pairs to f generated by renormalization (see §1.7). Let $\sigma = \{\sigma_n\}_{n=0}^{\infty}$ be the corresponding sequence of symbols $\pm$ (see §1.7), i.e., $\sigma_n = +$ if $\{(l_n, L_n); (r_n, R_n)\}$ is in Case $(+)$ and $\sigma_n = -$ if $\{(l_n, L_n); (r_n, R_n)\}$ is in Case $(-)$.

Let k_1 be the number of consecutive $+$ or $-$ starting from σ_0, k_2 be the number of consecutive $-$ or $+$ starting from σ_{k_1}, and so on. Then σ is k_1 $+$ or $-$ followed by k_2 $-$ or $+$ and followed by k_3 $+$ or $-$ and so on. Let us denote σ as $\sigma = (+)^{k_1}(-)^{k_2}(+)^{k_3}\ldots$ or $\sigma = (-)^{k_1}(+)^{k_2}(-)^{k_3}\ldots$.

Remark 3.16. We call the sequence $(k_1, k_2, \ldots)$ the combinatorics of f. It relates with the continuous fractional expansion of the rotation number $\rho(f)$.

If $\sigma_j = +$ for $0 \leq j \leq k_1 - 1$, then $L_j = L_0$ for $0 \leq j \leq k_1$ and $R_{k_1} \subset R_{k_1-1} \subset \cdots \subset R_1 \subset R_0$, and moreover, $\{(l_j, L_j); (r_j, R_j)\}$ is in Case $(+)$ for $0 \leq j \leq k_1 - 1$ and $\{(l_{k_1}, L_{k_1}); (r_{k_1}, R_{k_1})\}$ is in Case $(-)$. Similarly, If $\sigma_j = -$ for $0 \leq j \leq k_1 - 1$, then $L_{k_1} \subset L_{k_1-1} \subset \cdots \subset L_1 \subset L_0$ and $R_j = R_0$ for $0 \leq j \leq k_1$, and moreover, $\{(l_j, L_j); (r_j, R_j)\}$ is in Case $(-)$ for $0 \leq j \leq k_1 - 1$ and $\{(l_{k_1}, L_{k_1}); (r_{k_1}, R_{k_1})\}$ is in Case $(+)$. Let

$$fl_1 = l_{k_1-1} \qquad \text{and} \qquad fr_1 = r_{k_1-1}$$

and

$$FL_1 = L_{k_1-1} \qquad \text{and} \qquad FR_1 = R_{k_1-1}.$$

In general, for each $i \geq 1$, let $n_i = \sum_{j=1}^{i} k_j$. (Set $n_0 = 0$.) Then

$$\sigma_{n_i} \ldots \sigma_{n_{i+1}-1} = \underbrace{+\ldots+}_{k_{i+1}} \qquad \text{or} \qquad \underbrace{-\ldots-}_{k_{i+1}}.$$

In the case $\underbrace{+\ldots+}_{k_{i+1}}$, $L_j = L_{n_i}$ for $n_i \leq j \leq n_{i+1}$ and $R_{n_{i+1}} \subset R_{n_{i+1}-1} \subset \cdots \subset R_{n_i+1} \subset R_{n_i}$, and moreover, $\{(l_j, L_j); (r_j, R_j)\}$ is in Case $(+)$ for $n_i \leq j \leq n_{i+1} - 1$ and $\{(l_{n_{i+1}}, L_{n_{i+1}}); (r_{n_{i+1}}, R_{n_{i+1}})\}$ is in Case $(-)$. Similarly, in the case $\underbrace{-\ldots-}_{k_{i+1}}$, $L_{n_{i+1}} \subset L_{n_{i+1}-1} \subset \cdots \subset L_{n_i+1} \subset L_{n_i}$ and $R_j = R_{n_i}$ for

$n_i \leq j \leq n_{i+1}$, and moreover, $\{(l_j, L_j); (r_j, R_j)\}$ is in Case $(-)$ for $n_i \leq j \leq n_{i+1} - 1$ and $\{(l_{n_{i+1}}, L_{n_{i+1}}); (r_{n_{i+1}}, R_{n_{i+1}})\}$ is in Case $(+)$. Let

$$fl_{i+1} = l_{n_{i+1}-1} \qquad \text{and} \qquad fr_{i+1} = r_{n_{i+1}-1}$$

and

$$FL_{i+1} = L_{n_{i+1}-1} \qquad \text{and} \qquad FR_{i+1} = R_{n_{i+1}-1}.$$

We call the sequence

$$\Big\{\{(fl_i, FL_i); (fr_i, FR_i)\}\Big\}_{i=1}^{\infty}$$

the sequence of faster renormalizations of f. Note that

$$fl_i = f^{\circ a_i} - b_i \qquad \text{and} \qquad fr_i = f^{\circ A_i} - B_i$$

for some integers a_i, b_i, A_i, and B_i. Let $|I|$ mean the length of an interval I.

Definition 3.10. *Suppose f is an irrational circle mapping in $\mathcal{S}$. Let*

$$\Big\{\{(fl_{i,t}, FL_{i,t}); (fr_{i,t}, FR_{i,t})\}\Big\}_{i=1}^{\infty}$$

be the corresponding sequence of faster renormalizations of $f_t(x) = f(x+t)-t$. Then f is said to have bounded nearby geometry if there is a constant $C > 0$ such that

$$\left|\log\left(\frac{|FL_{i,t}|}{|FR_{i,t}|}\right)\right| \leq C$$

for all integers $i > 0$ and all $0 \leq t < 1$.

For any real number x, we can find a unique number $y(x)$ in $[f(0)-1, f(0))$ such that $x - y = m$ is an integer. Let $\chi : x \mapsto y(x)$. For every interval I such that $|I| < 1$, let $J = \chi(I) \subseteq [f(0)-1, f(0))$. We do not distinguish x and $y(x)$ and I and J if there is no confusion.

Theorem 3.15. *A C^{1+bv} irrational circle diffeomorphism has bounded nearby geometry.*

Proof. Let us prove it for $t = 0$ (for arbitrary $0 \leq t < 1$, the proof is similar). Suppose

$$\{(fl_i, FL_i); (fr_i, FR_i)\}$$

is in Case (+) (in Case (−), the argument is similar), where

$$fl_i = l_{n_i-1}, \qquad fr_i = r_{n_i-1}$$

and

$$FL_i = L_{n_i-1}, \qquad FR_i = R_{n_i-1}.$$

Then $\{(l_{n_i}, L_{n_i}); (r_{n_i}, R_{n_i})\}$ is in Case (−). There are positive integers A_i and B_i such that

$$fr_i = f^{\circ A_i} - B_i.$$

Let E_0 (see Fig. 3.6) be the preimage of FL_i under fr_i and $E_m = f^{\circ m}(E_0)$ (in $[f(0)-1, f(0)]$). Let $H_0 = FR_i \setminus E_0$. Then we have that $E_{A_i} = FL_i$ and $fr_i(H_0) \subseteq E_0$ (see Fig. 3.6). There is an integer $m_i \geq A_i$ such that $\cup_{m=0}^{m_i}(E_m \cup E_{A_i+m})$ covers $[f(0)-1, f(0)]$ but at most twice. Let m_i be the smallest such an integer. From the proofs of Lemma 1.12 and Lemma 1.3, there is a constant $C_1 > 0$ such that

$$\left|\log\left(\frac{(f^{\circ m})'(x)}{(f^{\circ m})'(y)}\right)\right| \leq C_1$$

for $0 \leq m \leq m_i$ and $x, y \in E_0 \cup E_{A_i}$ (or $x, y \in E_{A_i} \cup E_{2A_i}$).

Let $\tau_m = |E_m|/|E_{A_i+m}|$. Then

$$e^{-C_1}\tau_0 \leq \tau_m \leq e^{C_1}\tau_0 \qquad \text{and} \qquad e^{-C_1}\tau_{A_i} \leq \tau_{m+A_i} \leq e^{C_1}\tau_{A_i}$$

for $0 \leq m \leq m_i$. This implies that for $0 \leq m \leq m_i$,

$$|E_m| + |E_{A_i+m}| = (\tau_m + 1)\tau_{A_i+m}|E_{2A_i+m}| \leq e^{3C_1}(\tau_0 + 1)\tau_0|E_{2A_i+m}|.$$

Moreover,

$$1 \leq \sum_{m=0}^{m_i}\Big(|E_m| + |E_{A_i+m}|\Big) \leq e^{3C_1}(\tau_0+1)\tau_0 \sum_{m=0}^{m_i}|E_{2A_i+m}| \leq 2e^{3C_1}(\tau_0+1)\tau_0.$$

Hence there is a constant $C_2 > 0$ such that $\tau_0 \geq C_2^{-1}$ for all $i \geq 1$. By the similar argument and by considering $f^{-m}|E_{-A_i} \cup E_0 \cup E_{A_i}$, we can get $\tau_0^{-1} \geq C_2^{-1}$.

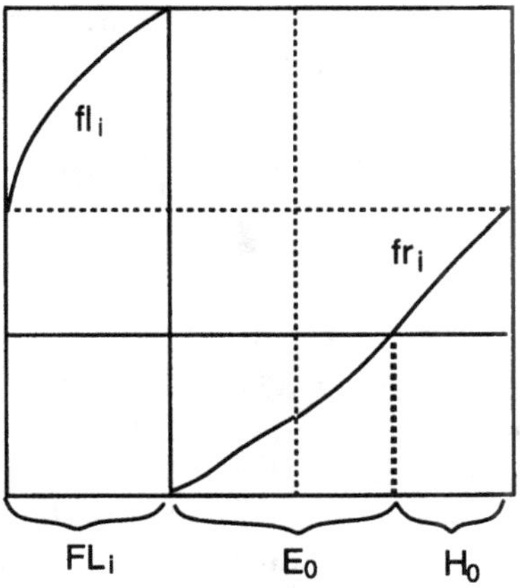

Fig. 3.6

Lemma 1.12 says that f has bounded distortion. Thus there is a constant $C_3 > 0$ such that

$$\left|\log\left(\frac{(fr_i)'(x)}{(fr_i)'(y)}\right)\right| \leq C_3$$

for any x and y in FR_i. Since $fr_i(H_0) \subseteq E_0$ and since $fr_i(E_0) = E_{A_i}$,

$$\frac{|H_0|}{|E_0|} = \frac{(fr_i)'(x)}{(fr_i)'(y)} \frac{|fr_i(H_0)|}{|E_{A_i}|} \leq e^{C_3}\tau_0 \leq C_4 = e^{C_3}C_2,$$

where x and y are in FR_i. Because $FL_i = E_{A_i}$ and $FR_i = E_0 \cup H_0$, we then have a constant $C > 0$ such that

$$C^{-1} \leq \frac{|FL_i|}{|FR_i|} \leq C$$

for all $i > 0$. ∎

Remark 3.17. Using the bounded nearby geometry property, we can discuss by using the method in §3.5, the quasisymmetric property of the conjugacy between two C^{1+bv}-irrational circle diffeomorphisms whose combinatorics $(k_1, k_2, \ldots)$ is bounded, i.e., $k_i \leq N_0$ for all $i > 0$ where N_0 is a constant integer.

Corollary 3.5. *Suppose f is a C^{1+bv} circle diffeomorphism and*

$$\Big\{\{(fl_{i,t}, FL_{i,t}); (fr_{i,t}, FR_{i,t})\}\Big\}_{i=1}^{\infty}$$

is the corresponding sequence of faster renormalizations of $f_t(x) = f(x+t)-t$. *Then there is a constant* $0 < \mu < 1$ *such that*

$$\frac{|FL_{i+2,t}|}{|FL_{i,t}|} \leq \mu \qquad \text{and} \qquad \frac{|FR_{i+2,t}|}{|FR_{i,t}|} \leq \mu$$

for all $i > 0$ *and all* $0 \leq t < 1$.

Proof. Let us prove it for $t = 0$ (for all $0 \leq t < 1$, the argument is similar). First let us prove it when $\{(fl_i, FL_i); (fr_i, FR_i)\}$ is in Case (+). In this case $fl_i = l_{n_i-1}$ and $fr_i = r_{n_i-1}$, $\{(l_{n_i}, L_{n_i}); (r_{n_i}, R_{n_i})\}$ is in Case (−), and $L_{n_i} = FL_i$ and $R_{n_i} = FR_{i+1} \subset FR_i$.

If $k_{i+1} = 1$, then $FL_{i+1} = FL_i$. Let E_0 be the preimage of FL_i under fr_i and $H_0 = FR_i \setminus E_0$. Then E_0 and H_0 is mapped onto $FL_{i+1} = FL_i$ and onto FR_{i+1} by fr_i (see Fig. 3.6). By Lemma 1.12, there is a constant $C_1 > 0$ such that

$$\left|\log\left(\frac{(fr_i)'(x)}{(fr_i)'(y)}\right)\right| \leq C_1$$

for all x and y in FR_i. Then, following Theorem 3.15, there is a constant $C_2 > 0$ such that

$$C_2^{-1} \leq \frac{|H_0|}{|E_0|} \leq C_2.$$

Since $FR_{i+1} \subseteq E_0$ and $FR_i = E_0 \cup H_0$, we have a constant $0 < \mu < 1$ so that

$$\frac{|FR_{i+1}|}{|FR_i|} \leq \mu.$$

If $k_{i+1} > 1$, then $FL_{i+1} = L_{n_{i+1}-1} \subset FL_i$. Consider $\{(l_{n_i}, L_{n_i}); (r_{n_i}, R_{n_i})\}$ which is in Case (−). Let H_2 be the preimage of FR_{i+1} under l_{n_i}. Let E_2 be the preimage of FL_{i+1} under l_{n_i} (see Fig. 3.7). By Lemma 1.12, there is a constant $C_3 > 0$ such that

$$\left|\log\left(\frac{(l_{n_i})'(x)}{(l_{n_i})'(y)}\right)\right| \leq C_3$$

for all x and y in L_{n_i}. This and Theorem 3.15 imply that there is a constant $C_4 > 0$ such that

$$C_4^{-1} \leq \frac{|H_2|}{|E_2|} \leq C_4.$$

Let E_1 and H_1 be intervals in FR_i (see Fig. 3.7) such that

$$l_{n_i}^{\circ(k_{i+1}-2)} \circ fr_i(E_1) = E_2$$

and

$$l_{n_i}^{\circ(k_{i+1}-2)} \circ fr_i(H_1) = H_2.$$

From the proof of Lemma 1.12 (refer to Fig. 3.7), there is a constant $C_5 > 0$ such that

$$\left|\log\left(\frac{(l_{n_i}^{\circ(k_{i+1}-2)} \circ fr_i)'(x)}{(l_{n_i}^{\circ(k_{i+1}-2)} \circ fr_i)'(y)}\right)\right| \leq C_5$$

for all $x, y \in E_1 \cup H_1$. This and the previous inequalities imply that

$$(e^{C_5}C_4)^{-1} \leq \frac{|H_1|}{|E_1|} \leq e^{C_5}C_4.$$

Since $FR_{i+1} \subseteq E_1$ and $E_1 \cup H_1 \subseteq FR_i$, we have a constant $0 < \mu < 1$ such that

$$\frac{|FR_{i+1}|}{|FR_i|} \leq \mu.$$

Similarly, if $\{(fl_i, FL_i); (fr_i, FR_i)\}$ is in Case $(-)$,

$$\frac{|FL_{i+1}|}{|FL_i|} \leq \mu.$$

Since the signs of the faster renormalizations of f are alternate, so we have that

$$\frac{|FL_{i+2}|}{|FL_i|} \leq \mu \qquad \text{and} \qquad \frac{|FR_{i+2}|}{|FR_i|} \leq \mu$$

for all $i \geq 1$.

Fig. 3.7

∎

Definition 3.11. *We call the number*

$$\mu_f = \limsup_{i\to\infty}\left\{\sup_{0\leq t<1}\left\{\max\left\{\sqrt{\frac{|FL_{i+2,t}|}{|FL_{i,t}|}}, \sqrt{\frac{|FR_{i+2,t}|}{|FR_{i,t}|}}\right\}\right\}\right\}$$

the geometric factor of f.

From Corollary 3.5, we have

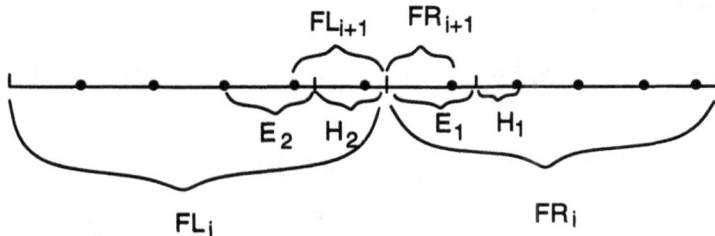

Corollary 3.6. *The geometric factor of a C^{1+bv} irrational circle mapping is strictly less than one.*

Definition 3.12. *Suppose f is an irrational circle mapping. Let $(k_1, k_2, \ldots)$ be the combinatorics of f and let μ_f be the geometric factor of f. We say that the geometry and the combinatorics of f are compatible if*

$$\sum_{i=1}^{\infty} k_i \left(\sqrt{\mu_f + \epsilon}\right)^i < \infty$$

for some $\epsilon > 0$

A sequence of integers $(k_1, k_2, \ldots,)$ is said to be exponentially controlled if $\sum_{i=1}^{\infty} k_i \mu^i$ converges for every $0 \leq \mu < 1$. For example, the sequence $(1^s, 2^s, \ldots i^s, \ldots,)$ is exponentially controlled where $s > 0$ is an integer.

Corollary 3.7. *The geometry and the combinatorics of a C^{1+bv} irrational circle mapping whose combinatorics $(k_1, k_2, \ldots)$ is exponentially controlled are compatible.*

Remark 3.18. Let f be a circle mapping in $\mathcal{S}$ and let $\{(l, L); (r, R)\}$ be the induced commuting pair. Let $\{(fl_i, FL_i); (fr_i, FR_i)\}$ be the ith faster renormalization of f. Let $\alpha_i = fl_i(0) - fr_i(0)$ and let $h_i(x) = \alpha_i x$ be the linear rescaling. Define

$$\mathcal{R}^{oi}(l) = h_i^{-1} \circ fl_i \circ h_i, \qquad \mathcal{R}^{oi}(r) = h_i^{-1} \circ fr_i \circ h_i$$

and

$$\mathcal{R}^{oi}(L) = h_i^{-1}(FL_i), \qquad \mathcal{R}^{oi}(R) = h_i^{-1}(FR_i).$$

Then

$$\{(\mathcal{R}^{oi}(l), \mathcal{R}^{oi}(L)); (\mathcal{R}^{oi}(r), \mathcal{R}^{oi}(R))\}$$

is a commuting pair with $|\mathcal{R}^{oi}(L) \cup \mathcal{R}^{oi}(R)| = 1$. Moreover, $\mathcal{R}^{oi}(l)(0) - 1 = \mathcal{R}^{oi}(r)(0)$. Therefore, we can define a circle mapping $\mathcal{R}^{oi}(f)$ in $\mathcal{S}$ by first defining

$$\mathcal{R}^{oi}(f)(x) = \begin{cases} \mathcal{R}^{oi}(r)(x) + 1, & \text{if } x \in [0, \mathcal{R}^{oi}(l)(0)]; \\ \mathcal{R}^{oi}(l)(x-1) + 1, & \text{if } x \in [\mathcal{R}^{oi}(l)(0), 1] \end{cases}$$

and then by defining $\mathcal{R}^{oi}(f)(x+n) = \mathcal{R}^{oi}(f)(x) + n$ for any $x \in [0,1]$ and any integer n. We call $\mathcal{R}^{oi}(f)$ the ith renormalization of f. If the combinatorics of

f is $(k_1, k_2, \ldots)$, then the combinatorics of $\mathcal{R}^{\circ i}(f)$ is $(k_{i+1}, k_{i+2}, \ldots)$. Suppose $\mathbf{k} = (k, k, \ldots)$ is a constant sequence where $k \geq 1$ is an integer. Let $\mathcal{S}_{\mathbf{k}}$ be the space of all circle mappings whose combinatorics is $\mathbf{k}$. Then $\mathcal{R}^{\circ i}(\mathcal{S}_{\mathbf{k}}) \subseteq \mathcal{S}_{\mathbf{k}}$. The renormalization operator $\mathcal{R}$ has a fixed point $R_\rho(x) = x + \rho$ which is the rigid rotation in $\mathcal{S}_{\mathbf{k}}$. Consider $\mathcal{S}^3_{\mathbf{k}}$ as the subspace of all C^3 circle diffeomorphisms in $\mathcal{S}_{\mathbf{k}}$. Lemma 3.13 in the next section will say that for every f in $\mathcal{S}^3_{\mathbf{k}}$, $\mathcal{R}^{\circ i}(f)$ (more precisely, the induced commuting pair from $\mathcal{R}^{\circ i}(f)$) converges exponentially to the rigid rotation R_ρ (more precisely, the induced commuting pair from R_ρ) in the C^1-topology.

3.12. Herman's Theorem

In this section, we prove that Herman's theorem holds for an irrational circle diffeomorphism whose geometry and combinatorics are compatible. To avoid technical difficulty, we discuss it for a C^3 irrational circle diffeomorphism.

Theorem 3.16 (Herman's theorem). *Every C^3 irrational circle diffeomorphism f whose geometry and combinatorics are compatible is C^1 conjugate to the rigid rotation $R_\rho(x) = x + \rho$ by the same rotation number ρ.*

We apply the estimate in §3.11. If h is a C^1 orientation-preserving diffeomorphism of $\mathbf{R}$ satisfying $h(x+1) = h(x) + 1$ and $h \circ f = R_\rho \circ h \pmod 1$, then

$$\frac{h'(f(x))}{h'(x)} = \frac{1}{f'(x)}$$

for all real numbers x where h' is a positive continuous periodic function of period one. Conversely, we have that

Lemma 3.12. *Suppose f is a C^1 circle diffeomorphism in $\mathcal{S}$ and suppose ϕ is a positive continuous periodic function of period one defined on the real line satisfying*

$$\frac{\phi(f(x))}{\phi(x)} = \frac{1}{f'(x)}.$$

Then f is C^1 conjugate to the rigid rotation R_ρ by the same rotation number ρ.

Proof. Let $h(x) = b\int_0^x \phi(\xi)\, d\xi$ where $b = 1/\int_0^1 \phi(\xi) d\xi$. Then h is a C^1 orientation-preserving diffeomorphism satisfying

$$h(x+1) = h(x) + 1 \qquad \text{and} \qquad h(f(x)) = h(x) + \rho$$

where $\rho = h(f(0))$ is the rotation number of f. ∎

Let f be a C^3-irrational circle diffeomorphism. For any real number x, we can find a unique number $y(x)$ in $[f(0)-1, f(0))$ such that $x - y = m$ is an integer. Let $\chi : x \mapsto y(x)$. For every interval I such that $|I| < 1$, let $J = \chi(I) \subseteq [f(0)-1, f(0))$. We do not distinguish x and $y(x)$, and I and J if there is no confusion.

Let $x_n = f^{\circ n}(0)$ be the image of 0 under $f^{\circ n}$ and $O = \{x_n\}_{n=0}^{\infty}$ be the orbit of x. Then O is dense in $[f(0)-1, f(0))$ by Lemmas 1.7, 1.11, and 1.12. Let us define $\phi_0(x_0) = 1$ and

$$\phi_0(x_n) = \prod_{i=0}^{n-1} \frac{1}{f'(x_i)}$$

for $n \geq 1$. Then it is easy to check that

$$\frac{\phi_0(f(x_n))}{\phi_0(x_n)} = \frac{1}{f'(x_n)}.$$

If ϕ_0 is uniformly continuous and is bigger than a positive constant on O, then it can be extended to a positive continuous function on $[f(0)-1, f(0)]$ and then to a positive continuous periodic function ϕ of period one on the real line $\mathbf{R}$ satisfying the condition in Lemma 3.12. So f is C^1 conjugate to the rigid rotation R_ρ by the same rotation number ρ from Lemma 3.12. Hence we have an equivalent statement of Theorem 3.16.

Theorem 3.17 (Herman's theorem). *Suppose f is a C^3 irrational circle diffeomorphism in $\mathcal{S}$ whose geometry and combinatorics are compatible. Then ϕ_0 is uniformly continuous on O and bigger than a positive constant.*

Suppose f is a C^1 irrational circle mapping and suppose

$$\Big\{\{(fl_{i,t}, FL_{i,t}); (fr_{i,t}, FR_{i,t})\}\Big\}_{i=1}^{\infty}$$

is the corresponding sequence of faster renormalizations of $f_t(x) = f(x+t)-t$. Let

$$\sigma = (+)^{k_1}(-)^{k_2}\ldots \qquad \text{or} \qquad (-)^{k_1}(+)^{k_2}\ldots$$

be the corresponding sequence of symbols $\pm$. Denote

$$d_i = \sup_{0\leq t<1} \Big\{\max\{\max_{x\in FL_{i,t}} |\log((fl_{i,t})'(x))|, \max_{x\in FR_{i,t}} |\log((fr_{i,t})'(x))|\}\Big\}.$$

The next lemma is crucial in the proof of Theorem 3.17.

Lemma 3.13. *Suppose f is a C^3 irrational circle diffeomorphism and $0 \leq \mu_f < 1$ is the geometric factor of f. Then*

$$\limsup_{i\to\infty}\{d_i^{\frac{1}{i}}\} \leq \sqrt{\mu_f}.$$

Proof. Recall $fl_{i,t} = f_t^{\circ a_i} - b_i$ and $fr_{i,t} = f_t^{\circ A_i} - B_i$, where a_i, b_i, A_i, and B_i are positive integers. Then

$$fl_{i,t}(x) = f^{\circ a_i}(x+t) - t - b_i \qquad \text{and} \qquad fr_{i,t}(x) = f^{\circ A_i}(x+t) - t - B_i,$$

and

$$FL_{i,t} = [fr_{i,t}(0), 0] \qquad \text{and} \qquad FR_{i,t} = [0, fl_{i,t}(0)].$$

Suppose $t_0(i)$ minimizes $|FR_{i,t}|$. Let us assume $t_0(i) = 0$. Then

$$\frac{d}{dt}(fl_{i,t}(0))|_{t=0} = (f^{\circ a_i})'(0) - 1 = 0.$$

Let $m_i > 0$ be the smallest positive integer such that $\cup_{m=0}^{m_i} f^{\circ m}(FL_i)$ covers $[f(0)-1, f(0)]$ but at most twice.

For every $x \in [f(0)-1, f(0)]$, let $m > 0$ be the first integer such that $z = f^{-m}(x)$ is in FL_i. For $f^{\circ(-m+a_i)}(x)$, we have

$$\log\left(\left(f^{\circ(-m+a_i)}\right)'(x)\right) = \log\left(\left(f^{-m}\right)'(f^{\circ a_i}(x))\right) + \log\left(\left(f^{\circ a_i}\right)'(x)\right),$$

and

$$\log\left(\left(f^{\circ(-m+a_i)}\right)'(x)\right) = \log\left(\left(f^{\circ a_i}\right)'(f^{-m}(x))\right) + \log\left(\left(f^{-m}\right)'(x)\right).$$

This implies that

$$\begin{aligned}\left|\log\left(\left(f^{\circ a_i}\right)'(x)\right)\right| &= \left|\log\left(\left(f^{\circ(-m+a_i)}\right)'(x)\right) - \log\left(\left(f^{-m}\right)'(f^{\circ a_i}(x))\right)\right| \\ &= \Big|\left(\log\left(\left(f^{\circ a_i}\right)'(f^{-m}(x))\right) - \log\left(\left(f^{\circ a_i}\right)'(0)\right)\right) \\ &\quad - \left(\log\left(\left(f^{-m}\right)'(f^{\circ a_i}(x))\right) - \log\left(\left(f^{-m}\right)'(x)\right)\right)\Big| \\ &\le \|N(f^{\circ a_i})\| \cdot |FL_i| + \|N(f^{\circ m})\| \cdot |FL_i \cup FR_i|\end{aligned}$$

where $N(\cdot)$ is the nonlinearity (see §2.1) and $\|N(\cdot)\|$ is the super norm over $[f(0)-1, f(0)]$. We note that $\log\left(\left(f^{\circ a_i}\right)'(0)\right) = 0$, that z and $fl_i(z)$ are in $FL_i \cup FR_i$, and that

$$\begin{aligned}&\left|\log\left(\left(f^{-m}\right)'(f^{\circ a_i}(x))\right) - \log\left(\left(f^{-m}\right)'(x)\right)\right| \\ &= \left|\log\left(\left(f^{\circ m}\right)'(f^{\circ a_i}(z))\right) - \log\left(\left(f^{\circ m}\right)'(z)\right)\right|.\end{aligned}$$

From Theorem 3.15, there is a constant $C > 0$ such that

$$\left|\log\left(\left(f^{\circ a_i}\right)'(x)\right)\right| \leq C\Big(||N(f^{\circ a_i})|| + ||N(f^{\circ m})||\Big)|FR_i|.$$

The condition that f is a C^3 diffeomorphism is used to estimate $||N(f^{\circ k})||$. (For a $C^{2+\alpha}$ irrational circle diffeomorphism for some $0 < \alpha \leq 1$, the reader may refer to [KAO,KHS] for the estimation of $||N(f^{\circ k})||$ and may consider the discrete form of the Schwarzian derivative.) Recall two relations between Schwarzian derivative and nonlinearity (see §2.1)

$$S(f^{\circ k}) = \sum_{j=0}^{k-1} S(f) \circ f^{\circ j} \cdot ((f^{\circ j})')^2$$

and

$$S(f^{\circ k}) = (N(f^{\circ k}))' - \frac{1}{2}(N(f^{\circ k}))^2.$$

Suppose $N(f^{\circ k})$ takes the maximum value at z_0. Then $\big(N(f^{\circ k})\big)'(z_0) = 0$ and

$$\begin{aligned} ||N(f^{\circ k})||^2 = |N(f^{\circ k})(z_0)|^2 &= 2\left|\sum_{j=0}^{k-1} S(f)(f^{\circ j}(z_0)) \cdot \left((f^{\circ j})'(z_0)\right)^2\right| \\ &\leq 2||S(f)|| \sum_{j=0}^{k-1} \left((f^{\circ j})'(z_0)\right)^2. \end{aligned}$$

Let $\nu_i = \max_{0 \leq t < 1} |FL_{i,t}|$. From the fact that $f^{\circ k}|FL_{i,z_0}$ for $k = a_i$ or m has bounded distortion (refer to Lemma 1.12 in §1.11 and Theorem 3.15), there are constants we denote all of them as $C > 0$ such that for $0 \leq j \leq k$,

$$|(f^{\circ j})'(z_0)| \leq C\frac{|f^{\circ j}(FL_{i,z_0})|}{|FL_{i,z_0}|} \leq C\frac{|FL_{i,f^{\circ j}(z_0)}|}{|FR_{i,z_0}|} \leq C\frac{\nu_i}{|FR_i|}.$$

Similarly,

$$\sum_{j=0}^{k-1} |(f^{\circ j})'(z_0)| \leq C\sum_{j=0}^{k-1} \frac{|f^{\circ j}(FL_{i,z_0})|}{|FL_{i,z_0}|} \leq \frac{C}{|FR_i|}$$

for $k = m$ or a_i because $\sum_{j=0}^{k-1} |f^{\circ j}(FL_{i,z_0})| \leq 2$. Therefore,

$$||N(f^{\circ k})||^2 \leq 2||S(f)|| \sum_{j=0}^{k-1} \left((f^{\circ j})'(z_0)\right)^2 \leq C||S(f)||\frac{\nu_i}{|FR_i|^2}$$

for $k = m$ or a_i. Hence,

$$\left|\log\left(\left(f^{\circ a_i}\right)'(x)\right)\right| \leq C\sqrt{\nu_i}$$

for any $x \in [f(0)-1, f(0)]$. Since $\nu_i = FL_{i,t}$ for some t in $[0,1)$,

$$\sqrt{\nu_i} = \sqrt{\frac{FL_{i,t}}{FL_{i-2,t}}\frac{FL_{i-2,t}}{FL_{i-4,t}}\cdots\frac{FL_{i-2[\frac{i}{2}]+2,t}}{FL_{i-2[\frac{i}{2}],t}}}$$

where $[\frac{i}{2}]$ is the maximum of integers $n \leq i/2$. Thus

$$\limsup_{i\to\infty}\left(\sup_{0\leq t<1}\left\{\max_{x\in FL_{i,t}}\left\{\left|\log\left(\left(fl_{i,t}\right)'(x)\right)\right|\right\}\right\}\right)^{\frac{1}{i}} \leq \sqrt{\mu_f}.$$

A similar argument can be used to prove that for every $x \in [f(0)-1, f(0)]$,

$$\left|\log\left(\left(f^{\circ A_i}\right)'(x)\right)\right| \leq C\sqrt{\nu_i'}$$

where $\nu_i' = \max_{0\leq t<1}|FR_{i,t}|$. Thus

$$\limsup_{i\to\infty}\left(\sup_{0\leq t<1}\left\{\max_{x\in FR_{i,t}}\left\{\left|\log\left(\left(fr_{i,t}\right)'(x)\right)\right|\right\}\right\}\right)^{\frac{1}{i}} \leq \sqrt{\mu_f}.$$

∎

Proof of Theorem 3.17. Suppose $(k_1, k_2, \ldots)$ is the combinatorics of f and suppose

$$\left\{\{(fl_{i,t}, FL_{i,t}); (fr_{i,t}, FR_{i,t})\}\right\}_{i=1}^{\infty}$$

is the corresponding sequence of faster renormalizations of f_t.

Suppose

$$\{(fl_{i,x_n}, FL_{i,x_n}); (fr_{i,x_n}, FR_{i,x_n})\}$$

is in Case $(-)$ (in Case $(+)$, the argument is similar). Remember that $fl_{i,x_n} = l_{n_i-1,x_n}$ and $fr_{i,x_n} = r_{n_i-1,x_n}$ and that

$$\{(l_{n_i,x_n}, L_{n_i,x_n}); (r_{n_i,x_n}, R_{n_i,x_n})\}$$

is in Case $(+)$. Suppose

$$l_{n_i,x_n} = fl_{i,x_n} = f_{x_n}^{\circ e_i} - d_i$$

and

$$r_{n_i,x_n} = fl_{i,x_n} \circ fr_{i,x_n} = f_{x_n}^{\circ E_i} - D_i.$$

Then

$$R_{n_i,x_n} = [0, x_{n+e_i} - x_n] \qquad \text{and} \qquad L_{n_i,x_n} = [x_{n+E_i} - x_n, 0].$$

The interval $FR_{i,x_n} = R_{n_i,x_n}$ is cut by k_{i+1} points $x_{n+e_i+jE_i} - x_n$ for $1 \leq j \leq k_{i+1}$ and

$$\frac{\phi_0(x_{n+e_i+jE_i})}{\phi_0(x_{n+e_i+(j-1)E_i})} = \frac{1}{(fl_{i,x_n} \circ fr_{i,x_n})'(x_{n+a_i+(j-1)E_i} - x_n)}.$$

We also have

$$\frac{\phi_0(x_{n+e_i})}{\phi_0(x_n)} = \frac{1}{(fl_{i,x_n})'(0)}$$

Since the geometry and the combinatorics of f are compatible, there is a small number $\epsilon > 0$ such that $\mu_f + \epsilon < 1$ and such that

$$\sum_{i=1}^{\infty} k_i \left(\sqrt{\mu_f + \epsilon}\right)^i$$

converges. Now Lemma 3.13 implies that there is an integer $N > 0$ such that for $i \geq N$,

$$\left|\log\left(\frac{\phi_0(x_{n+e_i+jE_i})}{\phi_0(x_{n+e_i+(j-1)E_i})}\right)\right| \leq 2\left(\sqrt{\mu_f + \epsilon}\right)^i$$

and

$$\left|\log\left(\frac{\phi_0(x_{n+e_i})}{\phi_0(x_n)}\right)\right| \leq \left(\sqrt{\mu_f + \epsilon}\right)^i.$$

Therefore,

$$\left|\log\left(\frac{\phi_0(x_{n+e_i+jE_i})}{\phi_0(x_n)}\right)\right| \leq \left(2k_{i+1} + 1\right)\left(\sqrt{\mu_f + \epsilon}\right)^i$$

for $0 \leq j \leq k_{i+1}$. Hence for every x_q in FR_{i,x_n},

$$\left|\log\left(\frac{\phi_0(x_q)}{\phi_0(x_n)}\right)\right| \leq \sum_{j\geq i}\left(2k_{j+1} + 1\right)\left(\sqrt{\mu_f + \epsilon}\right)^j.$$

Similarly, for every x_q in FL_{i,x_n},

$$\left|\log\left(\frac{\phi_0(x_q)}{\phi_0(x_n)}\right)\right| \leq \sum_{j\geq i}\left(2k_{j+1} + 1\right)\left(\sqrt{\mu_f + \epsilon}\right)^j.$$

Since

$$\lim_{i\to\infty} \sum_{j\geq i}\left(2k_{j+1} + 1\right)\left(\sqrt{\mu_f + \epsilon}\right)^j = 0,$$

we have that $\log|\phi_0|$ is uniformly continuous on O. So ϕ_0 is uniformly continuous and bigger than a positive constant on O. ∎

Chapter Four

The Renormalization Method and Folding Mappings

This chapter introduces some recent developments in the renormalization theory of real folding mappings.

Consider the real quadratic family $P_t(x) = t-(t+1)x^2$. One can calculate the period doubling bifurcation parameters t_n for $0 \leq n < \infty$, i.e., P_t has a unique attractive periodic orbit of period 2^n for $t_n < t < t_{n+1}$. Let $\delta_n = (t_n - t_{n-1})/(t_{n+1} - t_n)$. A remarkable discovery by Feigenbaum [FE1,FE2] (see also the independent paper of Coullet and Tresser [COT] for this discovery) is that

- The limit $\delta = \lim_{n\to\infty} \delta_n$ exists and is universal, that means that δ does not depend on any particular family like the real quadratic family. For example, from the family $f_t(x) = t\sin(\pi x)$, $1/2 \leq t \leq 1$ and $0 \leq x \leq 1$, one gets the same δ;
- the limiting map P_{t_∞} is unique in the real quadratic family and has no attractive periodic point, instead, there is an attractive Cantor set. The geometry of this attractive Cantor set is also universal;
- the map $f_n = \alpha_n^{-1} \circ P_{t_\infty}^{\circ 2^n} \circ \alpha_n$ defined on $[-1,1]$ converges exponentially to a map g in the C^1-topology as n goes to infinity where $\alpha_n(x)$ is the linear map such that $f_n|[-1,0]$ is increasing, $f_n|[0,1]$ is decreasing, and $f_n(-1) = f_n(1) = -1$; the map g is also universal.

The reader may refer to Cvitanović's book [CVI], to Vul, Sinai, and Khanin's survey article [VSK], to Collet and Eckmann's book [COE], and to De Melo and Van Strien's book [MV2] for the background of this discovery.

The discovery can be generalized for infinitely renormalizable folding mappings with arbitrary exponents and asymmetries. Let I be the interval $[-1,1]$. A continuous map f from I into itself is called a folding mapping if $f|[-1,0]$ is strictly increasing and $f|[0,1]$ is strictly decreasing (See Fig. 4.1). For a folding mapping f, 0 is a unique turning point. For a C^1 folding mapping f, we assume that 0 is the unique singular point and is power law type, $|x|^\gamma$ for $\gamma > 1$; moreover, we use A to denote the asymmetry of f at 0 (see §3.3). We call $\gamma > 1$ and $A < 0$ the exponent and the asymmetry of f, respectively.

A folding mapping f is renormalizable if there is a subinterval $J = [c,d] \subset I$ containing 0 in its interior and an integer $n \geq 2$ such that

1) $f^{\circ n}$ is strictly monotone when restricted to $[c,0]$ and to $[0,d]$,

2) $f^{\circ i}(J) \cap \mathring{J} = \emptyset$ for all $0 < i < n$, and

3) $f^{\circ n}(J) \subseteq J$.

We can normalize J to I by a Möbius transformation $\alpha(x) = ux/(vx+w)$ fixing 0 such that

$$\mathcal{R}(f) = \alpha \circ f^{\circ n} \circ \alpha^{-1}$$

is again a folding mapping. Here, $\mathcal{R}(f)$ is called a renormalization. To fix notation, we always assume that $n \geq 2$ is the smallest such integer and that $J \neq I$ is the biggest such interval. We say that f is once n-renormalizable, that $\mathcal{R}(f)$ is the renormalization of f, and that $\mathcal{R}$ is the renormalization operator.

A folding mapping f is twice (n_1, n_2)-renormalizable if f is once n_1-renormalizable and $\mathcal{R}(f)$ is once n_2-renormalizable; f is k-*times* $(n_1, n_2, \ldots, n_k)$-renormalizable if $\mathcal{R}^{\circ i}(f)$ is n_{i+1}-renormalizable for $0 \leq i < k$; f is infinitely $(n_1, n_2, \ldots, n_k, \ldots)$-renormalizable if $\mathcal{R}^{\circ i}(f)$ is n_{i+1}-renormalizable for every integer $i \geq 0$. An infinitely $(n_1, n_2, \ldots, n_k, \ldots)$-renormalizable folding mapping f is of bounded type if $\{n_k\}_{k=1}^{\infty}$ is a bounded sequence; otherwise, f is of unbounded type. In particular, if all $n_k = 2$, f is called a Feigenbaum-like mapping.

Let $FM(\gamma, A)$ be the set of C^1 folding mappings f, whose exponent and asymmetry are $\gamma > 1$ and $A < 0$, and which satisfy $f(-1) = f(1) = -1$ and $f(0) \geq 0$. A family of folding mappings $\mathcal{F} = \{f_t\}_{t \in [a,b]}$ in $FM(\gamma, A)$ is full (see Fig. 4.1) if

(1) $F(t,x) = f_t(x)$ is C^1 on $a \leq t \leq b$ and $-1 \leq x \leq 1$ and

(2) $f_a(0) = 0$ and $f_b(0) = 1$.

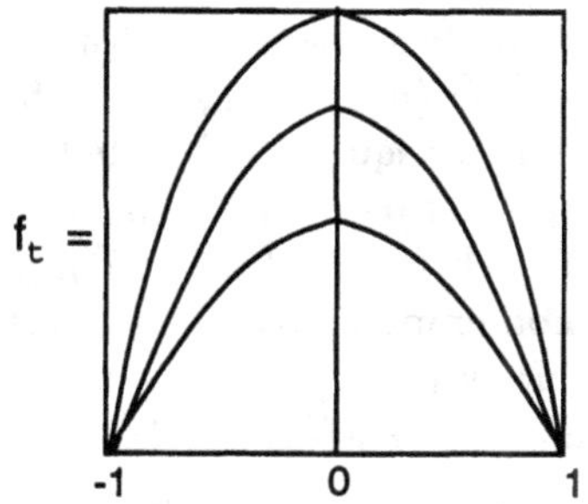

Graphs of folding mappings in a family.

Fig. 4.1

An example of a full family $\mathcal{F} = \{f_t\}_{t\in[0,1]}$ in $FM(\gamma, A)$ is

$$f_t(x) = \begin{cases} h_{1,t} \circ h(-|x|^\gamma) & \text{for } -1 \le x \le 0; \\ h_{2,t} \circ h(|x|^\gamma) & \text{for } 0 \le x \le 1 \end{cases}$$

where $h(x) = (x + a)/(ax + 1)$, $h_{1,t}(x) = (t + 1)(x + 1)/(a + 1) - 1$, and $h_{2,t}(x) = (t + 1)(x - 1)/(a - 1) - 1$ and where $a = (A + 1)/(A - 1)$. When $A = -1$,

$$f_t(x) = t - (t + 1)|x|^\gamma.$$

The topology of a folding mapping f can be described by its kneading sequence. The kneading sequence of a folding mapping f can be defined as $k_f = k_1k_2 \ldots k_n \ldots$ where

$$k_n = \begin{cases} L & \text{if } f^{\circ n}(0) < 0; \\ R & \text{if } f^{\circ n}(0) > 0; \\ C & \text{if } f^{\circ n}(0) = 0. \end{cases}$$

Let $\mathcal{K} = \{k_f \mid f \text{ is a folding mapping and } f(0) \ge 0, f(-1) = f(1) = -1\}$ be the set of all kneading sequences. It is known that $t \mapsto k_{f_t}$ is onto for a full family $\mathcal{F}$ in $FM(\gamma, A)$. We would like to note that $t \mapsto k_{f_t}$ is not onto for the family

$$\{\ f_t(x) = t - (1 + t)|x|\ \}_{0\le t\le 1}$$

because f_t is not C^1. For the theory of kneading sequences, see the paper of Milnor and Thurston [MIT] and the books of Collet and Eckmann [COE] and De Melo and Van Strien [MV2].

Let $\mathcal{F} = \{f_t\}_{t\in[0,1]}$ be a full family in $FM(\gamma, A)$ (where $\gamma > 1$ and $A < 0$). For every integer $n \ge 2$, there is an interval $B(n)$ of $[a, b]$ such that f_t is once n-renormalizable for each t in $B(n)$ and $\{\mathcal{R}(f_t)\}_{t\in B(n)}$ is a full family. For every sequence $\rho = (n_1, \ldots, n_k, \ldots)$ with entries in $\mathbf{N} \setminus \{1\}$, define $\rho_k = (n_1, \ldots, n_k)$. There is a sequence of nested intervals $\{B(\rho_k)\}_{k=1}^{\infty}$ such that $B(\rho_k) \subset B(\rho_{k-1})$, f_t is k-times $(n_1, \ldots, n_k)$-renormalizable for each t in $B(\rho_k)$, and such that for each $k \geq 1$, $\{\mathcal{R}^{\circ k}(f_t)\}_{t\in B(\rho_k)}$ is a full family. The set $B_\rho = \cap_{k=1}^{\infty} B(\rho_k)$ is non-empty, closed, and connected. Furthermore, for each $t \in B(\rho)$, f_t is infinitely $(n_1, \ldots, n_k, \ldots)$-renormalizable. One observes the following for a full family in $FM(\gamma, A)$ like the example after Fig. 4.1.

Observation 4.1. *Every set $B_\rho = \cap_{k=1}^{\infty} B(\rho_k)$ contains only one number. Consider $\rho_2 = (2, 2, \ldots)$. Let $\rho_{2,k} = (\overbrace{2, \ldots, 2}^{k\text{-times}})$, let $B(\rho_{2,k}) = [r_k, s_k]$, and let $\{t_\infty\} = \cap_{k=1}^{\infty} B(\rho_{2,k})$. Then*

$$\lim_{k\to\infty} \frac{r_k - t_\infty}{r_{k+1} - t_\infty} = \lim_{k\to\infty} \frac{s_k - t_\infty}{s_{k+1} - t_\infty} = \delta > 1$$

exists.

Observation 4.2. *The number δ is universal in the sense that it is independent of any particular full family in $FM(\gamma, A)$ like the example after Fig. 4.1. In other words, $\delta = \delta(\gamma, A)$ is a function of exponent $\gamma > 1$ and asymmetry $A < 0$.*

Observation 4.3. *Let $\Lambda(f_{t_\infty})$ be the closure of the critical orbit $\{f_{t_\infty}^{\circ k}(0)\}_{k=0}^{\infty}$. It is a Cantor set and can be treated as a non-escaping set of a $C^{1+\alpha}$ degree two expanding map σ for some $0 < \alpha < 1$. The scaling function s_σ of σ is universal (see §1.3).*

Observation 4.4. *Let $\{\mathcal{R}^{\circ k}(f_{t_\infty})\}_{k=0}^{\infty}$ be the sequence of renormalizations. Then*

$$\lim_{k\to\infty} \mathcal{R}^{\circ k}(f_{t_\infty}) = g$$

exists. The convergence is exponential in the C^1 topology. The limiting mapping g is again universal.

To avoid notational complication, we only discuss maps in $FM(\gamma, -1)$. The reader can refer to Remark 4.1 for a more general setting. A map h is $C^{k+\alpha}$ for $0 < \alpha < 1$ (or C^{k+1} or C^{k+Z}) if h is k-times differentiable and if the kth derivative $h^{(k)}$ is C^α (or Lipschitz or Zygmund). We say that a folding mapping $f(x) = h(-|x|^\gamma)$ is $C^{k+\alpha}$ for $0 < \alpha < 1$ (or C^{k+1} or

C^{k+Z}) if $h : [-1,0] \to [-1,h(0)]$ is a $C^{k+\alpha}$ (or C^{k+1} or C^{k+Z}) orientation-preserving diffeomorphism and fixes -1. Note that the asymmetry and the exponent of $f(x) = h(-|x|^{\gamma})$ is -1 and $\gamma > 1$. Let $FM^r(\gamma)$ be the set of C^r folding mappings $f(x) = h(-|x|^{\gamma})$, where $r = k + \alpha \geq 2$ and $0 < \alpha \leq 1$ (or $r = k + Z$). Let $FM^r_{\infty}(\gamma)$ be its subset of Feigenbaum-like mappings f, let $L^r_k(\gamma)$ be its subset of $(\overbrace{2,\ldots,2}^{k\text{-}times})$-renormalizable folding mappings f such that $\mathcal{R}^{\circ k}(f)(0) = 0$, and let $U^r_k(\gamma)$ be its subset of $(\overbrace{2,\ldots,2}^{k\text{-}times})$-renormalizable folding mappings f such that $\mathcal{R}^{\circ k}(f)(0) = 1$. One observes that the renormalization operator $\mathcal{R}$ on $FM^r(\gamma)$ has a fixed point g in $FM^r_{\infty}(\gamma)$. One also expects that every folding mapping f in $FM^r_{\infty}(\gamma)$ tends to g under forward iterations by $\mathcal{R}$.

Let $m \geq 2$ be an even integer and let U be a connected and simply connected domain in the complex plane $\mathbf{C}$ containing I. Let $FM^{\omega}(m,U)$ be the space of all analytic folding mappings f in $FM^2(m)$ which can be extended to U analytically, and to $\overline{U}$ continuously. Then $FM^{\omega}(m,U)$ equipped with the supremum norm is a Banach space, and the renormalization operator $\mathcal{R}$ on $FM^{\omega}(m,U)$ is differentiable. In this case, we can talk about the tangent map $T_g\mathcal{R}$ on the tangent space $T_g\big(FM^{\omega}(m,U)\big)$ of the renormalization operator $\mathcal{R}$ at its fixed point g. The linear operator $T_g\mathcal{R}$ is bounded and compact. Thus its spectrum except for 0 is discrete, and consists of eigenvalues. (An eigenvalue of an operator $\mathcal{L}$ is a number λ such that $\mathcal{L}v = \lambda v$ for a non-zero vector v.) Let $FM^{\omega}_{\infty}(m,U) = FM^2_{\infty}(m) \cap F^{\omega}(m,U)$.

Conjecture 4.1. *The renormalization operator $\mathcal{R}$ on $FM^{\omega}(m,U)$ has a unique hyperbolic fixed point g in $FM^{\omega}_{\infty}(m,U)$ with a codimension one stable manifold $W^s(g) = FM^{\omega}_{\infty}(m,U)$ and a dimension one unstable manifold $W^u(g)$. The eigenvalues of the tangent map $T_g\mathcal{R}$ on $T_g\big(FM^r(m,U)\big)$ are all real. One eigenvalue is $\delta > 1$ and the rest have absolute values strictly less than a positive number which is less than one. Here δ is the expanding rate of the restriction of $\mathcal{R}$ to $W^u(g)$ at g.*

Feigenbaum [FE1,FE2] (see also [COT]) explained the discovery by using Conjecture 4.1 for $FM^{\omega}(2,U)$ (refer to [COE,VSK,SU4,MV2]). Write $W^u(g)$ as a smooth curve $\{g_t\}_{t\in[0,1]}$ which is transversal to $W^s(g)$. Suppose $g_{r^0_k}$ is in $L^{\omega}_k(2)$ and $g_{s^0_k}$ is in $U^{\omega}_k(2)$. Suppose $\{t^0_k\}_{k=0}^{\infty}$ are the period doubling bifurcation parameters for $\{g_t\}_{t\in[0,1]}$. One claims that

$$\lim_{k\to\infty}\frac{t_k^0-t_\infty^0}{t_{k+1}^0-t_\infty^0}=\lim_{k\to\infty}\frac{r_k^0-t_\infty^0}{r_{k+1}^0-t_\infty^0}=\lim_{k\to\infty}\frac{s_k^0-t_\infty^0}{s_{k+1}^0-t_\infty^0}=\delta,$$

the expanding eigenvalue of $T_g\mathcal{R}$. Suppose $\mathcal{F}=\{f_t\}_{t\in[0,1]}$ is an analytic full family in $FM^\omega(2,U)$ transversal to $W^s(g)$. Then f_{r_k} is in $L_k^\omega(2)$ and f_{s_k} is in $U_k^\omega(2)$. Suppose $\{t_k\}_{k=0}^\infty$ are the period doubling bifurcation parameters for $\mathcal{F}$. By a known result in smooth dynamical systems (the λ-Lemma (see [HPS] and [PAM])), the family $\{\mathcal{R}(f_t)\}_{t\in[t_k,s_k]}$ converges to $W^u(g)$ exponentially as k goes to infinity. Thus

$$\lim_{k\to\infty}\frac{t_k-t_\infty}{t_{k+1}-t_\infty}=\lim_{k\to\infty}\frac{r_k-t_\infty}{r_{k+1}-t_\infty}=\lim_{k\to\infty}\frac{s_k-t_\infty}{s_{k+1}-t_\infty}=\delta.$$

In the space $FM^\omega(2,U)$, Lanford [LA1,LA2] gave the first complete proof of Conjecture 4.1 (except for the reality of the spectrum) using computer help. Later, Eckmann and Wittwer [ECW] gave a proof using less computer help. Campanino and Epstein [CAE], Campanino, Epstein and Ruelle [CER], and Epstein [EPS] proved the existence of the fixed point g of $\mathcal{R}$ in $FM_\infty^\omega(2,U)$. The proof of Lanford [LA1,LA2] considers the Taylor expansion of a folding mapping g in $FM^\omega(2,U)$ and solves the Cvitanović-Feigenbaum functional equation

$$g(x)=-\alpha g\big(g(-\frac{1}{\alpha}x)\big)$$

where $\alpha=2.5\ldots$ is the universal number. In order to solve this functional equation, Lanford applied Newton's method. He truncated the Taylor expansion of g to a polynomial and estimated the coefficients of this truncated polynomial to get an approximate solution of the Cvitanović-Feigenbaum functional equation. He proved that if the degree of this approximate solution is sufficiently high, it lies in the attractive basin of the fixed point of Newton's method. By taking a limit, one gets a solution of the Cvitanović-Feigenbaum functional equation. The proof of Epstein [EPS] applies a fixed point theorem to a different functional equation which also gives the fixed point g of $\mathcal{R}$ in $FM_\infty^\omega(2,U)$.

Now we formulate a general conjecture for $\gamma>1$ and $A=-1$. For an arbitrary asymmetry $A<0$, the formulation will be similar (refer to Remark 4.1). Let U be a connected and simply connected domain containing I. Let $FM^\omega(\gamma,U)$, $\gamma>1$, be the space of all folding mappings $f(x)=h(-|x|^\gamma)$ where $h:[-1,0]\to[-1,h(0)]$ is a diffeomorphism with $h(-1)=-1$, and h^{-1}

can be extended to U analytically and to $\overline{U}$ continuously. Let $RFM^{\omega}(\gamma, U)$ be the space of once renormalizable folding mappings in $FM^{\omega}(\gamma, U)$. Then $FM^{\omega}(\gamma, U)$ equipped with the supremum norm on $h^{-1}|\overline{U}$ is a Banach space. Consider the renormalization operator

$$\mathcal{R} : RFM^{\omega}(\gamma, U) \to FM^{\omega}(\gamma, U).$$

Let

$$\Lambda^{\omega}(\mathcal{R})(U) = \cap_{k=1}^{\infty} \mathcal{R}^{-k}(FM^{\omega}(\gamma, U))$$

be the maximal invariant set of $\mathcal{R}$ in $RFM^{\omega}(\gamma, U)$. Let $RFM^{1+1}(\gamma)$ be the space of once renormalizable folding mappings in $FM^{1+1}(\gamma)$. Let

$$\Lambda^{1+1}(\mathcal{R}) = \cap_{k=1}^{\infty} \mathcal{R}^{-k}(FM^{1+1}(\gamma))$$

be the maximal invariant set of the operator

$$\mathcal{R} : RFM^{1+1}(\gamma) \to FM^{1+1}(\gamma)$$

in $RFM^{1+1}(\gamma)$.

Conjecture 4.2. *There is an open domain $U \supset I$ such that $\Lambda^{\omega}(\mathcal{R})(U)$ is a hyperbolic invariant set (see [HPS] and [PAM]) of the differentiable operator*

$$\mathcal{R} : RFM^{\omega}(\gamma, U) \to FM^{\omega}(\gamma, U),$$

and is the attractor in $\Lambda^{1+1}(\mathcal{R})$, i.e., $\mathcal{R}^{\circ k}(f) \to \Lambda^{\omega}(\mathcal{R})(U)$ (in a $C^{1+\alpha}$-topology for some $0 \leq \alpha < 1$) as $k \to \infty$ for every f in $\Lambda^{1+1}(\mathcal{R})$.

In this chapter, we discuss some developments, due mostly to Sullivan [SU4], in this direction. In §4.1 and §4.4, we discuss the *a priori* real bounds found by Sullivan [SU4] for an infinitely renormalizable folding mapping f. Using these *a priori* real bounds, we discuss, in §4.5, the distortion of renormalizations $\{\mathcal{R}^{\circ k}(f)\}_{k=1}^{\infty}$ for an infinitely renormalizable folding mapping f in $FM^{1+Z}(\gamma)$ and in $FM^{1+1}(\gamma)$.

We introduce in §4.2 an induced Markov map for a Feigenbaum-like folding mapping $f(x) = h(-|x|^{\gamma})$, where $\gamma > 1$, as worked out in [JI5,JI10]. We prove that this Markov map has bounded nearby geometry similar to that in §3.1 and §3.11. By using the method in §3.5, this property enables us to study the quasisymmetric property of the conjugacy between two Feigenbaum-like mappings $f(x) = h(-|x|^{\gamma})$. The bounded geometry of the attractor $\Lambda(f)$ of a Feigenbaum-like mapping f is discussed in §4.6.

4.1. Infinitely Renormalizable S-Unimodal Mappings

Let $I = [-1,1]$, let $I_- = [-1,0]$, and let $I_+ = [0,1]$. A folding mapping f is said to be unimodal if $f(x) = h(-|x|^\gamma)$, where $\gamma > 1$, and where h is an orientation-preserving diffeomorphism from I_- into $f(I_-)$ with $h(-1) = -1$. A unimodal mapping $f(x) = h(-|x|^\gamma)$ is said to be S-unimodal if h is a C^3-diffeomorphism with non-positive Schwarzian derivative and it is said to be $C^{1+\alpha}$-unimodal (or C^{1+Z}-unimodal) if h is a C^1-diffeomorphism and h' is α-Hölder (or Zygmund). In particular, C^{1+1} means that h' is Lipschitz.

Remark 4.1. For a more general theory, we can consider

$$f(x) = \begin{cases} h_-(-|x|^\gamma) & \text{for } x \in I_-; \\ h_+(-|x|^\gamma) & \text{for } x \in I_+ \end{cases}$$

where $h_\pm : I_- \to f(I_-)$ are both orientation-preserving diffeomorphisms. Then $A = -h_+(0-)/h_-(0-)$ and $\gamma > 1$ are the asymmetry and the exponent of f. Most arguments in this chapter apply to the renormalizations of this general setting. If f has the exponent $\gamma > 1$ and the asymmetry $A < 0$, then $f(x)$ can be written in the above form under an appropriate smooth coordinate on I (see, for example, [JI2]).

For a unimodal mapping $f(x) = h(-|x|^\gamma)$, an open subinterval T of $I = [-1,1]$ is called a homterval if $f^{\circ n}$ is injective when restricted to T for all integers $n > 0$. If T is a homterval, then either f has a (weak) attractive periodic point p of period k in some $f^{\circ i}(T)$ (i.e., $|(f^{\circ k})'(p)| \leq 1$ and $f^{\circ nk}(x) \to p$ as $n \to \infty$ for $x - p \leq 0$ small or $x - p \geq 0$ small) or else T is a wandering interval (i.e., $f^{\circ n}(T) \cap f^{\circ m}(T) = \emptyset$ for all $n \neq m$). Singer [SIN] proved that for an S-unimodal mapping f, if it has an attractive periodic orbit, then the orbit of the critical point 0 of f is attracted to the periodic orbit. Guckenheimer [GU2] proved that an S-unimodal mapping has no wandering interval (De Melo and Van Strien [MV1] proved a more general result). Hence an infinitely renormalizable S-unimodal mapping f has no attractive periodic point and has no wandering interval. A periodic point p of a folding mapping f of period k is called expanding if $|(f^{\circ k})'(p)| > 1$. Henceforth we always assume that (**) $f(x) = h(-|x|^\gamma)$ has only expanding periodic points and no wandering interval. Under this assumption, the structure of an infinitely renormalizable unimodal mapping can be described as follows.

Let $f(x) = h(-|x|^\gamma)$ be an infinitely $(n_1, n_2, \ldots, n_k, \ldots)$-renormalizable unimodal mapping. Let $m_k = \prod_{i=1}^k n_i$. Let $I_k = [-a_k, a_k]$ be the maximal interval containing 0 (set $m_0 = 0$ and $I_0 = I$) such that

(a) $f^{\circ m_k}$ is monotone when restricted to $[-a_k, 0]$ and to $[0, a_k]$,
(b) $f^{\circ m_k}(I_k) \subseteq I_k$,
(c) I_k, $f(I_k)$, ..., $f^{\circ(m_k-1)}(I_k)$ have disjoint interiors, and
(d) $f^{\circ m_k}$ has exactly two fixed points p_k and q_k in I_k, and $a_k = |p_k|$.

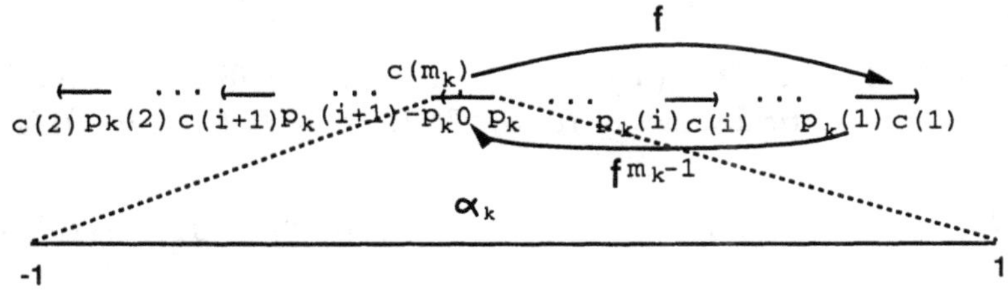

Fig. 4.2

The kth-renormalization of f is

$$\mathcal{R}^{\circ k}(f) = \alpha_k^{-1} \circ f^{\circ m_k} \circ \alpha_k$$

where $\alpha_k(x) = -p_k x$ is the linear rescaling from I_k to I. Suppose $c(i) = f^{\circ i}(0)$ is the ith critical value of f. For each $k \geq 1$, let $I_k(i) = f^{\circ i}(I_k)$ and let $p_k(i) = f^{\circ i}(p_k)$ for $0 \leq i < m_k$. Then $I_k(i)$ is an interval bounded by $p_k(i)$ and $c(i)$. We note that $I_k(0) = I_k$ is an interval bounded by $-p_k$ and p_k and that $I_k(m_k)$ is an interval bounded by p_k and $c(m_k)$. Except for $f : I_k \to I_k(1)$, all other mappings $f : I_k(i) \to I_k(i+1)$ are homeomorphisms for $1 \leq i < m_k$ (see Fig. 4.2). The kth-renormalization for $k \geq 1$ can be written as a unimodal mapping

$$\mathcal{R}^{\circ k}(f)(x) = h_k(-|x|^{\gamma})$$

where $h_k = \alpha_k^{-1} \circ f^{\circ(m_k-1)} \circ h \circ \tilde{\alpha}_k$ is a diffeomorphism from I_- into $h_k(I_-)$ and $\tilde{\alpha}_k(x) = |p_k|^{\gamma} x$ (set $h_0 = h$). Let $J_k(i)$ be the interval bounded by $c(i)$ and $c(m_k + i)$ for $k \geq 0$ and $0 \leq i < m_k$ (see Fig. 4.3).

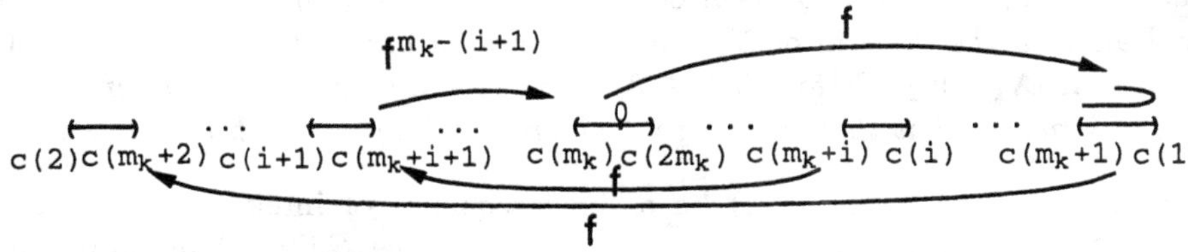

Fig. 4.3

Let $\xi_k = \{I_k(i)\}_{0 \leq i < m_k}$ and $\eta_k = \{J_k(i)\}_{0 \leq i < m_k}$. Because of the assumption $(**)$, both are composed of intervals with pairwise disjoint interiors.

Lemma 4.1. *For each $k > 0$ and $0 \le i < m_k$,*

$$\cup_{j=0}^{n_{k+1}-1} I_{k+1}(jm_k + i) \subseteq I_k(i), \qquad \cup_{j=0}^{n_{k+1}-1} J_{k+1}(jm_k + i) \subseteq J_k(i),$$

and $\Lambda(f) = \cap_{k=1}^{\infty} \cup_{i=0}^{m_k-1} J_k(i)$ is the closure of the critical orbit $\{c(i)\}_{i=0}^{\infty}$ and is a Cantor set.

Proof. Since $f^{\circ m_k}(I_k(0)) \subseteq I_k(0)$ and since $J_k(0)$ is bounded by the first and second critical values of $f_k = f^{\circ m_k}|I_k$, we have $f^{\circ m_k}(J_k(0)) \subseteq J_k(0)$. Because

$$I_{k+1}(jm_k) = f_k^{\circ j}(I_{k+1}(0)) \qquad \text{and} \qquad J_{k+1}(jm_k) = f_k^{\circ j}(J_{k+1}(0)),$$

we have

$$\cup_{j=0}^{n_{k+1}-1} I_{k+1}(jm_k) \subseteq I_k(0) \qquad \text{and} \qquad \cup_{j=0}^{n_{k+1}-1} J_{k+1}(jm_k) \subseteq J_k(0).$$

From these and the fact that $f^{\circ i}(I_{k+1}(jm_k)) = I_{k+1}(jm_k + i)$, we have

$$\cup_{j=0}^{n_{k+1}-1} I_{k+1}(jm_k + i) \subseteq I_k(i).$$

Similarly,

$$\cup_{j=0}^{n_{k+1}-1} J_{k+1}(jm_k + i) \subseteq J_k(i)$$

for $0 \le i < m_k$, which implies that the closure of the critical orbit $\{c(i)\}_{i=0}^{\infty}$ is contained in

$$\Lambda(f) = \cap_{k=1}^{\infty} \cup_{i=0}^{m_k-1} J_k(i).$$

The set $\Lambda(f)$ has no interior point, otherwise, f would have a wandering interval (refer to Fig. 4.3). By using similar arguments to those used in the proof of Theorem 1.1, $\Lambda(f)$ is the closure of the critical orbit, and is a Cantor set. ∎

Definition 4.1 (Bounded and eventually universally bounded). *Let $\mathcal{X}$ be a space of sequences $X = \{X_k\}_{k=0}^{\infty}$ of functions. We say $\mathcal{X}$ is beau if for each $X \in \mathcal{X}$, there is a constant $C_1 > 0$ such that $|X_k(x)| \le C_1$ for all $k \ge 0$ and all x in the domain of X_k, and if there is a universal constant $C_0 > 0$ such that $|X_k(x)| \le C_0$ for all $X \in \mathcal{X}$, large k, and all x in the domain of X_k (large k means that there is a big integer $K = K(X) > 0$ such that $k > K$).*

Let $\rho = (n_1, \ldots, n_k, \ldots)$ be a fixed sequence of integers $n \ge 2$ and let $\gamma > 1$ be a real number. Consider the space $\mathcal{SF}(\gamma, \rho)$ of sequences $X = \{\mathcal{R}^{\circ k}(f)(x) = h_k(-|x|^{\gamma})\}_{k=0}^{\infty}$ of renormalizations for all infinitely ρ-renormalizable S-unimodal mappings $f(x) = h(-|x|^{\gamma})$. Let $\mathcal{SN}(\gamma, \rho)$ be the space of sequences $Y = \{N(h_k^{-1})\}_{k=0}^{\infty}$ of non-linearities of $X = \{\mathcal{R}^{\circ k}(f)(x) = h_k(-|x|^{\gamma})\}_{k=0}^{\infty} \in \mathcal{SF}(\gamma, \rho)$.

Theorem 4.1 [SU4]. *The space $\mathcal{SN}(\gamma, \rho)$ is beau.*

Remark 4.2. The domain of h_k^{-1} for each $k \geq 0$ is $[-1, h_k(0)]$. Therefore, the domain of $N(h_k^{-1})$ in Theorem 4.1 is $[-1, h_k(0)]$. If h^{-1} can be extended to I as a C^3-diffeomorphism having non-negative Schwarzian derivative, then every h_k^{-1} can be extended to I as a C^3-diffeomorphism having non-negative Schwarzian derivative. In this case, one can think the domain of $N(h_k^{-1})$ in Theorem 4.1 is always I.

First we prove some lemmas. Let $\xi_k = \{I_k(i)\}_{0 \leq i < m_k}$, for $k = 1, 2, \ldots$, be the hierarchical system. For each interval $I_k(i)$, we use $LI_k(i)$ and $RI_k(i)$ to denote the unique intervals in ξ_k adjacent to $I_k(i)$ and on the left and right sides of $I_k(i)$, respectively. Let $LI_k^+(i)$ be the smallest interval containing $LI_k(i)$ and the left end-point of $I_k(i)$ and let $RI_k^+(i)$ be the smallest interval containing $RI_k(i)$ and the right end-point of $I_k(i)$ for $i = 0$ or $3 \leq i < m_k$ (see Fig. 4.4). Let $LI_k^+(2) = [-1, c(2)]$ and $RI_k^+(1) = [c(1), 1]$.

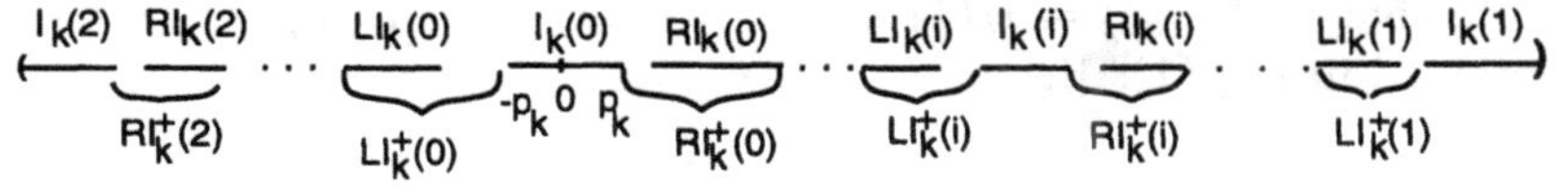

Fig. 4.4

Lemma 4.2. *There is a constant $C(\gamma) = \left((4/3)^{1/\gamma} - 1\right)/2$ and a sequence of positive real numbers $\{C(k, \gamma)\}_{k=1}^{+\infty}$ converging to $C(\gamma)$ such that*

$$\min\{|LI_k^+(0)|, |RI_k^+(0)|\} \geq C(k, \gamma)|I_k(0)|$$

for all $k > 0$.

Proof. For each k, let $I_k(i_k)$ be an interval with the minimum length in ξ_k. If $i_k = 0$, then we take $C(k, \gamma) = C(\gamma)$. Suppose $i_k \neq 0$. The relative positions of the boundary points of $LI_k(i_k)$, $I_k(i_k)$, and $RI_k(i_k)$ are shown in Fig. 4.5.

Suppose E_i and e_i are the sets of all the local maximum and minimum values of the ith-iterate of f. If $j \leq i$, then $E_j \subseteq E_i$ and $e_j \subseteq e_i$. This implies that any inverse branch of $f^{\circ i}$ can be defined homeomorphically on an interval bounded by a point in E_i and a point in e_i. By applying this fact, the inverse of $f^{\circ(i_k - 1)}|I_k(1)$ can be extended to a C^3 diffeomorphism G, with positive Schwarzian derivative, defined on an interval containing $LI_k(i_k) \cup I_k(i_k) \cup$

$RI_k(i_k)$ since $f^{\circ(i_k-1)}$ has no critical point in $I_k(1)$. From Lemma 2.6,

$$\begin{aligned} 4 &\geq \frac{|LI_k^+(i_k)\cup I_k(i_k)|\cdot|RI_k^+(i_k)\cup I_k(i_k)|}{|LI_k^+(i_k)|\cdot|RI_k^+(i_k)|} \\ &\geq \frac{|G(LI_k^+(i_k))\cup G(I_k(i_k))|\cdot|G(RI_k^+(i_k))\cup G(I_k(i_k))|}{|G(LI_k^+(i_k))|\cdot|G(RI_k^+(i_k))|}. \end{aligned}$$

Thus

$$\frac{|G(LI_k^+(i_k))\cup G(I_k(i_k))|}{|G(LI_k^+(i_k))|}\leq 4 \quad \text{and} \quad \frac{|G(LI_k^-(i_k))\cup G(I_k(i_k))|}{|G(LI_k^-(i_k))|}\leq 4.$$

So

$$|G(LI_k^+(i_k))|\geq \frac{|G(I_k(i_k))|}{3} \quad \text{and} \quad |G(LI_k^-(i_k))|\geq \frac{|G(I_k(i_k))|}{3}.$$

We note that $G(I_k(i_k)) = I_k(1)$ and that either $G(LI_k^+(i_k))$ or $G(RI_k^+(i_k))$ is contained in $LI_k^+(1)$. Hence

$$|LI_k^+(1)|\geq \frac{|I_k(1)|}{3}.$$

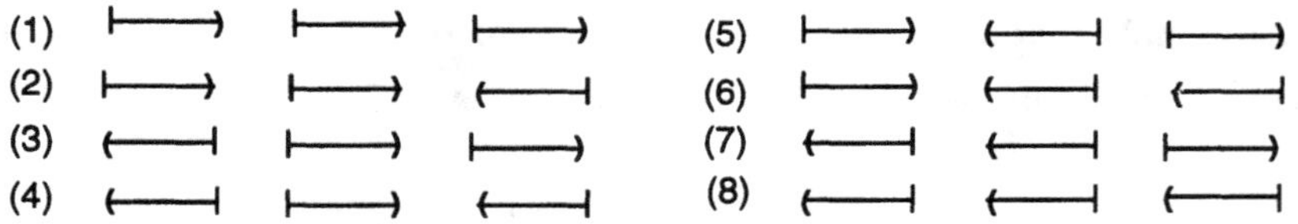

(means a local minimum value,) means a local maximum value, and | means a periodic point.

Fig. 4.5

Since $f(x) = h(-|x|^\gamma)$ maps $LI_k^+(0)\cup I_k(0)\cup RI_k^+(0)$ onto $LI_k^+(1)\cup I_k(1)$ (see Fig. 4.6) and since the length of $I_k(1)$ tends to zero as k goes to infinity, we can find a constant $C(k,\gamma)$ such that

$$C(k,\gamma)\to C(\gamma)=\frac{\left(\frac{4}{3}\right)^{\frac{1}{\gamma}}-1}{2}$$

as $k\to\infty$ and such that

$$\min\{|LI_k^+(0)|,|RI_k^+(0)|\}\geq C(k,\gamma)|I_k(0)|$$

for all $k > 0$.

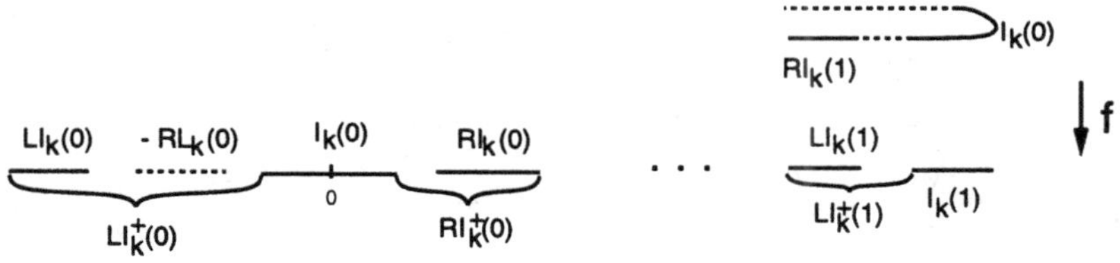

Fig. 4.6

■

Lemma 4.3. *For any $k \geq 1$, any $1 \leq i < m_k$, and any $x \in I_k(0)$,*

$$\max_{x \in I_k(i)} \left| N\left(\left(f^{\circ(m_k - i)} | I_k(i) \right)^{-1} \right)(x) \right| \leq \frac{2}{C(k,\gamma)|I_k(0)|}.$$

Proof. Since $f^{\circ(m_k-i)}|I_k(i)$ is a diffeomorphism from $I_k(i)$ to $I_k(0)$ and since its inverse can be extended to a diffeomorphism defined on an interval containing $LI_k^+(0) \cup I_k(0) \cup RI_k^+(0)$, the lemma follows from the previous lemma and Lemma 2.4. ■

Lemma 4.4. *For any $k \geq 1$, any $1 \leq i < m_k$, and any x and y in $I_k(i)$,*

$$\left| \log \left(\frac{|(f^{\circ(m_k-i)})'(x)|}{|(f^{\circ(m_k-i)})'(y)|} \right) \right| \leq \frac{2}{C(k,\gamma)} \frac{|x-y|}{|I_k(0)|}.$$

Proof. This is a corollary of Lemma 4.3 and Lemma 2.1. ■

Proof of Theorem 4.1. For each $k > 0$, the kth-renormalization is a unimodal map:

$$\mathcal{R}^{\circ k}(f)(x) = \alpha_k^{-1} \circ f^{\circ m_k} \circ \alpha_k(x) = h_k(-|x|^\gamma)$$

where

$$h_k = \alpha_k^{-1} \circ f^{\circ(m_k-1)} \circ h \circ \tilde{\alpha}_k : I_- \to h_k(I_-)$$

and

$$\alpha_k(x) = -p_k x \qquad \text{and} \qquad \tilde{\alpha}_k(x) = |p_k|^\gamma x.$$

Note that

$$\begin{aligned} \left| N(h_k^{-1}) \right| = & \Big| p_k \cdot \left(N(h^{-1}) \circ f^{-(m_k-1)} \circ \alpha_k \right) \cdot \left((f^{-(m_k-1)})' \circ \alpha_k \right) \\ & + p_k \cdot \left(N\left(\left(f^{\circ(m_k-1)} | I_k(1) \right)^{-1} \right) \circ \alpha_k \right) \Big|. \end{aligned}$$

From Lemma 4.3 and the fact $|I_k(0)| = 2|p_k|$, we have, for $x \in I$,

$$\max_{x\in[-1,1]}\left|p_k \cdot N\big((f^{\circ(m_k-1)}|I_k(1))^{-1}\big)((\alpha_k(x))\right| \le \frac{1}{C(k,\gamma)}.$$

By applying Lemma 4.4, one sees that for $x \in I$

$$\begin{aligned}|(f^{-(m_k-1)})'(\alpha_k(x))| &\le \exp\Big(\frac{2}{C(k,\gamma)}\Big)|(f^{-(m_k-1)})'(p_k)| \\ &= \exp\Big(\frac{2}{C(k,\gamma)}\Big)\frac{|f'(p_k)|}{|(f^{\circ m_k})'(p_k)|} \\ &\le \exp\Big(\frac{2}{C(k,\gamma)}\Big)|f'(p_k)|\end{aligned}$$

because $|(f^{\circ m_k})'(p_k)| > 1$. This implies that

$$p_k \cdot \Big(N(h^{-1}) \circ f^{-(m_k-1)} \circ \alpha_k\Big) \cdot \Big((f^{-(m_k-1)})' \circ \alpha_k\Big) \to 0 \qquad (4.1)$$

uniformly on $[-1, h_k(0)]$ as $k \to \infty$. (If h^{-1} can be extended to I as a C^3-diffeomorphism having non-negative Schwarzian derivative, then (4.1) holds on I.) The theorem now follows from the fact that $C(k,\gamma)$ tends to $C(\gamma)$ as k goes to infinity. ∎

Remark 4.3. For $X = \{\mathcal{R}^{\circ k}(f)(x) = h_k(-|x|^\gamma)\}_{k=0}^\infty \in \mathcal{SF}(\gamma,\rho)$, Theorem 4.1 implies that $\log(h_k^{-1})'$ is a Lipschitz function on $[-1, h_k(0)]$ for $k \ge 0$, and that the sequence of Lipschitz constants of $\log(h_k^{-1})'$ is bounded. Moreover, there is a universal constant $C_0 > 0$ such that the Lipschitz constant of $\log(h_k^{-1})'$ is less that C_0 for large k. Since $h_k([-1,0]) = [-1, h_k(0)]$, there is a point $x_k \in [-1,1]$ such that $1/2 \le (h_k^{-1})'(x_k) \le 1$, therefore, the sequence $\{\log(h_k^{-1})'\}_{k=0}^\infty$ is bounded and eventually universally bounded. From the Ascoli-Arzelà theorem, the sequence $\{h_k\}_{k=0}^\infty$ is precompact in $\mathcal{M}(1+\alpha)$ for any $0 \le \alpha < 1$ which is the Banach space of all orientation-preserving $C^{1+\alpha}$-diffeomorphisms $h : [-1,0] \to [-1, h(0)]$ with the norm

$$||h||_{1+\alpha} = \sup_{x\in[-1,0]} |h(x)| + \sup_{x\neq y\in[-1,0]} \frac{|h'(x) - h'(y)|}{|x-y|^\alpha}.$$

If h_0^{-1} can be extended to I as a C^3-diffeomorphism having non-negative Schwarzian derivative, then $\{h_k\}_{k=0}^\infty$ is precompact in $\tilde{\mathcal{M}}(1+\alpha)$ for any $0 \le \alpha < 1$ which is the Banach space of all orientation-preserving $C^{1+\alpha}$-diffeomorphisms $h : [-1,0] \to [-1, h(0)]$ such that h^{-1} can be extended to I as a $C^{1+\alpha}$-diffeomorphism with the norm

$$||h||_{1+\alpha} = \sup_{x\in I}|h^{-1}(x)| + \sup_{x\neq y\in I} \frac{|(h^{-1})'(x) - (h^{-1})'(y)|}{|x-y|^\alpha}.$$

In §4.5, we discuss a similar result under less assumptions.

4.2. Markov Maps Induced from Infinitely Renormalizable Mappings

In Chapter Three, we used the technique called Markov partitions to study the geometry of a geometrically finite one-dimensional mapping; we now use this technique to study infinitely renormalizable folding mappings, which can be treated as geometrically infinite one-dimensional maps.

Let $f(x) = h(-|x|^\gamma)$ be a Feigenbaum-like S-unimodal mapping (see the beginning of this chapter and see §4.1). The hierarchical system $\xi_k = \{I_k(i)\}_{0\leq i<m_k}$ for $k = 1, 2, \ldots$ induced from f is quite simple. For each $k > 0$, the interval $I_k(0)$ is bounded by p_k, the periodic point of f of period 2^{k-1} with the smallest absolute value, and by $-p_k$. The mapping $f^{\circ 2^k}|I_k(0)$ has two fixed points, p_k and p_{k+1}, where $I_{k+1}(0)$ is bounded by p_{k+1} and $-p_{k+1}$. Every interval $I_k(i) \in \xi_k$ contains two intervals $I_{k+1}(i)$ and $I_{k+1}(2^k + i)$ in ξ_{k+1}. The intervals $I_{k+1}(i)$ and $I_{k+1}(2^k + i)$ have a common endpoint $p_{k+1}(i)$ (see Fig 4.7).

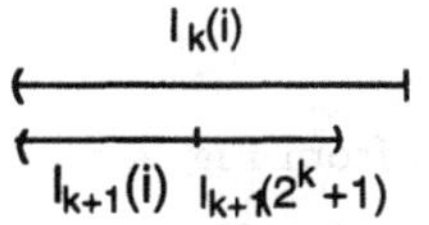

Fig. 4.7

Using the sequence of nested intervals $\{I_k(0)\}_{k=1}^{\infty}$, we construct an induced partition of $I = [-1, 1]$. Let P_{-0} and P_0 be the left and right connected components of $I \setminus I_1(0)$. Inductively, let P_{-k} and P_k be the left and right connected components of $I_k(0) \setminus I_{k+1}(0)$. Finally, set $P_\infty = \{0\}$. The collection

$$\beta_0 = \{P_{-0}, P_0, P_{-1}, P_1, \ldots P_{-k}, P_k, \ldots P_\infty\}$$

forms a partition of $I = [-1, 1]$ (see Fig. 4.8); that is, P_i and P_j have disjoint interiors for $i \neq j$ and $I = P_\infty \cup \cup_{k=0}^{\infty}(P_{-k} \cup P_k)$. Let F be the function defined as $F(0) = 0$ and

$$F(x) = \begin{cases} f(x), & x \in P_{-0} \cup P_0; \\ f^{\circ 2}(x), & x \in P_{-1} \cup P_1; \\ \vdots & \\ f^{\circ 2^i}(x), & x \in P_{-i} \cup P_i; \\ \vdots & \end{cases}.$$

The map F is continuous on I (see Fig. 4.8).

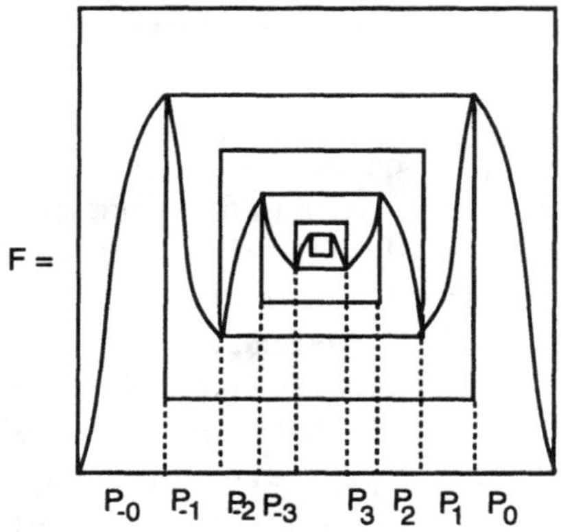

Fig. 4.8

Lemma 4.5. *For every even integer* $k = 2n \geq 0$, $F(P_{\pm k}) = \cup_{i=k}^{\infty} P_{-i} \cup \cup_{i=k+1}^{\infty} P_i$ *and for every odd integer* $k = 2n+1 > 0$, $F(P_{\pm k}) = \cup_{i=k+1}^{\infty} P_{-i} \cup \cup_{i=k}^{\infty} P_i$.

Proof. This can be seen from Fig. 4.8. ∎

From Lemma 4.5, the mapping F and the partition β_0 satisfy the Markov property in the sense that the image of every element in the partition β_0 is the union of some intervals in the partition. (But the number of intervals in the partition is infinite.) We call F an induced (infinite) Markov map.

Let $g_{\pm i} = (F|P_{\pm i})^{-1}$ be the inverse branches of F for $1 \leq i < \infty$. Suppose $w = i_0 i_1 \ldots i_{k-1}$ is a finite sequence of elements of $\overline{\mathbf{Z}} = \mathbf{Z} \cup \{-0\}$. We say w is admissible if the range P_{i_l} of g_{i_l} is contained in the domain $F_{i_{l-1}}(P_{i_{l-1}})$ of $g_{i_{l-1}}$ for $l = 1, \ldots, k-1$. For an admissible sequence $w = i_0 i_1 \ldots i_{k-1}$, we define the composition $g_w = g_{i_0} \circ g_{i_1} \circ \cdots \circ g_{i_{k-1}}$. We use $D(g_w)$ to denote the domain of g_w and use $|D(g_w)|$ to denote the length of the interval $D(g_w)$.

Definition 4.2. *We say the induced Markov map* F *has bounded nearby geometry if there is a constant* $C = C(f) > 0$ *such that*

(i) $C^{-1} \leq |P_k|/|\cup_{i=k+1}^{\infty} P_i| \leq C$ *and* $C^{-1} \leq |P_{-k}|/|\cup_{i=k+1}^{\infty} P_{-i}| \leq C$ *for all* $k \geq 0$, *and*

(ii) $|N(g_w)(x)| \leq C/|D(g_w)|$ *for all* x *in* $D(g_w)$ *and all finite admissible sequences* w *of* $\overline{Z}$.

Remark 4.4. Condition **(ii)** implies that the distortion $|\log(|g_w(x)|/|g_w(y)|)|$ of g_w at any x and y in $D(g_w)$ is bounded by C. Condition **(i)** is an analogous to bounded nearby geometry for a geometrically finite one-dimensional map (see §3.5).

Theorem 4.2 [JI10]. *Suppose $f(x) = h(-|x|^\gamma)$ is a Feigenbaum-like S-unimodal mapping. Then the induced Markov map F has bounded nearby geometry.*

Before proving Theorem 4.2, we prove some useful lemmas.

Lemma 4.6. *Let $h : [-1,0] \to \mathbf{R}$ be a C^3 orientation-preserving diffeomorphism such that $S(h)(x) \le 0$ for all x in $[-1,0]$. Suppose ϕ is a Möbius transformation satisfying*
a) *$\phi(a) = h(a)$ for $a = 0$ and -1, and*
b) *$N(h^{-1})(-1) \ge N(\phi^{-1})(-1)$. Then $\phi(x) \le h(x)$ for all x in $[-1,0]$.*

Proof. Let $Z = h^{-1} \circ \phi$. Then $Z(a) = a$ for $a = 0$ and -1. For x in $[-1,0]$, we have

$$S(Z)(x) = \left(\phi'(x)\right)^2 S(h^{-1})(\phi(x)) \ge 0.$$

We must show that $Z(x) \le x$ for all x in $[-1,0]$.

Using **b)**, one has $N(Z)(-1) \ge 0$. This implies that $Z''(-1) \ge 0$, and moreover, that

$$Z(x) \ge F(x) = -1 + Z'(-1)\cdot(x+1)$$

for small $x+1 \ge 0$. Since $S(Z)(x) \ge 0$ for all x in $[-1,0]$ and $Z'(-1) = F'(-1)$, we have that $Z(x) \ge F(x)$ for all $x \in [-1,0]$ (otherwise, $Z'(x)$ would have a local maximum point in $(-1,0)$ (see Remark 2.6)). In particular, $Z(0) \ge F(0)$. Hence $Z'(-1) \le 1$. Therefore, $Z(x) \le x$ for all x in $[-1,0]$ because $S(Z)(x) \ge 0$ for all x in $[-1,0]$ (refer to Fig. 2.1). So $\phi(x) \le h(x)$ for all x in $[-1,0]$. ■

Let $SF(\gamma, C)$ be the subspace of S-unimodal mappings $f(x) = h(-|x|^\gamma)$ such that

$$\min_{x\in h([-1,0])} \{N(h^{-1})(x)\} \ge -C.$$

Lemma 4.7. *There is a constant $C_1 = C_1(\gamma, C) > 0$ such that $f(0) \ge C_1$ for every infinitely renormalizable mapping f in $SF(\gamma, C)$.*

Proof. Suppose $f(x) = h(-|x|^\gamma)$ is a mapping in $SF(\gamma, C)$. Since h is a C^3 orientation-preserving diffeomorphism such that $S(h)(x) \le 0$ for all x in $[-1,0]$, one can compare h with some Möbius transformation ϕ. Let ϕ

be the Möbius transformation satisfying $\phi(a) = h(a)$ for $a = -1$ and 0, and $N(\phi^{-1})(-1) = -C$. Then

$$\phi(x) = \frac{x+1}{\frac{C}{2}(x+1) + \frac{1}{f(0)+1} - \frac{C}{2}} - 1.$$

From Lemma 4.6, $\phi(x) \le h(x)$ for all x in $[-1, 0]$.

Suppose, for the moment, that $c = f(0) > 0$ is a variable. Let $C_1 = C_1(\gamma, C) > 0$ be the smallest solution of $\phi(-|c|^\gamma) = 0$. Then for $0 < c < C_1$,

$$f^{\circ 2}(0) \ge \phi^{\circ 2}(0) = \phi(-|c|^\gamma) > 0.$$

This says that $f^{\circ 2}$ has an attractive fixed point and is thus not once renormalizable. Hence if f is infinitely renormalizable, then $f(0) > C_1$. ∎

For an S-unimodal mapping $f(x) = h(-|x|^\gamma)$ with $f(0) > 0$, let q_f be the fixed point of f in $(0,1)$.

Lemma 4.8. *There is a constant $C_2 = C_2(\gamma, C) > 0$ such that $q_f \ge C_2$ for all infinitely renormalizable map f in $SF(\gamma, C)$.*

Proof. Let

$$\phi_0(x) = \frac{x+1}{\frac{C}{2}(x+1) + \frac{1}{C_1+1} - \frac{C}{2}} - 1$$

and let C_2 be the fixed point of $\phi_0(-|x|^\gamma)$ in $(0,1)$. Define

$$\phi(x) = \frac{x+1}{\frac{C}{2}(x+1) + \frac{1}{f(0)+1} - \frac{C}{2}} - 1.$$

Then the fixed point q' of $\phi(-|x|^\gamma)$ in $(0,1)$ is greater than C_2 since $f(0) > C_1$. But $q_f \ge q' > C_2$. ∎

Proof of Theorem 4.2. Let $\mathcal{R}^{\circ k}(f) = h_k(-|x|^\gamma)$ be the kth-renormalization of f for $k \ge 0$. It is the rescaling of $f^{\circ 2^k}|I_k(0)$ to $[-1,1]$. From Theorem 4.1, there is a constant $C = C(f) > 0$ such that

$$\max_{x \in h_k([-1,0])} \left|(N(h_k^{-1}))(x)\right| \le C$$

for all $k \ge 0$ (set $h_0 = h$). Lemma 4.8 implies that there is a constant $C_2 = C_2(\gamma, C) > 0$ such that

$$\frac{|I_{k+1}(0)|}{|I_k(0)|} \ge C_2$$

for all $k \geq 0$ (set $I_0(0) = I$).

Let $k > 0$. From Lemma 4.2 and the fact that $I_{k+1}(2^k)$ is either $LI^+_{k+1}(0)$ or $RI^+_{k+1}(0)$, there is a constant $C_3 = C_3(f) > 0$ such that

$$\frac{|I_{k+1}(2^k)|}{|I_{k+1}(0)|} \geq C_3.$$

This implies that

$$\frac{|I_{k+1}(0)|}{|I_k(0)|} \leq C_4 = \frac{1}{2C_3 + 1}.$$

Now take $C_5 = \max\{C_2^{-1}, C_4\}$. Then

$$C_5^{-1} \leq \frac{|I_{k+1}(0)|}{|I_k(0)|} \leq C_5.$$

This proves **(i)** of Definition 4.2 because

$$P_{-k} \cup P_k = \overline{I_k(0) \setminus I_{k+1}(0)}, |P_{-k}| = |P_k|, \text{ and } I_{k+1}(0) = \overline{\cup_{i=k+1}^{\infty}(P_{-i} \cup P_i)}.$$

Now we prove **(ii)** of Definition 4.2. For an integer $i \neq 0$, g_i can be extended to the interval

$$\Omega_{|i|} = I_{|i|-1}(2^{|i|-1}) \cup D(g_i) \cup I_{|i|}(2^{|i|})$$

as a C^3-diffeomorphism with $S(g_i)(x) \geq 0$ for all x in $\Omega_{|i|}$. Without loss of generality, we may assume that each of g_0 and g_{-0} can be extended to the interval

$$\Omega_0 = (-\infty, -1] \cup D(g_0) \cup I_1(1)$$

with $S(g_0)(x) \geq 0$ and $S(g_{-0})(x) \geq 0$ for all x in Ω_0.

Let $w = i_0 i_1 \ldots i_{k-1}$ be an admissible sequence of elements of $\overline{Z} = \mathbf{Z} \cup \{-0\}$ and let $g_w = g_{i_0} \circ g_{i_1} \circ \cdots \circ g_{i_{k-1}}$. By the definition of an admissible sequence, one can check that

$$|i_0| \leq |i_1| \leq \cdots \leq |i_{k-1}|.$$

Hence g_w can be extended to the domain $\Omega_{|i_{k-1}|}$ as a C^3-diffeomorphism with $S(g_w)(x) \geq 0$ for all x in $\Omega_{|i_{k-1}|}$. We note that $D(g_w) = D(g_{i_{k-1}})$ and that the intervals $\Omega_{|i|}$ are nested for $|i| = 0, 1, \ldots$. Then **(ii)** of Definition 4.2 now follows from Lemmas 2.4 and 4.2. ∎

Remark 4.5. Starting with any infinitely renormalizable folding mapping f, we can construct an induced Markov map F. For such a mapping f of bounded type, we can prove Theorem 4.2 (see [JI5]).

4.3. Conjugacies between Infinitely Renormalizable Mappings

It is known that two Feigenbaum-like S-unimodal mappings are topologically conjugate. The proof of this depends on two deep facts, the kneading theory developed by Milnor and Thurston [MIT] and the non-wandering interval theorem proved by Guckenheimer [GU2]. Using the Markov partitions constructed in §4.2, we can obtain a topological model for Feigenbaum-like mappings as did Sinai [SI1] and Bowen [BO1,BO4] for hyperbolic dynamical systems. Using Theorem 4.2, we give a simple proof of the result that two Feigenbaum-like S-unimodal mappings are topologically conjugate; moreover, we prove that the conjugacy between two Feigenbaum-like S-unimodal mappings is quasisymmetric (see Definition 2.6).

Let f be a Feigenbaum-like mapping, let $\beta_0 = \{P_{\pm k}\}_{k=0}^{\infty} \cup \{P_\infty\}$ be the induced partition, and let F be the induced Markov map. Let $A = (a_{ij})$ be the bi-infinite matrix such that $a_{ij} = 1$ if $P_j \subseteq F(P_i)$ and $a_{ij} = 0$ otherwise. From the construction of F, we can see that

1. for $i = \pm 2n$ with $n \geq 0$, $a_{ij} = 1$ if and only if $|j| > |i|$ or $j = -2n$;
2. for $i = \pm(2n+1) > 0$ with $n > 0$, $a_{ij} = 1$ if and only if $|j| > |i|$ or $j = 2n+1$.

A sequence $w_k = i_0 i_1 \ldots i_k$ of $\overline{\mathbf{Z}}$ is admissible if $a_{i_l i_{l+1}} = 1$ for $0 \leq l < k$. For such an admissible sequence w_k, define $g_{w_k} = g_{i_0} \circ g_{i_1} \circ \cdots \circ g_{i_k}$ and $P_{w_k} = g_{w_k}(F(P_{i_k}))$. Consider the symbolic space

$$\Sigma_A = \{w = i_0 i_1 \ldots i_k i_{k+1} \ldots \mid i_k \in \overline{Z} \cup \{\infty\},\ a_{i_k i_{k+1}} = 1,\ k = 0, 1, \ldots\}$$

with the product topology and the shift map

$$\sigma_A(i_0 i_1 \ldots i_k i_{k+1} \ldots) = i_1 \ldots i_k i_{k+1} \ldots .$$

Lemma 4.9. *Let f be a Feigenbaum-like S-unimodal mapping, and let F be the induced Markov map. Then F is semi-conjugate to σ_A; that is, there is a continuous surjective map $H : \Sigma_A \to I$ such that $F \circ H = H \circ \sigma_A$.*

Proof. For any $w = i_0 i_1 \ldots i_k i_{k+1} \ldots$ in Σ_A, let $w_k = i_0 \ldots i_k$ for $k > 0$. Then w_k is admissible. Applying Theorem 4.2, we see that $\cap_{k=0}^{\infty} P_{w_k}$ contains the single point x_w. Set $H(w) = x_w$. The map $H : \Sigma_A \to I$ is continuous. Since $\cup_{w_k} P_{w_k} = I$, where w_k runs over all admissible sequences of length $k+1$, H is surjective. Moreover, every point $x \in I$ has at most two preimages in Σ_A; and only a boundary point x of P_{w_k}, for some admissible w_k, can have two preimages. It is now easy to see that

$$F \circ H(w) = H \circ \sigma_A(w).$$

∎

Theorem 4.3. *Any two Feigenbaum-like mappings f and g are topologically conjugate.*

Proof. Suppose F and G are the induced Markov maps from f and g. Let H_1 and H_2 be the semi-conjugacies from F and G to σ_A. From the proof of Lemma 4.9, $H = H_1 \circ H_2^{-1}$ can be defined as a homeomorphism of I such that $F \circ H = H \circ G$. Hence F and G are topologically conjugate. Furthermore, H is the conjugacy between f and g. ∎

From Theorem 4.2 and the proof of Theorem 3.2, we have

Theorem 4.4. *Let f and g be two Feigenbaum-like mappings and let H be the conjugacy between them. Then H is quasisymmetric.*

Proof. Suppose that $\beta_{0,f} = \{P_{\pm k,f}\}_{k=0}^{\infty} \cup \{P_{\infty,f}\}$ is the Markov partition induced by f and let $\beta_{0,g} = \{P_{\pm k,g}\}_{k=0}^{\infty} \cup \{P_{\infty,g}\}$ be the Markov partition induced by g. Define

$$\beta_{k,f} = \{P_{w_k,f} \mid w_k \text{ is an admissible sequence of } \overline{Z} \cup \{\infty\} \text{ of length } k+1\}$$

and

$$\beta_{k,g} = \{P_{w_k,g} \mid w_k \text{ is an admissible sequence of } \overline{Z} \cup \{\infty\} \text{ of length } k+1\}.$$

These are called the kth-partitions of $I = [-1,1]$ induced by f and g. From Theorem 4.2, there is a constant $C > 0$ so that

$$C^{-1} \le \frac{|P_{\pm l w_k,f}|}{|\cup_{i=l+1}^{\infty} P_{\pm i w_k,f}|} \le C$$

and

$$C^{-1} \le \frac{|P_{\pm l w_k,g}|}{|\cup_{i=l+1}^{\infty} P_{\pm i w_k,g}|} \le C$$

for all l in $\overline{Z}$ and all admissible sequences w_k of $\overline{Z}$ of length $k+1$. This means that the hierarchical systems $\{\beta_{k,f}\}_{k=0}^{\infty}$ and $\{\beta_{k,g}\}_{k=0}^{\infty}$ satisfy a property similar to bounded nearby geometry (see Definition 3.3). Using an argument similar to the proof of Theorem 3.2, we can now prove that H is quasisymmetric as follows.

First we construct a sequence of nested partitions $\{\eta_{k,f}\}_{k=0}^{\infty}$ of $I = [-1,1]$ a little different from $\beta_{0,f}$. Let $\eta_{0,f}$ consist of one interval I. Cut I into three intervals $L_0 = P_{-0,f}$, $M_0 = \overline{\cup_{i=1}^{\infty}(P_{-i,f} \cup P_{i,f})}$, and $R_0 = P_{0,f}$ (see Fig. 4.9). Then $\eta_{1,f} = \{L_0,\, M_0,\, R_0\}$.

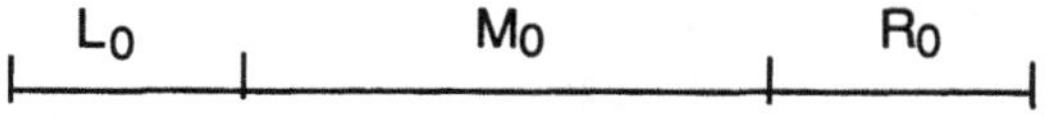

Fig. 4.9

The map F is a diffeomorphism when restricted to L_0 or to R_0, and $F(L_0) = F(R_0) = L_0 \cup M_0$. Cut L_0 (respectively, R_0) into the two intervals L_0L_0 and L_0M_0 (respectively, R_0L_0 and R_0M_0) which are preimages of L_0 and M_0 under $F|L_0$ (respectively, $F|R_0$). Cut M_0 into the three intervals $L_1 = P_{-1,f}$, $M_1 = \cup_{i=2}^{\infty}(P_{-i,f} \cup P_{i,f})$, and $R_1 = P_{1,f}$ (see Fig. 4.10). Then

$$\eta_{2,f} = \{L_0L_0, L_0M_0, L_1, M_1, R_1, R_0L_0, R_0M_0\}.$$

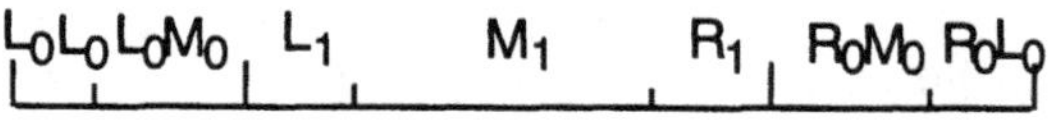

Fig. 4.10

Now we inductively define $\eta_{n,f}$ for $n \geq 3$. Suppose $\eta_{n,f}$ has been defined for some $n \geq 2$ and contains $L_{n-1} = P_{-(n-1),f}$, $M_{n-1} = \cup_{i=n}^{\infty}(P_{-i,f} \cup P_{i,f})$, and $R_{n-1} = P_{n-1,f}$. Cut M_{n-1} into the three intervals $L_n = P_{-n,f}$, $M_n = \cup_{i=n+1}^{\infty}(P_{-i,f} \cup P_{i,f})$, and $R_n = P_{n,f}$. For an interval $J \neq M_{n-1}$ in $\eta_{n,f}$, there is a maximum integer $i \geq 1$ such that $F^{\circ i}|J$ is a diffeomorphism. Then $F^{\circ i}(J)$ is one of **i)** M_{n-i-1}, **ii)** $L_{n-i} \cup M_{n-i}$, and **iii)** $R_{n-i} \cup M_{n-i}$. In **i)**, cut J into the three intervals $\mathcal{J} = \{JL_{n-i}, JM_{n-i}, JR_{n-i}\}$ which are the preimages of L_{n-i}, M_{n-i} and R_{n-i} under $F^{\circ i}|J$. In **ii)**, cut J into the two intervals $\mathcal{J} = \{JL_{n-i}, JM_{n-i}\}$ which are the preimages of L_{n-i} and M_{n-i} under $F^{\circ i}|J$. In **iii)**, cut J into the two intervals $\mathcal{J} = \{JM_{n-i}, JR_{n-i}\}$ which are the preimages of M_{n-i} and R_{n-i} under $F^{\circ i}|J$. Then

$$\eta_{n+1,f} = \cup_{J\in\eta_{n,f}}\mathcal{J} \cup \{L_n, M_n, R_n\}.$$

We have thus defined a sequence $\eta_f = \{\eta_{n,f}\}_{n=0}^{\infty}$ of nested partitions from $\beta_{0,f}$. Similarly, one can define a sequence $\eta_g = \{\eta_{n,g}\}_{n=0}^{\infty}$ of nested partitions from $\beta_{0,g}$.

From the construction of η_f, and from Theorem 4.2, η_f has bounded and bounded nearby geometry which is defined in §3.1. More precisely, there is a constant $C > 0$ such that [**(BG)** (*bounded geometry*)] for any pair of intervals $J \subseteq T$ with J in η_{n+1} and T in η_n, $n \geq 0$,

$$\frac{|J|}{|T|} \geq C,$$

and such that [**(BNG)** (*bounded nearby geometry*)] for any pair of intervals J_1 and J_2 in η_n with a common endpoint, $n \geq 1$,

$$\frac{|J_1|}{|J_2|} \geq C.$$

The statement **(BG)** follows directly from Theorem 4.2 and the construction of η_f. To prove the statement **(BNG)**, one need to check when the common endpoint point q of J_1 and J_2 is a preimage of a fixed point p_k of F under some iterate of F. In this case, let $J_{1,i} = F^{\circ i}(J_1)$ and $J_{2,i} = F^{\circ i}(J_2)$ for $i \geq 0$. Then there is a biggest integer $j \geq 0$ such that $F^{\circ j}|J_1 \cup J_2$ is a diffeomorphism. So $F^{\circ j}(q) = p_{\pm k}$ (see Fig. 4.11). Therefore there exists another integer $m \geq j$ such that both of $F^{\circ m}|J_1$ and $F^{\circ(m+1)}|J_2$ are diffeomorphisms and $J_{1,m} = J_{2,m+1} = P_k$ or P_{-k}. This implies that $J_{1,l} = J_{2,l+1}$ for all $j < l \leq m$. In particular, $J_{1,j} = J_{2,j+1}$. So $J_{1,j} = F(J_{2,j})$. From Theorem 4.2, there is a constant $C_0 > 0$ such that $C_0^{-1} \leq |F'(x)| \leq C_0$ for all x in $P_{\pm k}$ and all $k \geq 0$. Hence

$$C_0^{-1} \leq \frac{|J_{1,j}|}{|J_{2,j}|} \leq C_0.$$

Applying Theorem 4.2 again, there is a constant $C_1 > 0$ such that

$$C_1^{-1} \leq \frac{|J_1|}{|J_2|} \leq C_1.$$

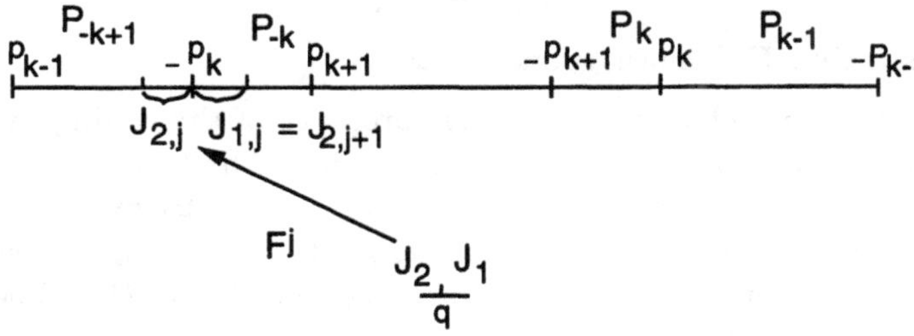

Fig. 4.11

Similarly, one can prove that η_g has bounded and bounded nearby geometry (**(BG)** and **(BNG)**).

From the bounded and bounded nearby geometry (**(BG)** and **(BNG)**) of η_f and of η_g, we now prove that h is quasisymmetric by using the exactly same argument as that in the proof of Theorem 3.2. Let us repeat the argument here for the convenience of the reader. For any $x < y$ in I, let $z = (x+y)/2$ and let $N > 0$ be the smallest integer such that there is an interval J in $\eta_{N,f}$ contained in $[x,y]$. Let $\tilde{J}$ be the interval in $\eta_{N-1,f}$ containing J. Then the union of $\tilde{J}$ and JJ, one of its adjacent intervals in $\eta_{N-1,f}$ contains $[x,y]$ (see Figs. 3.1, 3.2, and 3.3). Due to the bounded geometry **(BG)** and to the bounded nearby geometry **(BNG)** of η_g (and refer to Figs. 3.1, 3.2, and 3.3), there is a constant $C_2 > 0$ such that

$$\frac{|H(J)|}{|H([x,z])|} \geq C_2 \quad \text{and} \quad \frac{|H(J)|}{|H([z,y])|} \geq C_2.$$

Since η_f has bounded geometry (**(BG)**), the maximum length of intervals in $\eta_{n,f}$ tends to zero exponentially, that is, there are constants $C_3 > 0$ and $0 < \lambda < 1$ such that

$$\max_{J \in \eta_{n,f}} |J| \leq C_3 \lambda^n$$

for all $n \geq 0$. Thus, there is a constant integer $N_1 > 0$ (which does not depend on N) such that there are intervals J_1 and J_2 in η_{N+N_1}, contained in $[x,z]$ and $[z,y]$, respectively. This implies that $H(J_1)$ and $H(J_2)$ are contained in $H([x,z])$ and $H([z,y])$ respectively because H is the conjugacy.

Because of the bounded geometry **(BG)** and of the bounded nearby geometry **(BNG)** of η_g, there is again a constant $C_4 > 0$ (see Figs. 3.1, 3.2, and 3.3) such that

$$C_4^{-1} \leq \frac{|H(x) - H(z)|}{|H(z) - H(y)|} \leq C_4,$$

which means that H is quasisymmetric. ∎

Remark 4.6. The quasisymmetric property of the conjugacy between two conjugate infinitely renormalizable mappings is first proved in [SU4] for quadratic-like Feigenbaum-like maps (see §5.1 and §5.9) by using complex method. The proof of Theorem 4.4 presented here is for more general unimodal mappings; it is a real method developed in [JI5,JI7,JI10]. The method in the proof of Theorem 4.4 can be applied to the conjugacy between two conjugate infinitely renormalizable S-unimodal mappings $f(x) = h(-|x|^\gamma)$ for $\gamma > 1$ of bounded type (see [JI5]). Pałuba [PAL] developed another approach. The recent work of Lyubich [LYU] and of Graczyk and Świątek [SW2,GRS] promotes the study of the quasisymmetric property of the conjugacy between two conjugate infinitely renormalizable real quadratic polynomials of unbounded type. Their work if it is completed, combining with the work of Milnor and

Thurston [MIT], Thurston (see [DH4]), Douady and Hubbard [DH3] (see §5.3), Sullivan [SU4], McMullen [MC1], and Yoccoz (refer to §5.6), would give a proof of the conjecture that hyperbolic real quadratic polynomials are open and dense in the space of all real quadratic polynomials. The program to attack this conjecture was outlined by Sullivan [SU4] based on the work of Milnor and Thurston [MIT], Thurston (see [DH4]), and Douady and Hubbard [DH4] as follows. Consider the real quadratic family $\{P_t(x) = t - (1+t)x^2\}_{0 \leq t \leq 1}$. A point p is called a periodic point of P_t of period k if $P_t^{\circ i}(p) \neq p$ for all $0 \leq i < k$ but $P_t^{\circ k}(p) = p$. Let $\lambda_p = (P_t^{\circ k})'(p)$ be the eigenvalue of P_t at a periodic point p of period k. If $0 < |\lambda_p| < 1$, p is called attractive and if $\lambda_p = 0$, p is called super-attractive. From the work of Singer [SIN], every P_t can have at most one attractive or super-attractive periodic point. If P_t has an attractive or super-attractive periodic point, it is called hyperbolic, otherwise it is called non-hyperbolic. A conjugacy class in the real quadratic family is a subset of all parameters t such that for every two parameters t_1 and t_2 in this subset, P_{t_1} and P_{t_2} are topologically conjugate. A conjugacy class is hyperbolic if it contains a hyperbolic quadratic polynomial, otherwise it is called non-hyperbolic. From the work of Milnor and Thurston [MIT] in the kneading theory (see also [COE,MV2]) and the work of Thurston (see [DH4]) in combinatorial rigidity, every conjugacy class is either an interval or a point. The conjecture is then equivalent to another conjecture that every non-hyperbolic conjugacy class consists of only one point. Consider P_{t_1} and P_{t_2} as quadratic-like maps (see §5.1), from the work of Douady and Hubbard [DH3], if they are hybrid equivalent (see §5.3), then $t_1 = t_2$. This implies if a non-hyperbolic conjugacy class is actually a hybrid equivalent class, then it consists of only one point. From the work of Sullivan [SU4] and McMullen [MC1], if two real infinitely renormalizable polynomials P_{t_1} and P_{t_2} are quasisymmetrically conjugate on the interval $[-1, 1]$, then they are hybrid equivalent when we consider them as two quadratic-like maps. Sullivan [SU4] also proved that an infinitely renormalizable conjugacy class of bounded type is actually a hybrid equivalent class (see the beginning of §5.9). A consequence of the work of Yoccoz (see §5.6) is that if a non-hyperbolic conjugacy class is not infinitely renormalizable, then it is actually hybrid equivalent class and consists of only one point. The study of the quasisymmetric property of the conjugacy between P_{t_1} and P_{t_2} in an infinitely renormalizable conjugacy class of unbounded type would be established in the work of Lyubich [LYU] and of Graczyk and Świątek [SW2,GRS]. McMullen [MC1] approached this problem from no invariant line field point of view. The methods in the work of Sullivan [SU4], McMullen [MC1], Lyubich [LYU], Graczyk and Świątek [SW2,GRS] and Yoccoz (refer to §5.6) are complex. An interesting research problem is to prove that the conjugacy between two conjugate infinitely renormalizable real quadratic polynomials of unbounded type is quasisymmetric by a real method as we did in the proof of Theorem 4.4. The reason is that a real method may also give the proof of the conjecture that the conjugacy between two conjugate infinitely renormalizable folding mappings of unbounded type in the family $\{f_t(x) = t - (1+t)|x|^\gamma\}_{0 \leq t \leq 1}$, $\gamma > 1$, is quasisymmetric. The proof of this conjecture is completely open.

4.4. Infinitely Renormalizable C^{1+Z}-Unimodal Mappings

Let $f(x) = h(-|x|^\gamma)$, $\gamma > 1$, be an infinitely $(n_1, n_2, \ldots)$-renormalizable C^{1+Z}-unimodal mapping. Set $m_k = \prod_{i=1}^k n_i$. Let $\eta_k = \{J_k(i)\}_{1\leq i<m_k}$ and $\xi_k = \{I_k(i)\}_{1\leq i<m_k}$ be the hierarchical systems as in §4.1. Then $J_k(i) \subseteq I_k(i)$, and the map f permutes the intervals in η_k and in ξ_k.

Let $LI_k(i)$ and $RI_k(i)$ be the intervals in ξ_k which are adjacent to $I_k(i)$ and on the left and right sides of $I_k(i)$. Let $T_k(i)$ be the smallest interval containing $LI_k(i)$, $I_k(i)$, and $RI_k(i)$. Let $LI_k^+(i)$ and $RI_k^+(i)$ be the connected components of $T_k(i) \setminus I_k(i)$ containing $LI_k(i)$ and $RI_k(i)$ (see Fig. 4.4). For each $I_k(i)$ in ξ_k and each i, $1 \leq i < m_k$, $f^{\circ j}|I_k(i)$ is a homeomorphism from $I_k(i)$ onto $I_k(i+j)$ for $1 \leq j \leq m_k - i$. From the proof of Lemma 4.2, the inverse of $f^{\circ j}|I_k(i)$ can be extended to a homeomorphism $g_{k,ij}$ defined on an interval containing $T_k(i+j)$. Let $M_k(i) = g_{k,ij}(T_k(i+j))$ (it depends on j too). Since each $f^{\circ l}(M_k(i))$ intersects with at most three intervals in ξ_k, for $0 \leq l \leq j$, the intersection multiplier of $\{f^{\circ l}(M_k(i))\}_{1\leq l\leq j}$ is at most three. We thus have the following lemma.

Lemma 4.10. *There is a constant $C_1 = C_1(f) > 0$ such that for any i, $1 \leq i < m_k$, and any j, $1 \leq j \leq m_k - i$, the cross-ratio distortion $D(T, M; g_{k,ij})$ of $g_{k,ij}$ on any $M \subset T \subseteq T_k(i+j)$, is bounded by $-C_1$ from below, that is,*

$$-C_1 \leq D(T, M; g_{k,ij}).$$

Proof. Let $Q_\gamma(x) = -|x|^\gamma$. For any i, $1 \leq i < m_k$, and any j, $1 \leq j \leq m_k - i$, let $M_k(i+l) = f^{\circ l}(M_k(i))$. Then

$$g_{k,ij}|T_k(i+j) = S_0 \circ K_0 \circ \cdots \circ S_l \circ K_l \circ \cdots \circ S_{j-1} \circ K_{j-1},$$

where

$$S_l = \big(Q_\gamma|M_k(i+l)\big)^{-1} : Q_\gamma\big(M_k(i+l)\big) \to M_k(i+l)$$

is a C^3-diffeomorphism with positive Schwarzian derivative and where

$$K_l = \big(h|Q_\gamma(M_k(i+l))\big)^{-1} : M_k(i+l+1) \to Q_\gamma(M_k(i+l))$$

is a C^{1+Z}-diffeomorphism (see Fig. 4.12).

From Lemma 2.11, S_l increases the cross-ratio. So $D(T, M; S_l) \geq 0$ for any $M \subset T \subseteq Q_\gamma(M_k(i+l))$. From Lemma 2.17, there is a constant $C_1 = C(f) > 0$ such that for any $M \subset T \subseteq M_k(i+l+1)$

$$-\frac{C_1}{6}|M_k(i+l+1)| \leq D(T, M; K_l).$$

Hence

$$-C_1 \le -\frac{C_1}{6}\sum_{l=0}^{j-1} |M_k(i+l+1)| \le D(T, M; g_{k,ij})$$

for any $M \subset T \subseteq T_k(i+j)$ because the intersection multiplier of $\{M_k(i+l+1)\}_{l=0}^{j-1}$ is at most three.

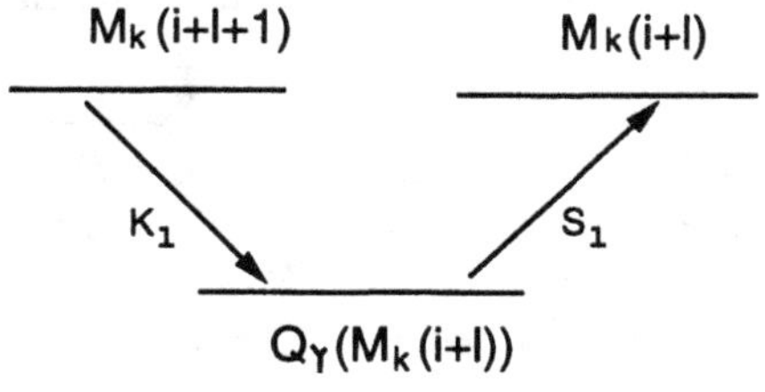

Fig. 4.12

■

Remark 4.7. Let $C'_{1,k} = C_1 \sum_{i=1}^{m_k-1} |I_k(i)|$. By more careful arguments, one can show that the cross-ratio distortion $D(T, M; g_{k,ij})$ is bounded from below by $-C'_{1,k}$ for any $M \subset T \subseteq T_k(i+j)$. Hence, if $\sum_{i=1}^{m_k-1} |I_k(i)|$ tends to zero as k goes to infinity (as it will be proven in Theorem 4.5), then the lower bound of the cross-ratio distortion will tends to zero as k goes to infinity. This is discussed later.

Lemma 4.11. *There is a constant $C_2 = C_2(f) > 0$ such that*

$$\min\{|LI_k^+(i)|, |RI_k^+(i)|\} \ge C_2|I_k(i)|$$

for all $k > 0$ and all i, $0 \le i < m_k$.

Proof. For each $k > 0$, let $I_k(i_k)$ be an interval in ξ_k with minimum length. Then

$$\min\{|LI_k^+(i_k)|, |RI_k^+(i_k)|\} \ge |I_k(i_k)|.$$

Let $i = 1$ and let $j = i_k - 1$. Let a and b be the midpoints of $LI_k^+(i_k)$ and $RI_k^+(i_k)$. Let $I = T_k(i_k)$ and $I' = [a, b]$. Then $Cr(I, I') \ge C_0$ where $C_0 > 0$ is a constant. From Lemma 2.13 and Lemma 4.10, $g_{k,ij}|I'$ is C-quasisymmetric and C depends only on C_0 and C_1 (see Lemma 2.13). Since $g_{k,ij}(I_k(i_k)) = I_k(1)$ and since one of the connected components of $g_{k,ij}(I' \setminus I_k(i_k))$ is contained in $LI_k^+(1)$, we can find a constant $C_3 > 0$, depending on C_0 and C_1, such that

$$|LI_k^+(1)| \ge C_3|I_k(1)|.$$

By using arguments similar to those in the proof of Lemma 4.2, one shows that there is a constant $C_4 = C_4(C_3, \gamma) > 0$ such that

$$\min\{|LI_k^+(0)|, |RI_k^+(0)|\} \geq C_4|I_k(0)|.$$

For any i, $1 \leq i < m_k$, let $j = m_k - i$ and let a and b be the midpoints of $LI_k^+(0)$ and $RI_k^+(0)$. Take $I = T_k(0) = T_k(m_k)$ and $I' = [a, b]$. Consider $g_{k,ij}$ defined on I. By repeating the above argument, one finds a constant $C_5 > 0$ depending only on C_0 and C_1 such that

$$\min\{|LI_k^+(i)|, |RI_k^+(i)|\} \geq C_5|I_k(i)|;$$

this is because $g_{k,ij}|I'$ is C-quasisymmetric, because $g_{k,ij}(I_k(0)) = I_k(i)$, and because the two connected components of $g_{k,ij}(I' \setminus I_k(0))$ are contained in $LI_k^+(i)$ and in $RI_k^+(i)$, respectively. ∎

Theorem 4.5 [GU1,SU4]. *Let f be an infinitely $(n_1, n_2, \ldots, \ldots)$-renormalizable C^{1+Z}-unimodal mapping and let Λ_f be the closure of the critical orbit of f. Then Λ_f is a Cantor set whose Lebesgue measure is zero.*

Proof. Each $J_k(i)$ in η_k contains $J_{k+1}(jm_k+i)$, $0 \leq j \leq n_{k+1}-1$, which are pairwise disjoint. Let $G_k(i)$ be the complement of $\cup_{j=0}^{n_{k+1}-1} J_{k+1}(jm_k+i)$ in $J_k(i)$. Suppose $G_{k+1}(i,1)$, …, $G_{k+1}(i, n_{k+1}-1)$ are the connected components of $G_k(i)$ arranged in real line order. Each of them is called a k-gap.

The interval $J_k(0)$ is sent to $J_k(1)$ by f (see Fig. 4.6). The preimage of $LJ_{k+1}^+(1)$ under f thus contains two intervals. If $n_{k+1} > 2$, then one of them is contained in a k-gap G_0 adjacent to $J_{k+1}(0)$. If $n_{k+1} = 2$, then one of them is contained in a $(k-1)$-gap G_0 adjacent to $J_k(0)$. By Lemma 4.11,

$$|G_0| \geq C_2|J_{k+1}(0)|$$

in all cases.

If $n_{k+1} > 2$, apply Lemma 4.10 and refer to the proof of Lemma 4.11; for every $J_{k+1}(jm_k)$, $2 < j < n_{k+1}$, there is a k-gap G_j adjacent to it such that

$$|G_j| \geq C_2|J_{k+1}(jm_k)|.$$

For $j = 1$, assume $LJ_{k+1}(m_k) \subseteq J_k(0)$ (similarly if $RJ_{k+1}(m_k) \subseteq J_k(0)$). Since

$$|LJ_{k+1}^+(m_k)| \geq C_2|J_{k+1}(m_k)|$$

and since $LJ^+_{k+1}(m_k)$ is the union of $LJ_{k+1}(m_k)$ and a k-gap, we can find a k-gap G_1 adjacent to either $J_{k+1}(m_k)$ or $LJ_{k+1}(m_k)$ such that

$$|G_1| \geq 2C_2|J_{k+1}(m_k)|.$$

Similarly, there is a k-gap G_2 such that

$$|G_2| \geq 2C_2|J_{k+1}(2m_k)|.$$

If $n_{k+1} = 2$, we consider $J_{k+2}(jm_k)$ for $0 \leq j < n_{k+1}n_{k+2}$. A similar argument implies that there is a k-gap or $(k+1)$-gap G_j adjacent to each J_{jm_k} such that

$$|G_j| \geq C_2|J_{k+2}(jm_k)|$$

for $2 < j < n_{k+1}n_{k+2}$. Also there are two k-gaps G_1 and G_2 such that

$$|G_1| \geq 2C_2|J_{k+2}(m_k)| \qquad \text{and} \qquad |G_2| \geq 2C_2|J_{k+2}(2m_k)|.$$

Let us arrange $\{J_{k+1}(jm_k + i)\}$ in the real line order as $J_{k+1}(i,1)$, $\ldots$, $J_{k+1}(i,n_k)$. Then $\{J_{k+1}(i,j)\}_{j=1}^{n_{k+1}}$ and $\{G_{k+1}(i,j)\}_{j=1}^{n_{k+1}-1}$ lie alternately on the real line, and for each $J_{k+1}(i,j)$, $1 < j < n_{k+1}$, there are two k-gaps adjacent to it. Hence there is a constant $C_3 > 0$ such that if $n_{k+1} > 2$, then

$$\begin{aligned}|J_k(0)| &= \sum_{j=1}^{n_{k+1}-1} \Big(|J_{k+1}(0,j)| + |G_{k+1}(0,j)|\Big) + |J_{k+1}(0,n_{k+1})| \\ &\leq C_3 \sum_{j=1}^{n_{k+1}-1} |G_{k+1}(0,j)|,\end{aligned}$$

and if $n_{k+1} = 2$, then

$$\begin{aligned}|J_k(0)| = &\sum_{i=1}^{n_{k+2}-1} \Big(|J_{k+2}(0,i)| + |G_{k+2}(0,i)|\Big) + |J_{k+2}(0,n_{k+2}| + |G_{k+1}(0,1)| \\ &+ \sum_{i=1}^{n_{k+2}-1} \Big(|J_{k+2}(m_k,i)| + |G_{k+2}(m_k,i)|\Big) + |J_{k+2}(m_k,n_{k+2})| \\ &\leq C_3 \sum_{i=1}^{n_{k+2}-1} \Big(|G_{k+2}(0,i)| + |G_{k+2}(m_k,i)|\Big) + |G_{k+1}(0,1)|.\end{aligned}$$

Thus if $n_{k+1} > 2$,

$$|J_k(0)| \leq C_3|G_k(0)|,$$

and if $n_{k+1} = 2$, then

$$|J_k(0)| \leq C_3\Big(|G_k(0)| + |G_{k+1}(0)| + |G_{k+1}(m_{k+1})|\Big).$$

In either case, we can find a constant $0 < C_4 < 1$ such that

$$\sum_{j=0}^{n_{k+1}n_{k+2}-1} |J_{k+2}(jm_k)| \leq C_4|J_k(0)|$$

for all $k > 0$. Now applying Lemma 4.10 and Lemma 4.11, we obtain a constant $0 < C_5 < 1$ such that

$$\sum_{j=0}^{n_{k+1}n_{k+2}-1} |J_{k+2}(jm_k + i)| \leq C_5|J_k(i)|$$

for all $k > 0$ and all $0 \leq i < n_k$. Hence there is a constant $C_6 > 0$ such that

$$\sum_{i=0}^{m_k-1} |J_k(i)| \leq C_6(\sqrt{C_5})^k$$

for all $k > 0$. Therefore,

$$\Lambda_f = \cap_{k=1}^{\infty} \cup_{i=0}^{m_k-1} J_k(i)$$

has zero Lebesgue measure. ∎

Remark 4.8. Theorem 4.5 is proved in [GU1] for a Feigenbaum-like S-unimodal map and is proved in [SU4] for any infinitely renormalizable C^{1+Z}-unimodal map. The proof presented here is due to Sullivan [SU4].

Let $f(x) = h(-|x|^\gamma)$ be an infinitely $(n_1, n_2, \ldots)$-renormalizable C^{1+Z}-unimodal mapping, and set $m_k = \prod_{i=1}^k n_i$. If we reconsider the proofs of Lemmas 4.10 and 4.11 after Theorem 4.5 (see Remark 4.7), we can obtain stronger results as follows.

Lemma 4.12. *The lower bound of the cross-ratio distortion $D(T, M; g_{k,ij})$ tends to zero, as $k \to \infty$, uniformly on $M \subset T \subseteq T_k(i+j)$ and on i and j where $1 \leq i < m_k$ and $1 \leq j \leq m_k - i$, more precisely,*

$$\inf\{D(T, M; g_{k,ij}) \mid M \subset T \subseteq T_k(i+j),\ 1 \leq i < m_k,\ 1 \leq j \leq m_k - i\} \to 0$$

as $k \to \infty$.

Lemma 4.13. *There is a universal constant $C > 0$, depending only on $\gamma > 1$ and on $\rho = (n_1, n_2, \ldots)$, such that*

$$\min\{|LI_k^+(i)|, |RI_k^+(i)|\} \geq C|I_k(i)|$$

for large $k > 0$ and all i, $0 \leq i < m_k$.

Lemma 4.14. *There is a universal constant $0 < \mu < 1$, depending only on $\gamma > 1$ and on $\rho = (n_1, n_2, \ldots)$, such that*

$$\sum_{j=0}^{n_k-1} |J_k(i + jm_{k-1})| \leq \mu |J_{k-1}(i)|$$

and

$$\sum_{j=0}^{n_k-1} |I_k(i + jm_{k-1})| \leq \mu |I_{k-1}(i)|$$

for large k and all i, $0 \leq i < m_k$.

Let $\gamma > 1$ and let $\rho = (n_1, n_2, \ldots)$ be a fixed sequence of integers $n \geq 2$. For an infinitely ρ-renormalizable C^{1+Z}-unimodal map $f(x) = h(-|x|^\gamma)$. Let $Q(f) = \{Q_{kij}\}$ be the sequence of quasisymmetric constants of $g_{k,ij}|I_k(i+j)$ where $1 \leq k < \infty$, $1 \leq i < m_k$, $1 \leq j \leq m_k - i$. Let $\mathcal{Q}$ be the space of sequences $Q(f)$ for all infinitely ρ-renormalizable C^{1+Z}-unimodal mappings f.

Lemma 4.15. *The space $\mathcal{Q}$ is beau.*

Let $FM^{1+Z}(\gamma, \rho)$ be the space of all infinitely ρ-renormalizable C^{1+Z}-unimodal mappings $f(x) = h(-|x|^\gamma)$. Let $\mathcal{R}^{\circ k}(f)(x) = h_k(-|x|^\gamma)$ be the kth-renormalization of $f \in FM^{1+Z}(\gamma, \rho)$ for $k \geq 0$. Let Q_k be the quasisymmetric constant of h_k on $[-1, 0]$. In particular, Lemma 4.15 implies

Theorem 4.6 [SU4]. *For each $f \in FM^{1+Z}(\gamma, \rho)$, there is a constant $C = C(f) > 0$ such that $Q_k \leq C$ for all $k > 0$. And there is a universal constant $C_0 > 0$ such that $Q_k \leq C_0$ for all $f \in FM^{1+Z}(\gamma, \rho)$ and large k.*

The reader may give the proofs of Lemmas 4.12, 4.13, 4.14 and Theorem 4.6 as exercises by going over the proofs of Lemmas 4.10 and 4.11 and Theorem 4.5. A result stronger than Theorem 4.6 will be proved in the next section.

4.5. The Distortion of an Infinitely Renormalizable Folding Mapping

We use the same notations as those in the previous section. For each $f \in FM^{1+Z}(\gamma,\rho)$, let $\xi_k = \{I_k(i)\}_{1\le i<m_k}$ be the hierarchical system as in §4.1. For each $k \geq 1$, $f^{\circ(m_k-1)} : I_k(1) \to I_k(m_k)$ is a homeomorphism. Its inverse can be extended to a homeomorphism g_k defined on an interval containing $T_k(0)$. Let $M_k(1) = g_k(T_k(0))$ and let $M_k(i+1) = f^{\circ i}(M_k(1))$ for $0 \le i \le m_k - 1$. We note that $M_k(m_k) = T_k(0)$. Let $Q_\gamma(x) = -|x|^\gamma$ and $W_k(i) = Q_\gamma(M_k(i))$ for $1 \le i \le m_k - 1$. Then $h(W_k(i)) = M_k(i+1)$ for $1 \le i \le m_k - 1$ (refer to Fig. 4.12). Let $K_i : M_k(i+1) \to W_k(i)$ denote the inverse of $h|W_k(i)$ and let $S_i : W_k(i) \to M_k(i)$ denote the inverse of $Q_\gamma|M_k(i)$ for $1 \le i \le m_k - 1$. Then

$$g_k = S_1 \circ K_1 \circ \cdots \circ S_i \circ K_i \circ \cdots \circ S_{m_k-1} \circ K_{m_k-1}.$$

For any x and y in $I_k(0)$, let $x_0 = g_k(x)$ and $y_0 = g_k(y)$, let $x_i = f^{\circ i}(x_0)$ and $y_i = f^{\circ i}(y_0)$ for $1 \le i \le m_k - 1$, and let $z_{i+1} = Q_\gamma(x_i)$ and $w_{i+1} = Q_\gamma(y_i)$, for $0 \le i \le m_k - 2$. The distortion

$$X_k(x,y) = \left|\log\left(\frac{|g_k'(x)|}{|g_k'(y)|}\right)\right|$$

of g_k at x and y can be estimated as

$$X_k(x,y) \le \left|\sum_{i=1}^{m_k-1}\Big(\log|K_i'(x_i)| - \log|K_i'(y_i)|\Big)\right| + \left|\sum_{i=1}^{m_k-1}\Big(\log|S_i'(z_i)| - \log|S_i'(w_i)|\Big)\right|.$$

Because h is a C^{1+Z}-diffeomorphism, the modulus of continuity of $\log(h^{-1})'$ is $|x-y|\log|x-y|$ (see Remark 2.12). Hence $\log(h^{-1})'$ is α-Hölder continuous for any $0 < \alpha < 1$. Let C_α be the α-Hölder constant of $\log(h^{-1})'$ for a fixed α. Then

$$X_k(x,y) \le C_\alpha \sum_{i=1}^{m_k-1} |x_i - y_i|^\alpha + \left|\sum_{i=1}^{m_k-1}\Big(\log|S_i'(z_i)| - \log|S_i'(w_i)|\Big)\right|.$$

Let L_i be the linear map which maps x_i to z_i, and y_i to w_i, for $1 \le i \le m_k - 1$. Then

$$\tilde{g}_k = S_1 \circ L_1 \circ S_2 \circ L_2 \circ \cdots \circ S_{m_k-1} \circ L_{m_k-1}$$

is a C^3-diffeomorphism with positive Schwarzian derivative defined on $T_k(0)$. By Lemma 2.4, there is a constant $C > 0$, obtained from Lemma 4.11, such that

$$|(N(\tilde{g}_k))(x)| \leq \frac{C}{|I_k(0)|}$$

for $x \in I_k(0)$. Hence

$$\left|\log\left(\frac{|\tilde{g}_k'(x)|}{|\tilde{g}_k'(y)|}\right)\right| \leq C\frac{|x-y|}{|I_k(0)|}$$

for any x and y in $I_k(0)$. Since

$$\log\left(\frac{|\tilde{g}_k'(x)|}{|\tilde{g}_k'(y)|}\right) = \sum_{i=1}^{m_k-1}\Big(\log|S_i'(z_i)| - \log|S_i'(w_i)|\Big),$$

we have

$$X_k(x,y) \leq C_\alpha \sum_{i=1}^{m_k-1} |x_i - y_i|^\alpha + C\frac{|x-y|}{|I_k(0)|}. \tag{4.2}$$

For each $I_k(i)$ in ξ_k, where $1 \leq i < m_k$, $f^{\circ j} : I_k(i) \to I_k(i+j)$ is a homeomorphism for $1 \leq j \leq m_k - i$. Its inverse can be extended to a homeomorphism $g_{k,ij}$ defined on an interval containing $T_k(i+j)$. By similar arguments, the distortion

$$X_{k,ij}(x,y) = \left|\log\left(\frac{|g_{k,ij}'(x)|}{|g_{k,ij}'(y)|}\right)\right|$$

of $g_{k,ij}$ at x and y in $I_k(i+j)$ can be estimated as

$$X_{k,ij}(x,y) \leq C_\alpha \sum_{l=1}^{j} |x_{ij,l} - y_{ij,l}|^\alpha + C\frac{|x-y|}{|I_k(i+j)|} \tag{4.3}$$

where $x_{ij} = g_{k,ij}(x)$ and $y_{ij} = g_{k,ij}(y)$, and $x_{ij,l} = f^{\circ l}(x_{ij})$ and $y_{ij,l} = f^{\circ l}(y_{ij})$.

Let $\gamma > 1$ and let $\rho = (n_1, n_2, \ldots)$ be a fixed sequence of integers $n \geq 2$. Let $FM^{1+1}(\gamma, \rho)$ be the space of all infinitely ρ-renormalizable C^{1+1}-unimodal mappings $f(x) = h(-|x|^\gamma)$. For each $f(x) = h(-|x|^\gamma) \in FM^{1+1}(\gamma, \rho)$, let $C_1 > 0$ be the Lipschitz constant of $\log(h^{-1})'$ on $[-1, h(0)]$. We can get the following stronger estimates

$$X_k(x,y) \leq C_1 \sum_{i=1}^{m_k-1} |x_i - y_i| + C\frac{|x-y|}{|I_k(0)|}, \tag{4.4}$$

and

$$X_{k,ij}(x,y) \le C_1 \sum_{l=1}^{j} |x_{ij,l} - y_{ij,l}| + C\frac{|x-y|}{|I_k(i+j)|}, \tag{4.5}$$

where C is a constant obtained via Lemma 4.11.

For f in $FM^{1+1}(\gamma,\rho)$, let $\{\mathcal{R}^{\circ k}(f)(x) = h_k(-|x|^\gamma)\}_{k=0}^{\infty}$ be the sequence of renormalizations of f (set $h_0 = h$). Let

$$Lip_k = \sup_{x\neq y\in h_k([-1,0])} \left\{ \frac{|\log(h_k^{-1})'(x) - \log(h_k^{-1})'(y)|}{|x-y|} \right\}$$

be the Lipschitz constant of $\log(h_k^{-1})'$ for $k \ge 0$.

Theorem 4.7 [SU4]. *For each $f \in FM^{1+1}(\gamma,\rho)$, there is a constant $C > 0$ such that $Lip_k \le C$ for all $k \ge 1$. Moreover, there is a universal constant $C_0 > 0$ such that $Lip_k \le C_0$ for all $f \in FM^{1+1}(\gamma,\rho)$ and large k.*

Proof. From Lemma 4.14 and from Inequality (4.5), the space of the sequences $X(f) = \{X_{k,ij}(x,y)$ defined on $I_k(i+j)\times I_k(i+j)$, for all $k \ge 0$, all $1 \le i < m_k$, and all $1 \le j \le m_k - i\}$ of distortions for all $f \in FM^{1+1}(\gamma,\rho)$ is beau. This implies that in (4.4),

$$|x_i - y_i| \le B_k |I_k(i)| \frac{|x-y|}{|I_k(0)|}$$

for $1 \le i < m_k$, where the space of the sequences $B(f) = \{B_k\}_{k=1}^{\infty}$ for all $f \in FM^{1+1}(\gamma,\rho)$ is beau. Hence we can bound the distortion $X_k(x,y)$ of g_k at x and y in $I_k(0)$ by

$$X_k(x,y) \le \Big(C_1 B_k \sum_{i=1}^{m_k-1} |I_k(i)| + C\Big)\frac{|x-y|}{|I_k(0)|}.$$

The theorem now follows from Lemma 4.14 and the fact that

$$h_k = \alpha_k^{-1} \circ f^{\circ(m_k-1)} \circ h \circ \tilde{\alpha}_k.$$

(Refer to Theorem 4.1.) ∎

Remark 4.9. For $f(x) = h(-|x|^\gamma) \in FM^{1+1}(\gamma,\rho)$, let

$$\mathcal{R}^{\circ k}(f)(x) = h_k(-|x|^\gamma)$$

be the kth-renormalization of f, for $k \geq 0$, where $h_k : [-1,0] \to [-1, h_k(0)]$ is a C^{1+1}-diffeomorphism. Since $1/2 \leq (h_k^{-1})'(x_k) \leq 1$ for some $x_k \in [-1,1]$, Theorem 4.7 implies that the sequence $\{\log(h_k^{-1})'\}_{k=0}^{\infty}$ is bounded and eventually universally bounded. Furthermore, the sequence of the Lipschitz constants of h_k', for $1 \leq k < \infty$ is bounded and eventually universal bounded. From Ascoli-Arzelà theorem, the sequence $\{h_k\}_{k=0}^{\infty}$ is precompact in $\mathcal{M}(1+\alpha)$ for any $0 \leq \alpha < 1$ (see Remark 4.3). If h_0^{-1} can be extended to $I = [-1,1]$ as a C^{1+1}-diffeomorphism, then $\{h_k\}_{k=0}^{\infty}$ is precompact in $\tilde{\mathcal{M}}(1+\alpha)$ (see Remark 4.3).

Suppose $f(x) = h(-|x|^{\gamma})$ is in $FM^{1+Z}(\gamma,\rho)$. Let Λ_f be the closure of the critical orbit of f.

Theorem 4.8 [SU4]. *If $\rho = (n_1, n_2, \ldots)$ is a bounded sequence of integers $n \geq 2$ and $\gamma > 1$ is a real number, then there is a universal constant $0 < \alpha_0 < 1$ depending only on $\gamma > 1$ and ρ such that for every $f \in FM^{1+Z}(\gamma,\rho)$, the Hausdorff dimension of Λ_f is less than or equal to α_0.*

Proof. Let $n_0 > 0$ be the least upper bound of $\{n_i\}_{i=1}^{\infty}$. From Lemma 4.14,

$$\sum_{j=0}^{n_k-1} |J_k(i+jm_{k-1})|^{\alpha} \leq n_k \Big(\frac{\sum_{j=0}^{n_k-1} |J_k(i+jm_{k-1})|}{n_k} \Big)^{\alpha} \leq n_k^{1-\alpha}\mu^{\alpha}$$

for large k. This implies that

$$\sum_{i=0}^{m_k-1} |J_k(i)|^{\alpha} \leq 2\Big(n_0^{1-\alpha}\mu^{\alpha}\Big)^{k-1}$$

for large k. Let $\alpha_0 = \log n_0/(\log n_0 - \log\mu)$. Then

$$\sum_{i=0}^{m_k-1} |J_k(i)|^{\alpha_0} \leq 2$$

for large k. Thus, the Hausdorff dimension of Λ_f is less than or equal to α_0. ■

Remark 4.10. Similarly,

$$\sum_{i=0}^{m_k-1} |I_k(i)|^{\alpha_0} < \infty$$

for large k.

For f in $FM^{1+Z}(\gamma,\rho)$, let $\{\mathcal{R}^{\circ k}(f)(x) = h_k(-|x|^{\gamma})\}_{k=0}^{\infty}$ be the sequence of renormalizations of f (set $h_0 = h$). Let $Hol_{k,\alpha}$ be the α-Hölder constant of $\log(h_k^{-1})'$ for $k \geq 0$. Let α_0 be the constant in Theorem 4.8.

Theorem 4.9 [SU4]. *Let $\rho = (n_1, n_2, \ldots)$ be a bounded sequence of integers $n \geq 2$ and let $\gamma > 1$ be a real number. For each $f \in FM^{1+Z}(\gamma, \rho)$ and each $\alpha_0 < \alpha < 1$, there is a constant $C > 0$ such that $Hol_{k,\alpha} \leq C$ for all $k \geq 0$. Moreover, for each $\alpha_0 < \alpha < 1$, there is a universal constant $C_0 > 0$ such that $Hol_{k,\alpha} \leq C_0$ for all $f \in FM^{1+Z}(\gamma, \rho)$ and large k.*

Proof. Since $\sum_{i=0}^{m_k-1} |I_k(i)|^\alpha$ tends to zero as k goes to infinity for $\alpha_0 < \alpha < 1$ and since Inequality (4.3), the space of the sequences $X(f) = \{X_{k,ij}(x, y)$ defined on $I_k(i+j) \times I_k(i+j)$ for all $k \geq 1$, all $1 \leq i < m_k$, and all $1 \leq j \leq m_k - i\}$ of distortions for all $f \in FM^{1+Z}(\gamma, \rho)$ is beau. This implies that in (4.2),

$$|x_i - y_i| \leq B_k |I_k(i)| \frac{|x - y|}{|I_k(0)|},$$

for $1 \leq i < m_k$, where the space of the sequences $B(f) = \{B_k\}_{k=0}^\infty$ for all $f \in FM^{1+Z}(\gamma, \rho)$ is beau. We can further bound the distortion $X_k(x, y)$ by

$$X_k(x, y) \leq \Big(C_\alpha B_k \sum_{i=1}^{m_k-1} |I_k(i)|^\alpha + C\Big) \frac{|x - y|^\alpha}{|I_k(0)|^\alpha}.$$

The theorem now follows from the end of the proof of Theorem 4.7. ∎

Remark 4.11. For $f(x) = h(-|x|^\gamma) \in FM^{1+Z}(\gamma, \rho)$, let

$$\mathcal{R}^{\circ k}(f)(x) = h_k(-|x|^\gamma)$$

be the kth-renormalization of f for $k \geq 0$, where $h_k : [-1, 0] \to [-1, h_k(0)]$ is a C^{1+Z}-diffeomorphism. Since $1/2 \leq (h_k^{-1})'(x_k) \leq 1$ for some $x_k \in [-1, 1]$, Theorem 4.9 implies that the sequence $\{\log(h_k^{-1})'\}_{k=0}^\infty$ is bounded and eventually universally bounded. Furthermore, the sequence of the α-Hölder constants of h'_k, for $1 \leq k < \infty$ is bounded and eventually universally bounded. From Ascoli-Arzelà theorem, the sequence $\{h_k\}_{k=0}^\infty$ is precompact in $\mathcal{M}(1+\alpha)$ for any $0 \leq \alpha < 1$ (see Remark 4.3). If h_0^{-1} can be extended to $I = [-1, 1]$ as a C^{1+Z}-diffeomorphism, then $\{h_k\}_{k=0}^\infty$ is precompact in $\tilde{\mathcal{M}}(1+\alpha)$ (see Remark 4.3).

4.6. The Attractor of an Infinitely Renormalizable Folding Mapping

Let $f(x) = h(-|x|^\gamma)$ be an infinitely $(n_1, n_2, \ldots)$-renormalizable C^{1+Z}-unimodal mapping, let $c(i) = f^{\circ i}(0)$ for $i \geq 0$, and let Λ_f be the closure of the critical orbit $\{c(i)\}_{i=0}^{\infty}$. The set Λ_f is the attractor of f in the sense of Milnor [MI1]. In this section, we discuss the geometry of Λ_f for a Feigenbaum-like mapping. (The discussion works as well for infinitely renormalizable C^{1+Z}-unimodal mapping of bounded type.)

Let $\gamma > 1$ be a real number and let $m_k = 2^k$ for $k \geq 0$. Suppose $f(x) = h(-|x|^\gamma)$ is a C^{1+Z} Feigenbaum-like unimodal mapping. Let $\eta_k = \{J_k(i)\}_{i=0}^{m_k-1}$, for $k = 1, 2, \ldots$, be the hierarchical system constructed in §4.1. Every $J_k(i)$ in η_k is bounded by $c(i)$ and $c(m_k + i)$ and contains only two intervals $J_{k+1}(i)$ and $J_{k+1}(m_k + i)$ in η_{k+1}. The complement

$$G_k(i) = J_k(i) \setminus \big(J_{k+1}(i) \cup J_{k+1}(m_k + i)\big)$$

is a subinterval of $J_k(i)$. Thus the hierarchical system $\{\eta_k\}_{k=1}^{\infty}$ gives a Cantor system $\mathcal{CS}(f)$ (see Definition 2.1), where $\mathcal{CS}(f) = \{\mathcal{I}, \mathcal{G}\}$ where $\mathcal{I} = \{\mathcal{I}_k\}_{k=0}^{\infty}$ and $\mathcal{G} = \{\mathcal{G}_k\}_{k=1}^{\infty}$ where $\mathcal{I}_k = \eta_{k+1}$ and $\mathcal{G}_k = \{G_k(i)\}_{i=1}^{m_k-1}$.

Definition 4.3. *The Cantor system $\mathcal{CS}(f)$ has bounded geometry if there is a constant $C > 0$ such that*

$$\frac{|J_{k+1}(i)|}{|J_k(i)|}, \ \frac{|J_{k+1}(m_k + i)|}{|J_k(i)|}, \ \frac{|G_k(i)|}{|J_k(i)|} \geq C$$

for all $k \geq 1$ and all $0 \leq i < m_k$.

Theorem 4.10 [SU4]. *Let $f(x) = h(-|x|^\gamma)$ be a C^{1+Z} Feigenbaum-like unimodal mapping. Then $\mathcal{CS}(f)$ has bounded geometry.*

Define

$$a_k = |J_{k+1}(0)|/|J_k(0)|,$$

$$b_k = |J_{k+1}(m_k)|/|J_k(0)|,$$

and

$$c_k = |G_k(0)|/|J_k(0)|.$$

Lemma 4.16. *If there is a constant* $C_1 > 0$ *such that*

$$a_k, b_k, c_k \geq C_1$$

for all $k > 0$*, then we can find a constant* $C_2 > 0$ *such that*

$$\frac{|J_{k+1}(i)|}{|J_k(i)|}, \frac{|J_{k+1}(m_k+i)|}{|J_k(i)|}, \frac{|G_k(i)|}{|J_k(i)|} \geq C_2$$

for all $k > 0$ *and all* $0 \leq i < m_k$.

Proof. By Lemma 2.13, Lemma 4.10, and Lemma 4.11,

$$f^{\circ(m_k-i)} : J_k(i) \to J_k(0)$$

is C-quasisymmetric for any $J_k(i)$, where $k > 0$ and $1 \leq i < m_k$. Further, $f^{\circ(m_k-i)}$ maps $J_{k+1}(i)$, $G_k(i)$ and $J_{k+1}(m_k+i)$ homeomorphically to $J_{k+1}(0)$, $G_k(0)$ and $J_{k+1}(m_k)$. There is thus a constant $C_2 > 0$ depending on C_1 such that

$$\frac{|J_{k+1}(i)|}{|J_k(i)|}, \frac{|J_{k+1}(m_k+i)|}{|J_k(i)|}, \frac{|G_k(i)|}{|J_k(i)|} \geq C_2$$

for all $k > 0$ and all $0 \leq i < m_k$. ■

Lemma 4.17. *There is a constant* $C_3 > 0$ *such that for all* $k > 0$,

$$c_k > C_3.$$

Proof. Let $J'_{k+2}(0) = -J_{k+2}(0) \cup J_{k+2}(0)$. From the proof of Lemma 4.11, there is a constant $C_4 > 0$ such that the length of each connected component of $f^{-1}(LJ^+_{k+2}(1))$ is greater than $C_4|J'_{k+2}(0)|$ for $k > 0$. One of the connected components is contained in $G_k(0)$ and the other is $J_{k+1}(0) \setminus J'_{k+2}(0)$ (see Fig. 4.13).

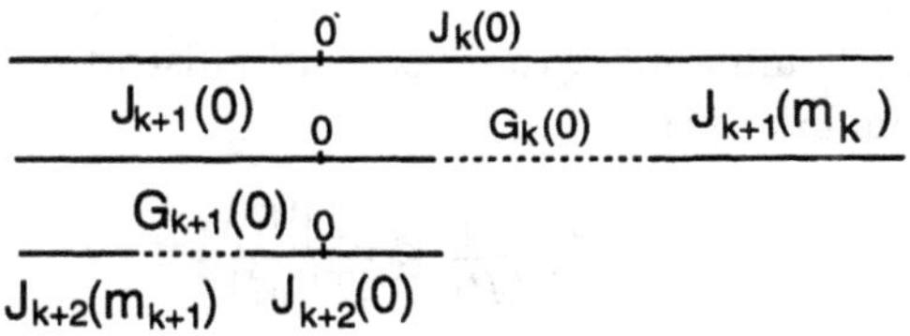

Fig. 4.13

Let $C_5 = C_4/(1 + C_4)$, then

$$|G_k(0)| \geq C_5 |J_{k+1}(0)|$$

for $k > 0$. From Lemma 4.11,

$$|J_{k+1}(0) \cup G_k(0)| \geq C_4 |J_{k+1}(m_k)|$$

because $|J_{k+1}(0) \cup G_k(0)|$ is either $LJ^+_{k+1}(m_k)$ or $RJ^+_{k+1}(m_k)$. Thus, there is a constant $C_3 > 0$ depending on C_5 such that $c_k > C_3$ for all $k > 0$. ∎

Proof of Theorem 4.10. Let $C > 0$ denote a constant (it may be different in different formulas). Let $T = J_k(1)$, $L = J_{k+1}(m_k+1)$, $M = G_k(1)$, and $R = J_{k+1}(1)$. Let $T' = f^{\circ(m_k-1)}(T) = J_k(0)$, $L' = f^{\circ(m_k-1)}(L) = J_{k+1}(0)$, $M' = f^{\circ(m_k-1)}(M) = G_k(0)$, and $R' = f^{\circ(m_k-1)}(R) = J_{k+1}(m_k)$. Then $f(T') = T$, $f(L') = R$, and $f(R') = L$ (see Fig. 4.14). Let $\tilde{L}' = L' \cup (-L')$ and $\tilde{M}' = T' \setminus (\tilde{L}' \cup R')$. Then $f(\tilde{M}') = M$.

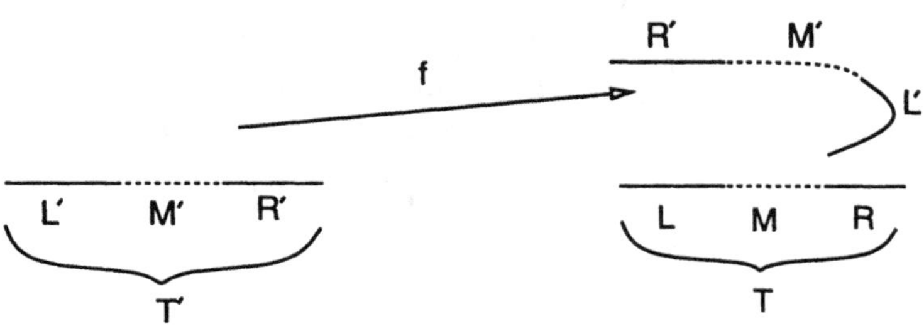

Fig. 4.14

From Lemma 4.17

$$C^{-1}\frac{|L'||R'|}{|T'||M'|} \leq \log\left(1 + \frac{|L'||R'|}{|T'||M'|}\right) \leq C\frac{|L'||R'|}{|T'||M'|},$$

and

$$C^{-1}\frac{|L||R|}{|T||M|} \leq \log\left(1 + \frac{|L||R|}{|T||M|}\right) \leq C\frac{|L||R|}{|T||M|}$$

for all $k > 0$. From Lemma 4.10, the cross-ratio distortion of $(f^{\circ(m_k-1)}|T)^{-1}$ is bounded, we thus have that

$$C\frac{|L'||R'|}{|T'||M'|} \leq \frac{|L||R|}{|T||M|}$$

for all $k > 0$.

Since $f(x) \sim -|x|^\gamma$ near 0, and since Lemma 4.17,

$$C^{-1}\frac{|L'|^\gamma}{|T'|^\gamma} \leq \frac{|R|}{|T|} \leq C\frac{|L'|^\gamma}{|T'|^\gamma},$$

$$C^{-1}\frac{|M'|}{|T'|} \le \frac{|M|}{|T|} \le C\frac{|M'|}{|T'|},$$

and

$$C^{-1}\frac{|R'|}{|T'|} \le \frac{|L|}{|T|} \le C\frac{|R'|}{|T'|},$$

for $k > 0$. Therefore,

$$a_k = \frac{|L'|}{|T'|} > C$$

for $k > 0$, and moreover,

$$\frac{|M|}{|T|} \ge C \qquad \text{and} \qquad \frac{|R|}{|T|} \ge C$$

for all $k > 0$.

Finally, we must prove that

$$b_k = \frac{|R'|}{|T'|} > C$$

for all $k \ge 1$. The proof is by contradiction. Suppose there is a subsequence $b_{k_i} \to 0$ as $k_i \to \infty$. Since the inverse of $f^{\circ(m_k-1)}|T$ can be extended to $LJ_k^+(0) \cup J_k(0) \cup RJ_k^+(0)$, from Lemma 4.10 and Lemma 4.11,

$$C \log\left(1 + \frac{|M'||D|}{|R'||E|}\right) \le \log\left(1 + \frac{|M||A|}{|R||B|}\right),$$

for all $k > 0$, where D is either $RJ_k^+(0)$ or $LJ_k^+(0)$ such that $E = M' \cup R' \cup D$ is an interval, and where A and B are the preimages of D and E under $f^{\circ(m_k-1)}$ (see Fig. 4.15). But

$$\log\left(1 + \frac{|M||A|}{|R||B|}\right)$$

is bounded for all $k > 0$ and

$$\log\left(1 + \frac{|M'||D|}{|R'||E|}\right) \to \infty$$

as $k_i \to \infty$. This is a contradiction.

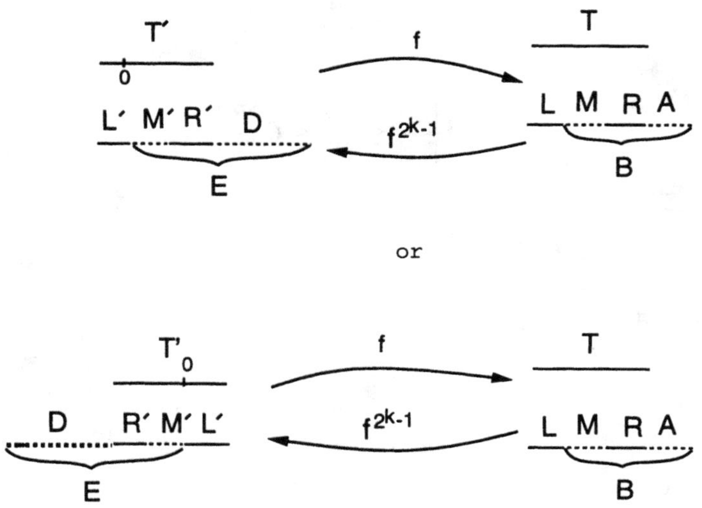

Fig. 4.15

Now Lemma 4.16 completes the proof. ∎

Remark 4.12. Let $FM_{\infty}^{1+Z}(\gamma)$ be the space of all C^{1+Z} Feigenbaum-like unimodal mappings $f(x) = h(-|x|^{\gamma})$ where $\gamma > 1$. Let $\mathcal{FCS}(\gamma) = \{\mathcal{CS}(f) \mid f \in FM_{\infty}^{1+Z}(\gamma)\}$. By using Lemmas 4.12 and 4.13, one can also discuss beau bounded geometry for $\mathcal{FCS}(\gamma)$ (see Definition 4.1).

4.7. Notes on the Limit Set of Renormalizations

Suppose $\rho = (n_1, n_2, \ldots)$ is a sequence of integers $n \geq 2$. Let $f(x) = h(-x^2)$ be an infinitely ρ-renormalizable C^{1+1}-unimodal mapping. Let

$$\mathcal{R}^{\circ k}(f)(x) = h_k(-x^2)$$

be the kth-renormalization of f for $k \geq 1$. (Set $h_0 = h$.) From Remark 4.9, $X = \{h_k\}_{k=0}^{\infty}$ is precompact in $\mathcal{M}(1+\alpha)$ for any $0 \leq \alpha < 1$. Hence every sequence in X has a convergent subsequence in $\mathcal{M}(1+\alpha)$. Let $\mathcal{LS}_f$ be the set of limits of all convergent subsequences of X in $\mathcal{M}(1+\alpha)$. When ρ is a bounded sequence, let f be an infinitely ρ-renormalizable C^{1+Z}-unimodal mapping, Remark 4.11 tells us that $X = \{h_k\}_{k=0}^{\infty}$ is precompact in $\mathcal{M}(1+\alpha)$ for any $0 \leq \alpha < 1$. Hence every sequence in X has a convergent subsequence in $\mathcal{M}(1+\alpha)$. In this case we still use $\mathcal{LS}_f$ to denote the set of limits of all convergent subsequences of X in $\mathcal{M}(1+\alpha)$.

Let $\mathbf{C}$ be the complex plane and let $\mathbf{R}$ be the real line. For a real number $\tau > 0$, let $I_\tau = (-1-\tau, 1+\tau)$ and define

$$\mathbf{C}_\tau = (\mathbf{C} \setminus \mathbf{R}) \cup I_\tau.$$

The Epstein class $\mathcal{E}_\tau$ is the set of all mappings $f(x) = h(-x^2)$ such that

1) $h : [-1, 0] \to [-1, h(0)]$ is a homeomorphism and fixes -1,
2) h^{-1} has a complex analytic extension $\phi : \mathbf{C}_\tau \to \phi(\mathbf{C}_\tau) \subseteq \mathbf{C}_\tau$ such that ϕ is one-to-one and such that $\phi(I_\tau) \subset I_\tau$.

Give $\mathcal{E}_\tau$ the topology of uniform convergence of ϕ on compact subsets of $\mathbf{C}_\tau$. More precisely, $f_k(x) = h_k(-x^2) \to f(x) = h(-x^2)$ in $\mathcal{E}_\tau$ as $k \to \infty$ if

$$\max_{z \in K} |\phi_k(z) - \phi(z)| \to 0 \qquad \text{as} \qquad k \to \infty$$

for every compact set $K \subset \mathbf{C}_\tau$. Then the Epstein class $\mathcal{E}_\tau$ is a compact topological space. Sullivan [SU4] (see also [MV2]) proved that there is a universal constant $\tau_0 > 0$ such that $\mathcal{LS}_f$ is contained in the Epstein class $\mathcal{E}_{\tau_0}$ for every infinitely ρ-renormalizable C^{1+1}-unimodal mapping f (or for every infinitely ρ-renormalizable C^{1+Z}-unimodal mapping f when ρ is bounded). In particular, for any C^{1+Z} Feigenbaum-like unimodal mapping $f(x) = h(-x^2)$, $\mathcal{LS}_f \subseteq \mathcal{E}_{\tau_0}$. A map in $\mathcal{E}_{\tau_0}$ is like a quadratic-like map (see §5.1).

Let $f_0(x) = h_0(-x^2) : U \to V$ be a real quadratic-like Feigenbaum-like map (see §5.1). Sullivan [SU4] developed the Teichmüller theory of Riemann surface laminations induced from f_0. Using this theory, he proved the convergence of renormalization operator for any quadratic-like Feigenbaum-like mapping f_0, i.e., $\mathcal{R}^{\circ k}(f_0) \to g$ as $k \to \infty$, where $\mathcal{R}(g) = g$. The idea of unrenormalization was given by McMullen [MC1,MC2] as follows. Consider the sequence of renormalizations $X = \{f_k = \mathcal{R}^{\circ k}(f_0)\}_{k=0}^{\infty}$ for a real quadratic-like Feigenbaum-like map f_0. Let $\mathcal{LS} = \mathcal{LS}_{f_0}$ be the limit set of X in $\mathcal{M}(1+\alpha)$ for any fixed $0 \leq \alpha < 1$. For any $g_0 \in \mathcal{LS}$, there is a subsequence $\{f_{k_i}\}_{i=1}^{\infty}$ such that $f_{k_i} \to g_0$. Consider another subsequence $\{f_{k_i-1}\}_{i=1}^{\infty}$. It also converges to a mapping $g_{-1} \in \mathcal{LS}$. It is clearly that $\mathcal{R}(g_{-1}) = g_0$ because $\mathcal{R}(f_{k_i-1}) = f_{k_i}$ for all $i \geq 1$. Inductively, one can build an induced tower from f_0 in $\mathcal{E}_{r_0}$:

$$\mathcal{T} = \{\cdots \stackrel{\mathcal{R}}{\to} g_{-j} \stackrel{\mathcal{R}}{\to} g_{-j+1} \stackrel{\mathcal{R}}{\to} \cdots \stackrel{\mathcal{R}}{\to} g_{-1} \stackrel{\mathcal{R}}{\to} g_0\}.$$

Two towers

$$\mathcal{T} = \{\cdots \stackrel{\mathcal{R}}{\to} g_{-j} \stackrel{\mathcal{R}}{\to} g_{-j+1} \stackrel{\mathcal{R}}{\to} \cdots \stackrel{\mathcal{R}}{\to} g_{-1} \stackrel{\mathcal{R}}{\to} g_0\}$$

and

$$\tilde{\mathcal{T}} = \{\cdots \stackrel{\mathcal{R}}{\to} \tilde{g}_{-j} \stackrel{\mathcal{R}}{\to} \tilde{g}_{-j+1} \stackrel{\mathcal{R}}{\to} \cdots \stackrel{\mathcal{R}}{\to} \tilde{g}_{-1} \stackrel{\mathcal{R}}{\to} \tilde{g}_0\}$$

in $\mathcal{E}_{r_0}$ are M-quasiconformally conjugate if there is a sequence of M-quasiconformal homeomorphisms $\{H_j\}_{j=0}^{\infty}$ such that

$$g_{-j} \circ H_j = H_j \circ \tilde{g}_{-j},$$

where $M \geq 1$ is a constant. Let f_0 and $\tilde{f}_0$ be two real quadratic-like Feigenbaum-like maps. Following the work of Sullivan [SU4] (see the beginning of §5.9), any two induced towers, $\mathcal{T}$ induced from f_0 and $\tilde{\mathcal{T}}$ induced from $\tilde{f}_0$, in $\mathcal{E}_{r_0}$ are M-quasiconformally conjugate. McMullen [MC1,MC2] further proved that an induced tower is rigid, that is, from the M-quasiconformal conjugacy, he proved that $\mathcal{T} = \tilde{\mathcal{T}}$. Thus $\mathcal{LS}$ consists of the universal mapping g for any f_0. This implies that $f_k \to g$ as $k \to \infty$ for any f_0.

About the exponential convergence of $\{f_k = \mathcal{R}^{\circ k}(f_0)\}_{k=0}^{\infty}$ to the fixed point g, there is an important development recently in [MC2] by McMullen. De Faria and De Melo [FAM] applied the work of McMullen [MC2] to the renormalization theory of critical circle mappings recently. Sullivan gave a program in [SU4] by Riemann surface laminations to the study of this problem. Kahn [KAH] gave another approach to this problem. Rand [RAN] and Pinto and Rand [PIR] developed a different approach to this problem by using Markov family.

Chapter Five

The Renormalization Method and Quadratic-Like Maps

The author's research in the study of dynamical systems generated by infinitely renormalizable quadratic polynomials and some other recent developments are introduced in this chapter.

Let $P_c(z) = z^2 + c$ be a quadratic polynomial where z is a complex variable and where c is a complex parameter. The filled-in Julia set K_c of P_c is, by definition, the set of points z which remain bounded under iterations of P_c. The Julia set J_c of P_c is the boundary of K_c. A central problem in the study of the dynamical system generated by P_c is to understand the topology and the geometry of K_c and of J_c. Douady and Hubbard [DH3] introduced a quadratic-like map $f : U \to V$ in this study, where f is proper, degree two, analytic branch cover and where U and V are domains isomorphic to a disc such that $\overline{U} \subset V$. The filled-in Julia set K_f of f is, similarly, the set of points which remain in U under iterations of f and the Julia set J_f of f is the boundary of K_f. In the first three sections, we introduce the work of Douady and Hubbard [DH3] which proves that a quadratic-like map with connected filled-in Julia set is hybrid equivalent to a unique quadratic polynomial. We also introduce some facts in complex dynamics and in the theory of quasiconformal mappings.

The Mandelbrot set $\mathcal{M}$ is the set of complex parameters c such that the Julia set J_c is connected. The Julia set J_c in the complement of $\mathcal{M}$ is a Cantor set. Douady and Hubbard [DH1], and independently, Sibony (see [CAG]), proved that $\mathcal{M}$ is connected. A complex number z is a periodic point of a quadratic-like map f of period $k \geq 1$ if $f^{\circ i}(z) \neq z$ for $1 \leq i < k-1$ but $f^{\circ k}(z) = z$. The set $O = \{z, f(z), \ldots, f^{\circ(k-1)}(z)\}$ is called a periodic orbit;

the number $\lambda_O = \left(f^{\circ k}\right)'(z)$ is called the multiplier of f at O; the periodic orbit O is attractive if $0 < |\lambda_O| < 1$, is super-attractive if $\lambda_O = 0$, and is repelling if $|\lambda_O| > 1$. For a quadratic polynomial P_c, let $c(i) = P_c^{\circ i}(0)$ be its ith critical value and let $CO(c) = \{c(i)\}_{i=0}^{\infty}$ be its critical orbit. It is known that if P_c has an either attractive or super-attractive periodic orbit O, then O must be the limit set of $CO(c)$. Therefore, P_c can have only one periodic orbit which is either attractive or super-attractive. A quadratic polynomial P_c for c in $\mathcal{M}$ is called hyperbolic if it has an either attractive or super-attractive periodic orbit in the complex plane $\mathbf{C}$. Let $\mathcal{HP}$ be the set of all c for which P_c is hyperbolic. One important conjecture in this direction is that $\mathcal{HP}$ is open and dense in $\mathcal{M}$. Douady and Hubbard [DH2] proved that this conjecture follows from another conjecture, that the Mandelbrot set $\mathcal{M}$ is locally connected. To understand the local connectivity of $\mathcal{M}$, it would be helpful to answer the question: for which c in $\mathcal{M}$ is the Julia set J_c locally connected ? Yoccoz made substantial progress in this direction, and proved that if P_c is non-renormalizable (or finitely renormalizable), then its Julia set J_c is locally connected. We prove this result in §5.6. (Using this result, Yoccoz proved further that $\mathcal{M}$ is locally connected at a non-renormalizable (or finitely renormalizable) point c (see [HUB]).) In this same section, we also discuss the two-dimensional Yoccoz puzzle of a quadratic-like map and its relation with the renormalizability of such a map. In §5.4, we discuss the quadratic family $\{P_c \mid c \in \mathbf{C}\}$ and prove the result due to Douady and Yoccoz (refer to [MI2,HUB]) about the landing of external rays at a repelling periodic point of a quadratic polynomial. In §5.5, we discuss some definitions of a renormalizable quadratic-like map.

There remain many points in $\mathcal{M}$ which are infinitely renormalizable. In the last five sections, we introduce the author's research on infinitely renormalizable quadratic-like maps. In §5.7, we prove that for a renormalizable quadratic-like map, the filled-in Julia set of any renormalization about the period of the two-dimensional Yoccoz puzzle does not depend on the choices of renormalization domains. In particular, the renormalized filled-in Julia set is the intersection of all critical pieces in the two-dimensional Yoccoz puzzle. We define an unbranched infinitely renormalizable quadratic-like map, and also define an infinitely renormalizable quadratic-like map having the *a priori* complex bounds. In the same section, we prove that for an unbranched infinitely renormalizable quadratic polynomial P_c having the *a priori* complex bounds, its Julia set J_c is locally connected. The proof uses the three-dimensional Yoccoz puzzle of an infinitely renormalizable quadratic polynomial P_c.

In §5.8, we discuss Sullivan's sector theorem and present a version discussed in [JI16]. We prove in §5.9 the result of Sullivan which says that the Feigenbaum polynomial has the *a priori* complex bounds. Following this, a corollary of the discussion in §5.7 is that the Julia set of the Feigenbaum polynomial is locally connected.

In §5.10, we discuss the local connectivity of the Mandelbrot set $\mathcal{M}$ at certain infinitely renormalizable points. We construct a subset Υ of the Mandelbrot set $\mathcal{M}$ which is dense on the boundary $\partial\mathcal{M}$ of the Mandelbrot set $\mathcal{M}$ such that for every point c in this subset, the corresponding quadratic polynomial P_c is infinitely renormalizable and such that the Mandelbrot set $\mathcal{M}$ is locally connected at c. In §5.11, we use the same idea to construct another subset $\tilde{\Upsilon}$ of the Mandelbrot set which is again dense on the boundary $\partial\mathcal{M}$ of the Mandelbrot set $\mathcal{M}$ such that for every point c in $\tilde{\Upsilon}$, the corresponding quadratic polynomial P_c is unbranched, infinitely renormalizable and has the *a priori* complex bounds. From the discussion in §5.7, the Julia set J_c for c in $\tilde{\Upsilon}$ is locally connected.

5.1. Quadratic-Like Maps

Let $P_c(z) = z^2 + c$ be a quadratic polynomial where z is a complex variable and c is a complex parameter. Let $V = \{z \in \mathbf{C} \mid |z| < r\}$ be a disk in the complex plane $\mathbf{C}$. For r large enough, $U = P_c^{-1}(V)$ is a simply connected domain, its closure is relatively compact in V, and $P_c : U \to V$ is a holomorphic, proper branch cover of degree two (see Fig. 5.1). This is a model of an object defined by Douady and Hubbard [DH3].

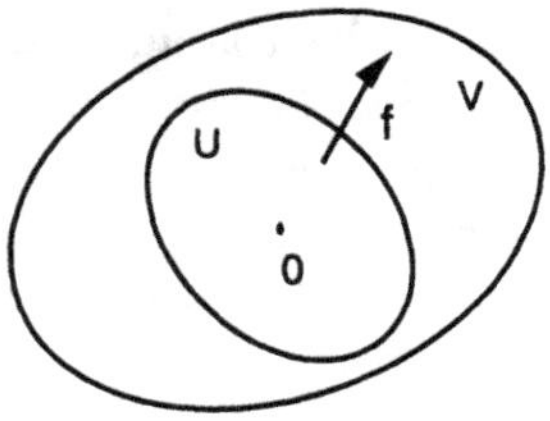

Fig. 5.1

Definition 5.1. *A quadratic-like map is a triple (U, V, f) where U and V are simply connected domains isomorphic to a disc with $\overline{U} \subset V$ and where $f : U \to V$ is a holomorphic, proper branch cover of degree two (see Fig. 5.1).*

Remark 5.1. A proper map f means that $f^{-1}(K)$ is compact for every compact set K. Douady and Hubbard [DH3] also defined a polynomial-like map. A quadratic-like map (U, V, f) has only one branched point b at which the derivative $f'(b)$ of f equals zero. We call b the critical point of f. Without loss of generality, we always assume that $b = 0$.

Suppose (U, V, f) is a quadratic-like map. Let $K_f = \cap_{n=0}^{\infty} f^{-n}(U)$ be the set of z such that $f^{\circ i}(z) \in U$ for all $i \geq 0$. It is called the filled-in Julia set of f and is compact. The Julia set J_f of f is the boundary of K_f. The proofs of the following theorem and corollary can be found in [BLA] and in [MI2].

Theorem 5.1. *Suppose (U, V, f) is a quadratic-like map. If the critical point of f is not in K_f, then $K_f = J_f$ is a Cantor set.*

Corollary 5.1. *Let (U, V, f) be a quadratic-like map. The filled-in Julia set K_f as well as the Julia set J_f is connected if and only if the critical point of f is in K_f.*

Suppose that Ω is a domain in $\mathbf{C}$ and that f is a self-map of Ω. A point z_0 in Ω is periodic with period $k \geq 1$ if $f^{\circ k}(z_0) = z_0$ but $f^{\circ i}(z_0) \neq z_0$ for all $1 \leq i < k$. For a periodic point z_0 of period k, $O = \{f^{\circ i}(z_0)\}_{i=0}^{k-1}$ is a periodic orbit of f. The number $\lambda = (f^{\circ k})'(z_0)$ is called the multiplier (or eigenvalue) of f at z_0 (or at O). A periodic point of period 1 is called a fixed point. A periodic point of period $k > 1$ of f is a fixed point of $f^{\circ k}$. A periodic point z_0 (or a periodic orbit O) of f is said to be repelling, super-attractive, attractive, or neutral if $|\lambda| > 1$, $\lambda = 0$, $0 < |\lambda| < 1$, or $|\lambda| = 1$. A neutral periodic point z_0 (or orbit O) is called parabolic if $\lambda^n = 1$ for some integer $n \geq 1$. The following theorem was proved by Kœnigs in 1884. A proof can be found in [MI2].

Theorem 5.2. *Suppose z_0 is a repelling or attractive periodic point of period k of a holomorphic function f defined on Ω. Then there is a neighborhood U of z_0 and a unique holomorphic diffeomorphism $h : U \to h(U)$ with $h(z_0) = 0$ and $h'(z_0) = 1$ such that*

$$h \circ f^{\circ k} \circ h^{-1}(z) = \lambda z$$

on $h(U)$, where $\lambda = (f^{\circ k})'(z_0)$ is the multiplier of f at z_0.

The proofs of the next two theorems can be found in [BLA].

Theorem 5.3. *Suppose z_0 is a super-attractive periodic point of period k of a holomorphic function f defined on Ω. Then there is a neighborhood U of z_0, a unique holomorphic diffeomorphism $h : U \to h(U)$ with $h(z_0) = 0$ and $h'(z_0) = 1$, and a unique integer $n > 1$ such that*

$$h \circ f^{\circ k} \circ h^{-1}(z) = z^n$$

on $h(U)$.

Suppose z_0 is a neutral fixed point of a holomorphic function f defined on Ω and let $\lambda = f'(z_0)$ be the multiplier of f at z_0. If we want to conjugate f to λz, we have to find a neighborhood U of z_0 and a holomorphic diffeomorphism $h : U \to h(U)$ such that $(h \circ f)(z) = \lambda h(z)$. This functional equation is called a Schröder equation. If z_0 is parabolic, then the Schrölder equation has no solution. But one has that

Theorem 5.4. *Let z_0 be a parabolic fixed point of a holomorphic function f defined on Ω. Either f is the identity or there are a neighborhood U of z_0, a unique holomorphic diffeomorphism $h : U \to h(U)$ with $h(z_0) = 0$ and $h'(z_0) = 1$, and a unique integer $n > 1$ such that*

$$h \circ f \circ h^{-1}(z) = \lambda z(1 + z^n)$$

on $h(U)$.

Let (U, V, f) be a quadratic-like map, let $c_f(i) = f^{\circ i}(0)$ be the ith critical value, and let $CO(f) = \{c_f(i)\}_{i=0}^{\infty}$ be the critical orbit of f. Suppose $O = \{z_0, f(z_0), \ldots, f^{\circ(k-1)}(z_0)\}$ is an attractive or super-attractive periodic orbit of period $k \geq 1$ of f. The set

$$B_O = \{z \in V \mid f^{\circ n}(z) \to O \text{ as } n \to \infty\}$$

is called the basin of O. Let $CB(f^{\circ i}(z_0))$ be the component of B_O containing $f^{\circ i}(z_0)$. The set $IB_O = \cup_{i=0}^{k-1} CB(f^{\circ i}(z_0))$ is called the immediate basin of O. One can refer to [BLA,MI2,SU2] for the following theorem.

Theorem 5.5. *Let (U, V, f) be a quadratic-like map and let $J_f = \partial K_f$ be its Julia set. Let E_f be the set of all repelling periodic points of f. Then*

(1) *J_f is completely invariant, i.e., $f(J_f) = J_f$ and $f^{-1}(J_f) = J_f$;*
(2) *J_f is perfect, i.e., $J'_f = J_f$, where J'_f is the set of limit points of J_f;*
(3) *E_f is dense in the Julia set J_f, i.e., $\overline{E_f} = J_f$;*
(4) *for any z in V, the limit set of $\{f^{-n}(z)\}_{n=0}^{\infty}$ is J_f;*
(5) *J_f has no interior point;*
(6) *If f has an attractive or super-attractive periodic orbit O, then the immediate basin IB_O contains the critical orbit $CO(f)$.*
(7) *If f has neither any attractive, nor any super-attractive, nor any neutral periodic points in V, then $K_f = J_f$.*

Suppose $\overline{\mathbf{C}} = \mathbf{C} \cup \{\infty\}$ is the extended complex plane. Then ∞ is a super-attractive fixed point of any quadratic polynomial P_c. The filled-in Julia set K_c of P_c is the set of all points which remain bounded under iterations of P_c. Let $\mathbf{D}_r = \{z \in \mathbf{C} \mid |z| < r\}$. As we mentioned in the beginning of this section, for $r > 1$ large enough, $U = P_c^{-1}(\mathbf{D}_r)$ is a simply connected domain and $\overline{U} \subset \mathbf{D}_r$. Thus (U, V, P_c) for $V = \mathbf{D}_r$ is a quadratic-like map.

When $c = 0$, $(\mathbf{D}_r, \mathbf{D}_{r^2}, P_0)$ is a quadratic-like map for every $r > 1$; the filled-in Julia set K_0 of P_0 is the closed unit disk $\overline{\mathbf{D}}$. For any c, Theorem 5.3 implies that there is a holomorphic diffeomorphism h_1 defined on a neighborhood $\overline{\mathbf{C}} \setminus \overline{\mathbf{D}}_r$ (for $r > 1$ large) about ∞ such that $h_1(\infty) = \infty$ and $h'_1(\infty) = 1$ and such that

$$h_1^{-1} \circ P_c \circ h_1(z) = z^2$$

on $\overline{\mathbf{C}} \setminus \overline{\mathbf{D}}_r$. Let $B_1(\infty) = h_1(\overline{\mathbf{C}} \setminus \overline{\mathbf{D}}_r)$ and let $B_n(\infty) = P_c^{-(n-1)}(B_1(\infty))$. If the filled-in Julia set K_c of P_c is connected, then all

$$P_c : B_n(\infty) \cap \mathbf{C} \to B_{n-1}(\infty) \cap \mathbf{C}$$

are unramified covering maps of degree two. In this case, we can inductively define holomorphic diffeomorphisms h_n on $\overline{\mathbf{C}} \setminus \overline{\mathbf{D}}_{r^{\frac{1}{2^n}}}$ such that

$$h_n^{-1} \circ P_c \circ h_n(z) = z^2$$

for z in $\overline{\mathbf{C}} \setminus \overline{\mathbf{D}}_{r^{\frac{1}{2^n}}}$ and for all $n > 1$. As n tends to infinity, we get a holomorphic diffeomorphism h_∞ defined on $\overline{\mathbf{C}} \setminus \overline{\mathbf{D}}$ such that

$$h_\infty^{-1} \circ P_c \circ h_\infty(z) = z^2 \tag{5.1}$$

for all z in $\overline{\mathbf{C}} \setminus \overline{\mathbf{D}}$. Moreover, $B_c(\infty) = h_\infty(\overline{\mathbf{C}} \setminus \overline{\mathbf{D}})$ is the basin of ∞ for P_c and $K_c = \overline{\mathbf{C}} \setminus B_c(\infty)$. Furthermore, for every $r > 1$ and for $U_r = h_\infty(\mathbf{D}_r)$, (U_r, U_{r^2}, P_c) is a quadratic-like map and its filled-in Julia set is always K_c.

5.2. Quasiconformal Mappings, Conformal Structures, and Moduli of Annuli

Let Ω be a domain of the complex plane $\mathbf{C}$ and let $h : \Omega \to h(\Omega)$ be a sense-preserving (or orientation-preserving) homeomorphism. Let $z = x + yi$ and $h(z) = u(x,y) + iv(x,y)$. If h is a C^1-diffeomorphism, then

$$h_z = \frac{1}{2}(u_x + v_y) + \frac{i}{2}(v_x - u_y),$$

and

$$h_{\overline{z}} = \frac{1}{2}(u_x - v_y) + \frac{i}{2}(v_x + u_y).$$

Since the Jacobian $Jac(h) = |h_z|^2 - |h_{\overline{z}}|^2 > 0$, we have $|h_{\overline{z}}| < |h_z|$. Let $w = h(z)$. Then

$$(|h_z| - |h_{\overline{z}}|)|dz| \leq |dw| \leq (|h_z| + |h_{\overline{z}}|)|dz|$$

where both equalities can be attained. The derivative $D(h)(z)$ maps a circle to an ellipse so that the ratio of the major to the minor axes of the ellipse is

$$M_h = \frac{|h_z| + |h_{\overline{z}}|}{|h_z| - |h_{\overline{z}}|}.$$

Suppose that

$$\mu_h = \frac{h_{\overline{z}}}{h_z}$$

is the complex dilatation of h and that $k_h(z) = |\mu_h(z)|$. Then

$$M_h = \frac{1 + k_h}{1 - k_h}, \qquad k_h = \frac{M_h - 1}{M_h + 1}.$$

The map h is holomorphic if and only if $h_{\overline{z}} \equiv 0$ (equivalently, if $M_h \equiv 1$ or if $k_h \equiv 0$). A holomorphic diffeomorphism is also called a conformal mapping (or analytic diffeomorphism or schlicht function or univalent function) whose derivative maps a circle to a circle. If $M = \sup_{z\in\Omega} M_h(z) \geq 1$ is a finite number, then h is called a quasiconformal mapping (or quasiconformal homeomorphism).

Actually, there is a more general definition of a quasiconformal mapping as follows. Let $h(z) = u(x,y) + iv(x,y)$ be a sense-preserving homeomorphism on a planar domain Ω. We say h is absolutely continuous on lines (ACL) in Ω if for every closed rectangle R in Ω with sides parallel to the x and y-axes, $u(x,y)$ and $v(x,y)$ are absolutely continuous on almost every horizontal line and on almost every vertical line in R. Of course, such a function has partial derivatives h_z and $h_{\overline{z}}$ for almost every (a.e.) point z in Ω.

Definition 5.2 (Analytic definition). *An orientation-preserving homeomorphism $h : \Omega \to h(\Omega)$ is quasiconformal if*

1) *h is ACL in Ω, and if*

2) *there is a constant $0 \le k < 1$ such that $|\mu_h| \le k$ a.e. on Ω where $\mu_h = h_{\overline{z}}/h_z$.*

Remark 5.2. Let $k = \sup_{z\in\Omega} |\mu_h(z)|$ and let $M = (1+k)/(1-k)$. Then h is called M-quasiconformal.

A quadrilateral $Q(z_1, z_2, z_3, z_4)$ is a Jordan domain, together with four ordered points z_1, z_2, z_3, z_4 on the boundary ∂Q so as to determine a positive orientation of ∂Q with respect to Q. Arcs (z_1, z_2) and (z_3, z_4) are called b-arcs and arcs (z_2, z_3) and (z_4, z_1) are called a-arcs. There is a unique sense-preserving conformal mapping sending Q to a rectangle R with sides a and b and sending (z_1, z_2) and (z_3, z_4) to b-sides and (z_2, z_3) and (z_4, z_1) to a-sides of R. The number $\mathrm{mod}(Q) = a/b$ is called the modulus of the quadrilateral $Q(z_1, z_2, z_3, z_4)$. Suppose that h is a sense-preserving homeomorphism and that $Q(z_1, z_2, z_3, z_4)$ is a quadrilateral. Then $h(Q)(h(z_1), h(z_2), h(z_3), h(z_4))$ is also a quadrilateral. Use $\mathrm{mod}(h(Q))$ to denote the modulus of $h(Q)(h(z_1), h(z_2), h(z_3), h(z_4))$.

Definition 5.3 (Geometric definition). *A sense-preserving homeomorphism $h : \Omega \to h(\Omega)$ is M-quasiconformal if for any quadrilateral $Q(z_1, z_2, z_3, z_4)$ in Ω,*

$$M^{-1} mod(Q) \le mod(h(Q)) \le M mod(Q).$$

The proofs of the next four theorems can be found in [AH1].

Theorem 5.6. *Definition 5.2 and Definition 5.3 are equivalent.*

Theorem 5.7. *If h is a quasiconformal mapping on a planar domain Ω, and $h_{\overline{z}} = 0$ a.e. on Ω, then h is a conformal mapping on Ω.*

Theorem 5.8. *If h is a quasiconformal mapping on a planar domain Ω, then for any Lebesgue measurable set $E \subset \Omega$,*

$$m(E) = \int_E Jac(h)\, dxdy$$

where $m(E)$ means the Lebesgue measure and $Jac(h)$ is the Jacobian of h. In particular, a quasiconformal mapping h maps null sets to null sets.

For a quasiconformal mapping h on a planar domain Ω, the complex dilatation $\mu_h = h_{\overline{z}}/h_z$ is a measurable function and $|\mu_h| \le k < 1$ a.e. on Ω.

In general, for a complex valued function μ with $|\mu| \leq k < 1$ a.e. on Ω,

$$h_{\overline{z}} = \mu h_z \tag{5.2}$$

is called a Beltrami equation. An important result in quasiconformal mappings is the Measurable Riemann Mapping Theorem as follows.

Theorem 5.9 (Measurable Riemann Mapping Theorem). *Suppose μ is a complex valued measurable function defined on $\mathbf{C}$ and $|\mu| \leq k < 1$ a.e. on $\mathbf{C}$. Then there is a unique quasiconformal mapping h^{μ} with complex dilatation μ that leaves 0, 1, ∞ fixed.*

Remark 5.3. The function μ is called the Beltrami coefficient of h^{μ}. In the theory of Beltrami equations, a solution h of $h_{\overline{z}} = \mu h_z$ can be expressed as a power series in μ (see [AH1,GAR]). Thus in Theorem 5.9, the normalized solution h^{μ} depends holomorphically on μ.

Suppose Ω_1, Ω_2, and Ω_3 are domains of $\mathbf{C}$ and suppose $f : \Omega_1 \to \Omega_2$ and $g : \Omega_2 \to \Omega_3$ are M_1- and M_2-quasiconformal mappings, respectively. Let $\eta = f(z)$. Then

$$(g \circ f)_z \overset{\text{a.e.}}{=} (g_\eta \circ f) f_z + (g_{\overline{\eta}} \circ f)(\overline{f})_z$$

and

$$(g \circ f)_{\overline{z}} \overset{\text{a.e.}}{=} (g_\eta \circ f) f_{\overline{z}} + (g_{\overline{\eta}} \circ f)(\overline{f})_{\overline{z}}$$

on Ω_1. One can now get a formula for composition. Let $\theta_f \overset{\text{a.e.}}{=} \overline{f_z}/f_z$. Then

$$\mu_{g \circ f} \overset{\text{a.e.}}{=} \frac{\mu_f + \mu_g \circ f \cdot \theta_f}{1 + \mu_g \circ f \cdot \overline{\mu}_f \cdot \theta_f}. \tag{5.3}$$

Therefore,

$$\mu_g \circ f \overset{\text{a.e.}}{=} \overline{\theta}_f \frac{\mu_{f \circ g} - \mu_f}{1 - \mu_{f \circ g} \overline{\mu}_f}.$$

For $g = f^{-1}$,

$$\mu_{f^{-1}} \overset{\text{a.e.}}{=} -\nu_f \circ f^{-1}$$

where $\nu_f \overset{\text{a.e.}}{=} \overline{\theta}_f \mu_f$ is the second complex dilatation of f. If g is conformal, then $\mu_g \overset{\text{a.e.}}{=} 0$ and then

$$\mu_{g \circ f} \overset{\text{a.e.}}{=} \mu_f. \tag{5.4}$$

If f is conformal, then $\mu_f \overset{\text{a.e.}}{=} 0$ and then

$$\nu_g \overset{\text{a.e.}}{=} \nu_{g \circ f} \circ f^{-1}. \tag{5.5}$$

If we set $h = g \circ f$, then

$$\mu_{h\circ f^{-1}} \circ f \overset{\text{a.e.}}{=} \overline{\theta}_f \frac{\mu_h - \mu_f}{1 - \mu_h \overline{\mu}_f}. \tag{5.6}$$

From the above calculations, we have

Theorem 5.10. *Suppose g and f are M_1- and M_2-quasiconformal mappings, respectively. Then $g \circ f$ is a $M_1 M_2$-quasiconformal mapping and f^{-1} is a M_2-quasiconformal mapping.*

The following theorem is due to Rickman.

Theorem 5.11 [RIC]. *Let $U \subset \mathbf{C}$ be a domain and $\Lambda \subset U$ be a compact set. Let ϕ and Φ be two homeomorphisms from U onto their images. Suppose that ϕ is quasiconformal on U, that Φ is quasiconformal on $U \setminus \Lambda$, and that $\phi = \Phi$ on Λ. Then Φ is quasiconformal, and $D(\Phi) = D(\phi)$ a.e. on Λ, where $D(\cdot)$ denotes the derivative of a function.*

Proof. Without loss of generality, we may assume $\Phi(U)$ and $\phi(U)$ are bounded. We must show that Φ is ACL and then find bounds for the eccentricity of its derivative a.e. on U. Since ϕ is ACL on U, it is enough to show that the real part $u = \Re(\Phi - \phi)$ and the imaginary part $v = \Im(\Phi - \phi)$ are ACL.

Let η_n be a continuous function such that

$$\eta_n(x) = \begin{cases} x - \frac{1}{n}, & x > \frac{2}{n}; \\ x + \frac{1}{n}, & x < -\frac{2}{n}; \\ 0, & -\frac{1}{2n} < x < \frac{1}{2n} \end{cases}$$

and such that $\eta_n'(x) < 1$ for x in $\mathbf{R}$. For a.e. horizontal line (respectively, vertical line) l, $u_n = \eta_n \circ u$ (respectively, $v_n = \eta_n \circ v$) is absolutely continuous on $l \cap U$ and is a Cauchy sequence in L^1, with limit u. Thus u (respectively, v) is absolutely continuous on $l \cap U$. Since $u_n = 0$ (respectively, $v_n = 0$) on a neighborhood of Λ for all n, $D(\Phi - \phi) = 0$ a.e. on Λ. Therefore, Φ is ACL on U and $|\mu_\Phi| \leq k < 1$ a.e. on U where k is a constant. Thus Φ is quasiconformal on U. ∎

A Riemann surface $\mathcal{S}$ is a connected one dimensional complex analytic manifold. This means that $\mathcal{S}$ is a connected Hausdorff topological space and there is a covering $\{U_\alpha\}$ of $\mathcal{S}$ by open sets and there are homeomorphisms $\{\phi_\alpha : U_\alpha \to \mathbf{C}\}$ such that if U_α and U_β overlap, the map

$$\phi_{\alpha\beta} = \phi_\alpha \circ \phi_\beta^{-1} : \phi_\beta(U_\alpha \cap U_\beta) \to \phi_\alpha(U_\alpha \cap U_\beta)$$

is conformal. The pair $\{U_\alpha, \phi_\alpha\}$ is called a system of charts of $\mathcal{S}$. Two systems of charts $\{U_\alpha, \phi_\alpha\}$ and $\{V_\beta, \psi_\beta\}$ are called compatible if whenever $U_\alpha \cap V_\beta$ is non-empty,

$$\phi_\alpha \circ \psi_\beta^{-1} : \psi_\beta(U_\alpha \cap V_\beta) \to \phi_\alpha(U_\alpha \cap V_\beta)$$

is conformal. The compatibility is an equivalence relation between systems of charts because the composition of conformal mappings is conformal (see Eq. (5.3)). The equivalence class σ of compatible systems of the system of charts $\{U_\alpha, \phi_\alpha\}$ of $\mathcal{S}$ is called the conformal structure. A homeomorphism $f : \mathcal{S} \to \mathcal{S}$ is called conformal (respectively, quasiconformal) if for any charts (U_α, ϕ_α) and (U_β, ϕ_β) in σ,

$$f_{\alpha\beta} = \phi_\beta \circ f \circ \phi_\alpha^{-1} : \phi_\alpha(U_\alpha \cap f^{-1}(U_\beta)) \to \phi_\beta(f(U_\alpha) \cap U_\beta)$$

is conformal (respectively, quasiconformal). A Beltrami coefficient μ on $\mathcal{S}$ is a complex valued function on $\mathcal{S}$ such that for every chart (U_α, ϕ_α) in σ, $\mu \circ \phi_\alpha^{-1}$ is measurable and $|\mu \circ \phi_\alpha^{-1}| \leq k$ a.e. on $\phi_\alpha(U_\alpha)$ for a fixed constant $0 \leq k < 1$.

Let $\mathcal{S}$ be a Riemann surface and let σ be the conformal structure of $\mathcal{S}$. Suppose μ is a Beltrami coefficient on $\mathcal{S}$. Applying Theorem 5.9, there is a quasiconformal mapping h_α on $\phi_\alpha(U_\alpha)$ with the complex dilatation $\mu \circ \phi_\alpha^{-1}$ for every chart (U_α, ϕ_α) in σ. Thus $H_\alpha = h_\alpha \circ \phi_\alpha : U_\alpha \to H_\alpha(U_\alpha)$ is a homeomorphism. If U_α and U_β overlap, then $H_\alpha \circ H_\beta^{-1}$ on $H_\beta(U_\beta \cap U_\alpha)$ is conformal (see Eq. (5.6)). We have a new system of charts $\{(U_\alpha, H_\alpha)\}$. Let σ_μ be the equivalent class of compatible systems of this new system of charts. It is a new conformal structure on $\mathcal{S}$. Every quasiconformal mapping of $\mathcal{S}$ whose complex dilatation is μ becomes a conformal mapping of $\mathcal{S}$ with this new conformal structure σ_μ.

Let $z = \phi_\alpha(p)$ be the coordinate of $\mathcal{S}$ where (U_α, ϕ_α) is a chart in σ and let $|dz|$ denote the metric on the tangent space $T_p\mathcal{S}$. Let μ be a Beltrami coefficient on $\mathcal{S}$. We use $\mu(z)$ to denote $\mu \circ \phi_\alpha^{-1}(z)$ on the chart (U_α, ϕ_α). Then

$$|dw| = |dz + \mu d\overline{z}|$$

is another metric on $T_p\mathcal{S}$ and is defined a.e. on U_α. Let $\tilde{z} = \phi_\beta(q)$ be the coordinate of $\mathcal{S}$ on another chart (U_β, ϕ_β) in σ and let

$$|d\tilde{w}| = |d\tilde{z} + \mu d\overline{\tilde{z}}|.$$

Let f be a quasiconformal mapping from an open set $U \subset U_\alpha$ onto another open set $V \subset U_\beta$ of $\mathcal{S}$. Let us still use f to denote $f_{\alpha\beta} = \phi_\beta \circ f \circ \phi_\alpha^{-1}$. Then

$$\begin{aligned} f^*(d\tilde{z}) &= f_z\, dz + f_{\overline{z}}\, d\overline{z} \\ &= f_z\Big(dz + \mu_f(z) d\overline{z}\Big) \end{aligned}$$

where $\mu_f = f_{\bar{z}}/f_z$ is the complex dilatation of f. So $|f^*(d\tilde{z})|$ is also a metric on T_pS for a.e. p in U. Now we consider the pull-back metric $|f^*(d\tilde{w})|$ on T_pS. Then

$$\begin{aligned}|f^*(d\tilde{w})| &= |f^*(d\tilde{z}) + \mu\circ f\cdot f^*(d\bar{\tilde{z}})| \\ &= \Big|\Big(f_z + \mu\circ f\cdot(\overline{f})_z\Big)\cdot\Big(dz + f^*(\mu)\,d\overline{z}\Big)\Big|\end{aligned}$$

where

$$f^*(\mu) = \frac{\mu_f + \mu\circ f\cdot\theta_f}{1+\mu\circ f\cdot\overline{\mu}_f\cdot\theta_f} \tag{5.7}$$

and $\theta_f = \overline{f_z}/f_z$. This calculation is independent of the choices of local coordinates. If $f: U\to V$ is a conformal mapping, then

$$f^*(\mu) = \theta_f\cdot\mu\circ f.$$

Definition 5.4. *Let U and V be two open sets of $\mathcal{S}$. Let $f: U\to V$ be a quasiconformal mapping. We say f preserves the conformal structure σ_μ on U and V if $f^*(\mu)=\mu$.*

From the above computations, we have

Theorem 5.12. *Let U and V be two open sets of $\mathcal{S}$. Let $f: U\to V$ be a quasiconformal mapping. let μ be a Beltrami coefficient on $\mathcal{S}$. Suppose h is the quasiconformal mapping of $\mathcal{S}$ whose complex dilatation is μ. Then $f_0 = h\circ f\circ h^{-1}: h(U)\to h(V)$ is conformal if and only if f preserves the conformal structure σ_μ on U and V.*

Proof. For any $p\in U$, let (U_α,ϕ_α), (U_β,ϕ_β), and (U_γ,ϕ_γ) be charts of $\mathcal{S}$ in σ such that $p\in U_\alpha$, $f(p)\in U_\beta$, and $h(p)\in U_\gamma$. Let $z=\phi_\alpha(p)$ be the coordinate. Let us still use h and f to denote $h_{\alpha\gamma}=\phi_\gamma\circ h\circ\phi_\alpha^{-1}$ and $f_{\alpha\beta}=\phi_\beta\circ f\circ\phi_\alpha^{-1}$. Because $h^*(dz)=h_z(dz+\mu d\overline{z})$ and because

$$\begin{aligned}(h\circ f)^*(dz) &= f^*\circ h^*(dz) \\ &= h_z\circ f\cdot\Big(f_z+\mu\circ f\cdot\overline{f_z}\Big)\cdot\Big(dz+f^*(\mu)d\overline{z}\Big),\end{aligned}$$

$$(f_0)^*(dz) = \Big((h^*)^{-1}\circ(h\circ f)^*\Big)(dz)$$

$$= h_z \circ f \cdot \left(f_z + \mu \circ f \cdot \overline{f_z}\right) \cdot \left((h^{-1})_z + f^*(\mu) \circ h^{-1} \cdot \overline{(h^{-1})_z}\right) \cdot \left(dz + \theta_{h^{-1}} \cdot \frac{-\mu + f^*(\mu)}{1 - f^*(\mu) \cdot \overline{\mu}}\, d\overline{z}\right).$$

Thus

$$\mu_{f_0} = \frac{-\mu + f^*(\mu)}{1 - f^*(\mu)\overline{\mu}} \theta_{h^{-1}}.$$

This calculation is independent of the choices of local coordinates. Therefore f_0 is conformal if and only if $f^*(\mu) = \mu$. ∎

We know that $\mathbf{C}$, $\overline{\mathbf{C}}$, and $\mathbf{D} = \{z \in \mathbf{C} \mid |z| < 1\}$ are simply connected Riemann surfaces. The following uniformization theorem (see [AH2]) says that up to conformal equivalence, they are the only simply connected ones.

Theorem 5.13 (Uniformization Theorem). *Let $\mathcal{S}$ be a simply connected Riemann surface. It is holomorphically diffeomorphic to exactly one of $\mathbf{C}$, $\overline{\mathbf{C}}$, or $\mathbf{D}$.*

Let $\mathcal{S}$ be a Riemann surface and let $(\tilde{\mathcal{S}}, \pi)$ be the universal cover of $\mathcal{S}$, where $\tilde{\mathcal{S}}$ is a simple connected Riemann surface and $\pi : \tilde{\mathcal{S}} \to \mathcal{S}$ is the universal covering map. From Theorem 5.13, we can identify $\tilde{\mathcal{S}}$ with one of $\overline{\mathbf{C}}$, $\mathbf{C}$, or $\mathbf{D}$. A Riemann surface $\mathcal{S}$ is hyperbolic if $\tilde{\mathcal{S}} = \mathbf{D}$.

Let $\mathcal{D}$ be the hyperbolic disk and let d_H be the hyperbolic distance (see §2.6). From Theorem 2.6, every one-to-one holomorphic mapping h from $\mathbf{D}$ into $\mathbf{D}$, which is not a Möbius transformation, strictly decreases the hyperbolic distance d_H, i.e,

$$d_H(h(z_1), h(z_2)) < d_H(z_1, z_2)$$

for all z_1 and z_2 in $\mathcal{D}$.

For a hyperbolic Riemann surface $\mathcal{S}$, let $d_{H,S}$ be the hyperbolic distance on $\mathcal{S}$ so that $\pi : \mathcal{D} \to \mathcal{S}$ is locally isometric. Any one-to-one holomorphic mapping h from $\mathcal{S}$ into $\mathcal{S}$, which is not an isometry with respect to $d_{H,\mathcal{S}}$, strictly decreases this hyperbolic distance $d_{H,\mathcal{S}}$, i.e., for any z_1 and z_2 in $\mathcal{S}$,

$$d_{H,\mathcal{S}}(h(z_1), h(z_2)) < d_{H,\mathcal{S}}(z_1, z_2).$$

A bounded domain Ω in $\mathbf{C}$ is a hyperbolic Riemann surface. An important family of hyperbolic Riemann surfaces is the family of bounded doubly connected domains in $\mathbf{C}$. A bounded doubly connected domain Ω is called an annulus. From complex analysis (see [BIE]), any annulus is holomorphically diffeomorphic to a unique round annulus $A_r = \{z \in \mathbf{C} \mid r < |z| < 1\}$ for $0 < r < 1$. The number

$$\operatorname{mod}(\Omega) = -\log r$$

is called the modulus of Ω. It is a conformal invariant, i.e., $\mathrm{mod}(h(\Omega)) = \mathrm{mod}(\Omega)$ whenever h is a conformal mapping from Ω onto $h(\Omega)$. Moreover, it is a quasi-invariant under quasiconformal mappings (see [AH1]), i.e.,

$$M^{-1}\mathrm{mod}(\Omega) \leq \mathrm{mod}(h(\Omega)) \leq M\mathrm{mod}(\Omega)$$

whenever h is M-quasiconformal.

Let Ω be a bounded doubly connected domain in $\mathbf{C}$. Then the complement $\mathbf{C}\setminus\Omega$ of Ω in $\mathbf{C}$ has two components. One is a connected and simply connected compact set E and the other is an unbounded set F. The bounded component E is a single point if and only if $\mathrm{mod}(\Omega) = \infty$ (see [BRH]).

Let Ω be an annulus. If $\Omega_1 \subseteq \Omega$ is a subannulus, then

$$\mathrm{mod}(\Omega_1) \leq \mathrm{mod}(\Omega). \tag{5.8}$$

If Ω_1, $\Omega_2 \subseteq \Omega$ are two disjoint subannuli, then

$$\mathrm{mod}(\Omega_1) + \mathrm{mod}(\Omega_2) \leq \mathrm{mod}(\Omega). \tag{5.9}$$

The proofs of these two inequalities can be found in [AH1].

Let E be a connected and simply connected compact subset of the open unit disk $\mathbf{D}$. Let

$$\mathrm{mod}(\mathbf{D}, E) = \sup_{\Omega}\{\mathrm{mod}(\Omega), \text{ where } \Omega \subseteq \mathbf{D}\setminus E \text{ is a round subannulus}\}.$$

Note that $\mathrm{mod}(\mathbf{D}, E) = \infty$ if E is a single point.

Let $\mathrm{diam}_H(E) = \sup_{z_1, z_2 \in E} d_H(z_1, z_2)$ be the hyperbolic diameter of E in $\mathbf{D}$. The following theorem can be found in [MC1].

Theorem 5.14. *The hyperbolic diameter $d_H(E)$ and the mod$(\mathbf{D}, E)$ are inversely related:*

$$\mathrm{diam}_H(E) \to 0 \qquad \Longleftrightarrow \qquad \mathrm{mod}(\mathbf{D}, E) \to \infty$$

and

$$\mathrm{diam}_H(E) \to \infty \qquad \Longleftrightarrow \qquad \mathrm{mod}(\mathbf{D}, E) \to 0.$$

More precisely, there is a constant $C > 0$ such that

$$C^{-1}\mathrm{diam}_H(E) \leq \exp(-\mathrm{mod}(\mathbf{D}, E)) \leq C\,\mathrm{diam}_H(E)$$

when $\mathrm{diam}_H(E)$ is small, while

$$\frac{C}{\mathrm{diam}_H(E)} \geq \mathrm{mod}(\mathbf{D}, E) \geq C^{-1}\exp(-\mathrm{diam}_H(E))$$

when $\mathrm{diam}_H(E)$ is large.

5.3. Internal and External Classes of Quadratic-Like Maps

Let (U, V, f) and (U', V', g) be two quadratic-like maps. They are topologically conjugate if there is a homeomorphism h from a neighborhood $K_f \subset X \subset U$ to a neighborhood $K_g \subset Y \subset U'$ such that $h \circ f = g \circ h$ on X, where K_f and K_g are the filled-in Julia sets of f and g. If h is quasiconformal (respectively, holomorphic), then they are quasiconformally (respectively, holomorphically) conjugate. If h can be chosen such that $h_{\overline{z}} = 0$ a.e. on K_f, then they are hybrid equivalent. The set

$$I(f) = \{g \mid g \text{ is hybrid equivalent to } f\}$$

is called the internal class of f.

Theorem 5.15 [DH3]. *If (U, V, f) is a quadratic-like map such that K_f is connected, then there is a unique quadratic polynomial $P(z) = z^2 + c_f$ (more precisely, (U_r, U_{r^2}, P) (see §5.1)) in $I(f)$*

Proof. Let γ in $V \setminus U$ be a C^ω curve isomorphic to a circle and let V' be the domain bounded by γ. Let $U' = f^{-1}(V') \subseteq V'$. Let

$$h_0 : \overline{V'} \to \overline{\mathbf{D}}_{r^2} = \{z \in \mathbf{C} \mid |z| \le r^2\}$$

for some $r > 1$ be a conformal mapping. Then (U_0, V_0, f_0) for $V_0 = \mathbf{D}_{r^2}$, for $U_0 = h_0(U')$, and for $f_0 = h_0 \circ f \circ h_0^{-1}$ is a quadratic-like map holomorphically conjugate to (U, V, f).

Suppose $P_0(z) = z^2$ and $\mathbf{S}^r = \{z \in \mathbf{C} \mid |z| = r\}$. The mapping $f_0 : \gamma' = f_0^{-1}(h_0(\gamma)) \to h_0(\gamma) = \mathbf{S}^{r^2}$ is a C^ω degree two covering map. The mapping $P_0 : \mathbf{S}^r \to \mathbf{S}^{r^2}$ is also a degree two C^ω covering map. There is a M-quasiconformal mapping h_1 from the closed annulus $\overline{A}_0 = \overline{V}_0 \setminus U_0$ to the closed annulus $\overline{A}_0' = \{z \in \mathbf{C} \mid r \le |z| \le r^2\}$ such that $h_1|\mathbf{S}^{r^2}$ is the identity and such that $P_0 \circ h_1 = h_1 \circ f_0$ on ∂U_0. We can define a degree two continuous branch (or ramified) cover of the extended complex plane $\overline{\mathbf{C}}$:

$$F(z) = \begin{cases} P_0(z), & \text{if } x \in \overline{\mathbf{C}} \setminus V_0; \\ (h_1(z))^2, & \text{if } x \in V_0 \setminus U_0; \\ f_0(z), & \text{if } x \in U_0. \end{cases}$$

We now define a measurable function μ a.e. on $\overline{\mathbf{C}}$ as follows: take $\mu(z) = 0$ for z in $\overline{\mathbf{C}} \setminus V_0$. Suppose $A_n = F^{-n}(A_0)$ for $n \ge 0$. Take $\mu = (F^{\circ(n+1)})^*(\mu)$ on A_n. In the composition $F^{\circ(n+1)}$, there is only one M-quasiconformal mapping

h_1 and all others are conformal. Thus, $|\mu(z)| \le k$ for $k = (M-1)/(M+1)$. Finally, take $\mu(z) = 0$ for z in K_f. Thus μ is a Beltrami coefficient on $\overline{\mathbf{C}}$. From the definition of μ, the map F preserves the conformal structure σ_μ on $\overline{\mathbf{C}}$. Following Theorem 5.9, there is a unique quasiconformal mapping h of $\overline{\mathbf{C}}$ which has complex dilatation μ and which fixes 0, 1, and ∞. Consider $P = h \circ F \circ h^{-1}$. From Theorem 5.7, P is conformal at a.e. point z. This means that P is a holomorphic degree two branch covering map. It is a rational function of $\overline{\mathbf{C}}$. But $P(\infty) = \infty$ and $P'(0) = 0$ imply that $P(z) = z^2 + c$ for some complex number c. This proves the existence of $P(z)$.

Suppose $P_1(z) = z^2 + c_1$ and $P_2(z) = z^2 + c_2$ are both in the internal class $I(f)$. Let K_1 and K_2 be their respectively filled-in Julia sets. Then there are two pairs of simply connected domains $K_1 \subset U \subset V$ and $K_2 \subset U' \subset V'$ and a quasiconformal mapping $\phi : V \to V'$ such that ϕ is conformal on K_1, such that (U, V, P_1) and (U', V', P_2) are quadratic-like maps, and such that $\phi \circ P_1 = P_2 \circ \phi$ on U. From §5.1, there is the unique conformal mapping $\psi : \overline{\mathbf{C}} \setminus K_1 \to \overline{\mathbf{C}} \setminus K_2$ such that $\psi(\infty) = \infty$, such that $\psi'(\infty) = 1$, and such that $\psi \circ P_1 = P_2 \circ \psi$ on $\overline{\mathbf{C}} \setminus K_1$. Define

$$\Phi(z) = \begin{cases} \phi(z), & \text{if } z \in K_1; \\ \psi(z), & \text{if } z \in \overline{\mathbf{C}} \setminus K_1. \end{cases}$$

We first prove that Φ is a continuous map. Actually, we need only check that $\psi(z) \to \phi(z)$, or equivalently, $\phi^{-1} \circ \psi(z) \to z$, as z in $V \setminus K_1$ tends to ∂K_1. Let $\alpha = \phi^{-1} \circ \psi$; it is a homeomorphism from $V \setminus K_1$ onto itself. Also, $\alpha \circ P_1 = P_1 \circ \alpha$ on $U \setminus K_1$. Let $A_1 = U \setminus K_1$ and $A_2 = V \setminus K_1$; each is a doubly connected domain. Let $\Pi_1 : \mathbf{D} \to A_1$ and $\Pi_2 : \mathbf{D} \to A_2$ be universal covering maps where $\mathbf{D}$ is the open unit disk. Let $\tilde{f}$ be the lift of $P_1 : A_1 \to A_2$, i.e., $\tilde{f}$ is a map of $\mathbf{D}$ and $\Pi_2 \circ \tilde{f} = P_1 \circ \Pi_1$. Then $\tilde{f}$ is bijective. Suppose τ_1 and τ_2 are automorphisms of $\mathbf{D}$ giving the generators of the fundamental groups $\pi_1(A_1)$ and $\pi_1(A_2)$ specified by orientation. Then $\tilde{f} \circ \tau_1 = \tau_2^2 \circ \tilde{f}$. Let $\tilde{\alpha}_1$ and $\tilde{\alpha}_2$ be lifts of $\alpha : A_1 \to A_1$ and $\alpha : A_2 \to A_2$. Then $\tau_1 \circ \tilde{\alpha}_1 = \tilde{\alpha}_1 \circ \tau_1$ and $\tau_2 \circ \tilde{\alpha}_2 = \tilde{\alpha}_2 \circ \tau_2$. Since $\tilde{\alpha}_2 \circ \tilde{f}$ and $\tilde{f} \circ \tilde{\alpha}_1$ are both lifts of $\alpha \circ P_1 = P_1 \circ \alpha$, there is an integer i such that

$$\tilde{f} \circ \tilde{\alpha}_1 = \tau_2^i \circ \tilde{\alpha}_2 \circ \tilde{f}$$

on $\mathbf{D}$. Since $\tau_1^{\pm 1} \circ \tilde{\alpha}_1$ and $\tau_2^{\pm 1} \circ \tilde{\alpha}_2$ are also lifts of $\alpha : A_1 \to A_1$ and $\alpha : A_2 \to A_2$, and since

$$\tilde{f} \circ (\tau_1^{\pm 1} \circ \tilde{\alpha}_1) = \tau_2^{i \pm 1} \circ (\tau_2^{\pm 1} \circ \tilde{\alpha}_2) \circ \tilde{f}$$

on $\mathbf{D}$, we can choose appropriate lifts $\tilde{\alpha}_1$ and $\tilde{\alpha}_2$ such that $i = 0$. Let $\tilde{g}$ be the inverse of $\tilde{f}$. Then

$$\tilde{\alpha}_1 \circ \tilde{g} = \tilde{g} \circ \tilde{\alpha}_2$$

on $\mathbf{D}$.

Consider the hyperbolic disk $\mathcal{D} = (\mathbf{D}, d_H)$. Since $\tilde{g}$ is holomorphic, it decreases the hyperbolic distance d_H, i.e.,

$$d_H(\tilde{g}(x), \tilde{g}(y)) \leq d_H(x, y)$$

for all x and y in $\mathbf{D}$.

Take a fundamental domain $\tilde{A} \subset \mathbf{D}$ of the fundamental group $\pi_1(A_1)$. Let $B = \overline{P_1^{-1}(U)} \setminus P_1^{-2}(U)$ and let $\tilde{B} = \Pi_2^{-1}(B)$ be the lift of B (where $\Pi_2 : \mathbf{D} \to A_2$ is the covering map). The intersection $\tilde{B}_0 = \tilde{B} \cap \tilde{A}$ is a compact subset of $\mathbf{D}$. Let

$$m = \sup_{x \in \tilde{B}_0} d_H(\tilde{\alpha}_2(x), x).$$

Let $\tilde{B}_1$ be the lift of $P_1^{-2}(U) \setminus K_1$ under the covering map Π_2. For any x in $\tilde{B}_1$, there is a point y in $\tilde{B}_0$ and integers $k \geq 0$ and $j \geq 0$ such that $x = \tilde{g}^{\circ j} \circ \tau_2^{\circ k}(y)$. Therefore

$$\begin{aligned} d_H(\tilde{\alpha}_1(x), x) &= d_H(\tilde{\alpha}_1 \circ \tilde{g}^{\circ j} \circ \tau_2^{\circ k}(y), \tilde{g}^{\circ j} \circ \tau_2^{\circ k}(y)) \\ &= d_H(\tilde{g}^{\circ j} \circ \tilde{\alpha}_2 \circ \tau_2^{\circ k}(y), \tilde{g}^{\circ j} \circ \tau_2^{\circ k}(y)) \\ &\leq d_H(\tilde{\alpha}_2 \circ \tau_2^{\circ k}(y), \tau_2^{\circ k}(y)) \\ &= d_H(\tau_2^{\circ k} \circ \tilde{\alpha}_2(y), \tau_2^{\circ k}(y)) \\ &= d_H(\tilde{\alpha}_2(y), y) \leq m. \end{aligned}$$

This means that the Euclidean distance $d(\tilde{\alpha}_1(x), x)$ tends to zero as x in $\tilde{B}_1$ approaches the boundary of $\mathbf{D}$. Hence the Euclidean distance $d(\alpha(z), z)$ tends to zero as z approaches to ∂K_1.

Similarly, Φ^{-1} is continuous. Thus Φ is a homeomorphism of $\overline{\mathbf{C}}$. From Theorem 5.11, Φ is a quasiconformal mapping of $\overline{\mathbf{C}}$. Since it is holomorphic on $\overline{\mathbf{C}} \setminus K_1$ and since $D(\Phi) = D(\phi)$ a.e. on K_1, where $D(\cdot)$ denotes the derivative of a function, from Theorem 5.7, Φ is a conformal mapping of $\overline{\mathbf{C}}$. It can only be a Möbius transformation. Furthermore, since Φ fixes 0 and ∞, and since Φ is the conjugacy between P_1 and P_2, then Φ is the identity map of $\overline{\mathbf{C}}$. Therefore, $c_1 = c_2$. ∎

Remark 5.4. Theorem 5.15 tells us that if a property of quadratic-like maps is invariant under quasiconformal conjugacy, then we can use quadratic polynomials to study it and vice versa.

Let $\mathbf{S}^1$ be the unit circle. Suppose (U, V, f) and (U', V', g) are two quadratic-like maps with connected filled-in Julia sets K_f and K_g. Then f

and g are externally equivalent if there exist connected open sets $K_f \subset U_1 \subset V_1 \subset U$ and $K_g \subset U_1' \subset V_1' \subset U'$, and a conformal mapping H from $U_1 \setminus K_f$ onto $U_1' \setminus K_g$ such that (U_1, V_1, f) and (U_1', V_1', g) are also quadratic-like maps and such that $H \circ f = g \circ H$ on $U_1 \setminus K_f$. From the end of §5.1, any quadratic polynomial P_c (more precisely, (U_r, U_{r^2}, P_c)) with connected filled-in Julia set K_c is externally equivalent to P_0 (more precisely, $(\mathbf{D}_r, \mathbf{D}_{r^2}, P_0)$). The set

$$E(f) = \{g \mid g \text{ is externally equivalent to } f\}$$

is called the external class of f. We associate to any $E(f)$ an analytic, degree two, expanding map of $\mathbf{S}^1$, which is unique up to analytic conjugacy, as follows.

Suppose (U, V, f) is a quadratic-like map whose filled-in Julia set K_f is connected. Since $V \setminus K_f$ is a doubly connected domain, there is a conformal mapping α from $V \setminus K_f$ onto an annulus $W_r = \{z \in \mathbf{C} \mid 1 < |z| < r\}$ such that $|\alpha(z)| \to 1$ when $d(z, K_f) \to 0$. Let $W_+ = \alpha(U \setminus K_f)$ and let $h_+ = \alpha \circ f \circ \alpha^{-1} : W_+ \to W_r$, which is a covering map. The map h_+ can be continuously extended to $\mathbf{S}^1$ since it is locally homeomorphic. Let $\tau(z) = 1/\overline{z}$ be the reflection with respect to the unit circle. Let $W_{1/r} = \tau(W_r)$ and $W_- = \tau(W_+)$. Let $W = W_{1/r} \cup \mathbf{S}^1 \cup W_r$ and $W_0 = W_+ \cup \mathbf{S}^1 \cup W_-$. By the Schwarz reflection theorem (see [BIE]), h_+ can be extended to an analytic map h from W_0 onto W so that it is a degree two covering map. Note that $W_0 \subset W$. If we consider W as a hyperbolic Riemann surface, any local inverse branch h^{-1} contracts the hyperbolic metric on W. Thus the restriction $h_f = h|\mathbf{S}^1$ strongly expands the hyperbolic metric. But the hyperbolic metric on $\mathbf{S}^1$ is just $a|dz|$. Thus h_f is a strongly expanding, holomorphic, degree two map of $\mathbf{S}^1$. The following theorem says that h_f is unique for $E(f)$ up to analytic conjugacy. Therefore, we can call h_f the external map of $E(f)$.

Theorem 5.16 [DH3]. *Suppose (U, V, f) and (U', V', g) are two quadratic-like maps with connected filled-in Julia sets K_f and K_g. Let h_f and h_g be maps constructed above for f and g. If f and g are externally equivalent, there is a real analytic diffeomorphism h of $\mathbf{S}^1$ such that $h \circ h_f = h_g \circ h$.*

Proof. Let $K_f \subset U_1 \subset V_1 \subset V$ be two domains such that (U_1, V_1, f) is a quadratic-like map. Let h_f and $\tilde{h}_f$ be maps constructed above for (U, V, f) and (U_1, V_1, f). Let α from $V \setminus K_f$ onto W_r and α_1 from $V_1 \setminus K_f$ onto W_{r_1} be the maps in the construction of h_f and $\tilde{h}_f$. Then $\beta_0 = \alpha \circ \alpha_1^{-1} : W_{r_1} \to W_r$ is a conformal mapping and can be continuously extended to $\mathbf{S}^1$. By the Schwarz reflection theorem, β can be extended to a conformal mapping from $W_{1/r_1} \cup \mathbf{S}^1 \cup W_{r_1}$ into $W_{1/r} \cup \mathbf{S}^1 \cup W_r$. The restriction $h_0 = \beta|\mathbf{S}^1$ is an analytic conjugacy from h_f to $\tilde{h}_f$, i.e., $h_f \circ h_0 = h_0 \circ \tilde{h}_f$.

Suppose $K_g \subset U_1' \subset V_1' \subseteq V$ are domains such that (U_1', V_1', g) is a quadratic-like map. Let h_g and $\tilde{h}_g$ be maps constructed above for (U', V', g) and (U_1', V_1', g). Similarly, there is an analytic conjugacy h_1 from h_g to $\tilde{h}_g$, i.e., $h_g \circ h_1 = h_1 \circ \tilde{h}_g$.

Suppose (U, V, f) and (U', V', g) are externally equivalent. Then there are domains $K_f \subset U_1 \subset V_1 \subset V$ and $K_g \subset U_1' \subset V_1' \subset V'$ such that (U_1, V_1, f) and (U_1', V_1', f) are quadratic-like maps, and there is a conformal mapping from $V_1 \setminus K_f$ onto $V_1' \setminus K_g$ such that $H \circ f = g \circ H$ on $U_1 \setminus K_f$. Let α_2 from $V_1 \setminus K_f$ onto W_{r_2} and α_3 from $V_1' \setminus K_g$ onto W_{r_3} be the maps in the construction of $\tilde{h}_f$ and $\tilde{h}_g$. Then $\beta_1 = \alpha_3 \circ H \circ \alpha_2^{-1} : W_{r_2} \to W_{r_3}$ is a conformal mapping and can be extended continuously to $\mathbf{S}^1$. By the Schwarz reflection theorem, β_1 can be extended to a conformal mapping β_2 from $W_{1/r_2} \cup \mathbf{S}^1 \cup W_{r_2}$ into $W_{1/r_3} \cup \mathbf{S}^1 \cup W_{r_3}$. The restriction $h_2 = \beta_2 | \mathbf{S}^1$ is an analytic conjugacy from $\tilde{h}_f$ to $\tilde{h}_g$, i.e., $\tilde{h}_f \circ h_2 = h_2 \circ \tilde{h}_g$. Now $h = h_0 \circ h_2 \circ h_1^{-1}$ is an analytic conjugacy from h_f to h_g. ∎

Similar to Theorem 5.15, we have the following theorem.

Theorem 5.17 [DH3]. *Suppose that (U, V, f) is a quadratic-like map and that its filled-in Julia set K_f is connected. Then (U, V, f) is holomorphically conjugate to a polynomial if and only if it is externally equivalent to $P_0(z) = z^2$.*

Proof. Suppose (U, V, f) is holomorphically conjugate to a quadratic polynomial P_c. There is a conformal mapping H on V such that $H \circ f \circ H^{-1} = P_c$ on $H(U)$. We know from the end of §5.1 that there is a conformal mapping H_1 from $U_{r^2} \setminus K_{P_c}$ onto $\mathbf{D}_{r^2} \setminus \overline{\mathbf{D}}$ such that $H_1 \circ P_c \circ H_1^{-1} = P_0$ on $\mathbf{D}_r \setminus \overline{\mathbf{D}}$. Let $V_1 = H^{-1}(U_{r^2})$ and $U_1 = H^{-1}(U_r)$, and let $H_2 = H_1 \circ H$. Then (V_1, U_1, f) is a quadratic-like map and

$$H_2 \circ f = P_0 \circ H_2$$

on $U_1 \setminus K_f$. This proves the "only if" part.

Suppose (U, V, f) is externally equivalent to $P_0(z) = z^2$. Then there are domains $U_1 \subset V_1 \subset V$ such that (U_1, V_1, f) is a quadratic-like map and there is a conformal mapping $H_2 : V_1 \setminus K_f \to \mathbf{D}_{R^2} \setminus \overline{\mathbf{D}}$ for $R > 1$ such that $H_2^{-1} \circ P_0 \circ H_2 = f$ on $U_1 \setminus K_f$.

Take $\mathbf{D}_{R_1^2}$ for $\sqrt{R} < R_1 < R$. Then (U_1', V_1', f) is a quadratic-like map where $V_1' = H_2^{-1}(\mathbf{D}_{R_1^2})$ and $U_1' = H_2^{-1}(\mathbf{D}_{R_1})$. We take two half spheres V_1' and $\overline{\mathbf{C}} \setminus \overline{\mathbf{D}}_{R_1}$. We glue these two half spheres, by identifying $z \in \overline{V'}_1 \setminus U_1'$ and $H_2(z)$, to get a Riemann surface. From Theorem 5.13, this Riemann surface is conformally equivalent to the extended complex plane $\overline{\mathbf{C}}$. We define a continuous map

$$F(z) = \begin{cases} P_0(z), & \text{if } x \in \overline{\mathbf{C}} \setminus U_1'; \\ f(z), & \text{if } x \in \overline{U'}_1. \end{cases}$$

Let $\mu(z) = 0$ for z in $\overline{\mathbf{C}} \setminus V_1'$, $\mu(z) = H_0^*(0) = 0$ for z in $V_1' \setminus U_1$, and $\mu(z) = 0$ for z in U_1'. The function μ is measurable. From Theorem 5.9, there is a unique conformal mapping h whose complex dilatation is μ and which fixes 0, 1, and ∞. Consider $P = h \circ F \circ h^{-1}$. From Theorem 5.7, P is conformal at every point $z \neq 0$, and $\neq \infty$. This means that P is a holomorphic branch covering map of degree two. It is a rational function of $\overline{\mathbf{C}}$. But $P(\infty) = \infty$ and $P'(0) = 0$ imply that $P(z) = z^2 + c$ for some complex number c. ∎

5.4. Quadratic Polynomials

Suppose that $P_c(z) = z^2 + c$ is a quadratic polynomial and that its filled-in Julia set K_c is connected. Let h_∞ be the holomorphic diffeomorphism in Eq. (5.1). Let $\mathbf{S}^R = \{z \in \mathbf{C} \mid |z| = R\}$ and let $s_R = h_\infty(\mathbf{S}^R)$ for $R > 1$. Then

$$P_c(s_R) = s_{R^2}. \tag{5.10}$$

The topological circle s_R for every $R > 1$ is called an equipotential curve of P_c. A curve

$$e_\theta = h_\infty(\{z \in \mathbf{C} \mid |z| > 1, \arg(z) = \theta\})$$

for $0 \leq \theta < 2\pi$ is called an external ray of P_c. Then

$$P_c(e_\theta) = e_{2\theta}. \tag{5.11}$$

Remark 5.5. Let

$$G(z) = \max\{0, \lim_{n\to\infty} \frac{1}{2^n} \log |P_c^{\circ n}(z)|\}$$

be the Green's function of K_c in $\overline{\mathbf{C}}$. Then $G(P_c(z)) = 2G(z)$. For any $R > 1$, the equipotential curve $s_R = G^{-1}(\log R)$ is a level curve of G.

Let $\mathbf{S}^1$ be the unit circle and let $\mathbf{D} = \{z \in \mathbf{C} \mid |z| < 1\}$ be the open unit disk. If h_∞ can be extended continuously to $\mathbf{S}^1$, then we have a unique continuous map $H : \overline{\mathbf{C}} \setminus \mathbf{D} \to \overline{\mathbf{C}} \setminus \mathring{K}_f$ such that $H|(\overline{\mathbf{C}} \setminus \overline{\mathbf{D}}) = h_\infty$. Using H, we can define an equivalence relation on $\mathbf{S}^1$: $z_1 \sim z_2$ if and only if $H(z_1) = H(z_2)$. Let $[z]$ be the equivalent class of z. Then $\tilde{P}_0([z]) = [P_0(z)]$ defines a map of the quotient space $X = \mathbf{S}^1/\sim$, since $z_1^2 \sim z_2^2$ if $z_1 \sim z_2$. The dynamical system $(\tilde{P}_0, X)$ is topologically conjugate to (P_c, J_c) by $\tilde{H}([z]) = H(z)$.

Question 5.1. *For which c can h_∞ be extended continuously to $\mathbf{S}^1$?*

A connected set X in $\mathbf{C}$ is locally connected if for any point p in X and any neighborhood V about p, there is another neighborhood $U \subset V$ about p such that $U \cap X$ is connected. The following classical theorem proved by Carathéodory in one complex variable gives a sufficient and necessary condition to extend h_∞ continuously to $\mathbf{S}^1$.

Theorem 5.18 [CAR]. *Let h be a Riemann mapping from $\mathbf{D}$ onto a simply connected domain Ω. Then h can be extended continuously to $\mathbf{S}^1$ if and only if the boundary $\partial\Omega$ (as well as Ω) is locally connected.*

Remark 5.6. If $\partial\Omega$ is a Jordan curve, then h can be extended to a homeomorphism from $\overline{\mathbf{D}}$ onto $\overline{\Omega}$. Moreover, if $\partial\Omega$ is made of finite number of analytic curves, then the extension restricted to $\mathbf{S}^1$ has non-zero derivative at every point other than a corner (see [BIE]).

The proof of this theorem can be found in [MI2]. We have an equivalent question by just concerning the topology of a Julia set.

Question 5.2. *For which c is J_c (as well as K_c) locally connected ?*

An external ray e_θ lands at K_c if e_θ has only one limit point at K_c. An external ray is periodic with period m if $e_\theta \cap P_c^{\circ i}(e_\theta) = \emptyset$ for all $1 \leq i < m$ but $P_c^{\circ m}(e_\theta) = e_\theta$. The following theorem is proved by Douady and Yoccoz (refer to [MI2,HUB]).

Theorem 5.19. *Let $P_c(z) = z^2 + c$ be a quadratic polynomial and suppose its filled-in Julia set K_c is connected. Then every repelling periodic point of P_c is a landing point of finitely many periodic external rays with the same period.*

To prove the theorem, we need the following lemmas.

Lemma 5.1. *If there is a periodic external ray e_θ landing at a repelling periodic point z_0, then there are only finitely many external rays landing at z_0. All these external rays are periodic with the same period.*

Proof. Suppose e_θ is a periodic external ray of period n landing at z_0. Then $P_c^{\circ n}(e_\theta) = e_\theta$. This implies that $P_c^{\circ n}(z_0) = z_0$. Suppose $e_{\theta'}$ for $\theta' \neq \theta$ is another external ray landing at z_0. Let

$$\theta_i'' = 2^{in}\theta' (\mathrm{mod} 2\pi) \qquad \text{and} \qquad \theta_i' = \theta_i'' + \theta.$$

Then $P_c^{\circ n}$ sends $e_{\theta_i''}$ onto $e_{\theta_{i+1}''}$ for $i = 0, 1, \ldots$. Assume that $P_c^{\circ n}$ is orientation-preserving around z_0 (otherwise, consider $P_c^{\circ 2n}$). If $\theta < \theta_0' < \theta_1' < \theta + 2\pi$, then $\theta < \theta_i' < \theta_{i+1}' < \theta + 2\pi$ for all $i > 0$. Thus θ_i' has a limit θ_∞'. The angle $\theta_\infty'' = \theta_\infty' - \theta$ is a fixed point of the map $\theta \mapsto 2^n\theta(\mathrm{mod} 2\pi)$ and attracts θ_i''. But this map has only expanding fixed points. The contradiction implies that $\theta_0' = \theta_1'$. Hence $e_{\theta'}$ is periodic and its period is less than or equal to n. If the period of $e_{\theta'}$ is less than n, by using a similar argument, the period of

e_θ is less than n. This is impossible. So the period of $e_{\theta'}$ is also n. Since $\theta'' = 2^n\theta''(\text{mod}2\pi)$, we have

$$\theta'' = \frac{2j\pi}{2^n - 1}$$

for some $0 < j < 2^n - 1$. Hence there are only finitely many external rays landing at z_0 and all of them have the same period n. ∎

Without loss of generality, we assume that z_0 is a repelling fixed point of P_c in Theorem 5.19. If z_0 is a periodic point of period n, we replace P_c with $P_c^{\circ n}$ in following arguments.

From Theorem 5.2, there is a holomorphic diffeomorphism h defined on a neighborhood U about z_0 such that $h(z_0) = 0$ and such that $h \circ P_c \circ h^{-1}(z) = \lambda z$ on $h(U)$, where $\lambda = P_c'(0)$ is the multiplier of P_c at 0. Let $\mathbf{D}_r = \{z \in \mathbf{C} \mid |z| < r\}$ be a disk such that $\overline{\mathbf{D}}_r \subset h(U)$. Then $\mathbf{C} = \cup_{n=0}^{\infty}\lambda^n\mathbf{D}_r$. Define $H(z) = P_c^{\circ n}(h^{-1}(\lambda^{-n}z))$ for any z in $\lambda^n\mathbf{D}_r \setminus \lambda^{n-1}\mathbf{D}_r$, for $n = 1, 2, \ldots$. Then $H : \mathbf{C} \to \mathbf{C}$ is analytic and $P_c \circ H(z) = H(\lambda z)$ on $\mathbf{C}$. Let $\tilde{K}_c = H^{-1}(K_c)$ be the inverse image of K_c under H. Since K_c is closed and P_c-invariant, then $\tilde{K}_c$ is closed and $\lambda^{\pm}\tilde{K}_c = \tilde{K}_c$.

Lemma 5.2. *Let V be a connected component of $\mathbf{C} \setminus \tilde{K}_c$. Then $H : V \to \mathbf{C} \setminus K_c$ is a universal cover.*

Proof. The domain V is simply connected. Otherwise, the complement of V in $\mathbf{C}$ would have at least two connected components and at least one of them, W, would be bounded. Let $n \geq 0$ be the smallest integer such that $W_n = \lambda^{-n}W \subset \mathbf{D}_r$. Since $W \cap \tilde{K}_c \neq \emptyset$, then $K_n = W_n \cap \tilde{K}_c \neq \emptyset$. Since H on $\mathbf{D}_r$ is a conjugacy, then $H(K_n)$ is a subset of K_c and is separated from other points of K_c by $H(\lambda^{-n}V)$. This contradicts the fact that K_c is connected.

Let us prove that $H : V \to \mathbf{C} \setminus K_c$ is a covering map. From Eq. (5.1), $\cup_{n=0}^{\infty} P_c^{\circ n}(H(\mathbf{D}_r))$ covers $\mathbf{C} \setminus K_c$. For any small open disk B in $\mathbf{C} \setminus K_c$. There is an integer $n \geq 0$ such that $P_c^{-n}(B)$ is in $H(\mathbf{D}_r)$ and such that $P_c^{\circ n}$ restricted to any connected component B' of $P_c^{-n}(B)$ is a homeomorphism. Let n be the smallest such integer and let B' be any connected component of $P_c^{-n}(B)$. Then $\tilde{B} = \lambda^n H^{-1}(B')$ is a domain in V and $H : \tilde{B} \to B$ is a homeomorphism. Therefore $H : V \to \mathbf{C} \setminus K_c$ is a covering map. ∎

Consider $\tilde{G} = G \circ H$. It is the Green's function G on $\mathbf{C} \setminus K_c$ lifting to $\mathbf{C} \setminus \tilde{K}_c$. So $\tilde{G}(\lambda z) = 2\tilde{G}(z)$. Let $\tilde{\phi}_V = \log \circ h_\infty^{-1} \circ H : V \to \mathbf{RH}$ be the lift of the mapping $h_\infty^{-1} : \mathbf{C} \setminus K_c \to \mathbf{C} \setminus \overline{\mathbf{D}}$ to a connected component V of $\mathbf{C} \setminus \tilde{K}_c$ (see Fig. 5.2), where $\mathbf{RH}$ is the right-half plane. Then $\tilde{\phi}_V$ is unique up to addition of a multiple of $2\pi i$. Thus $\tilde{G} = \Re(\tilde{\phi}_V)$ and $\tilde{G}(\tilde{\phi}_V^{-1}(w)) = \Re(w)$ where $\Re$ means the real part.

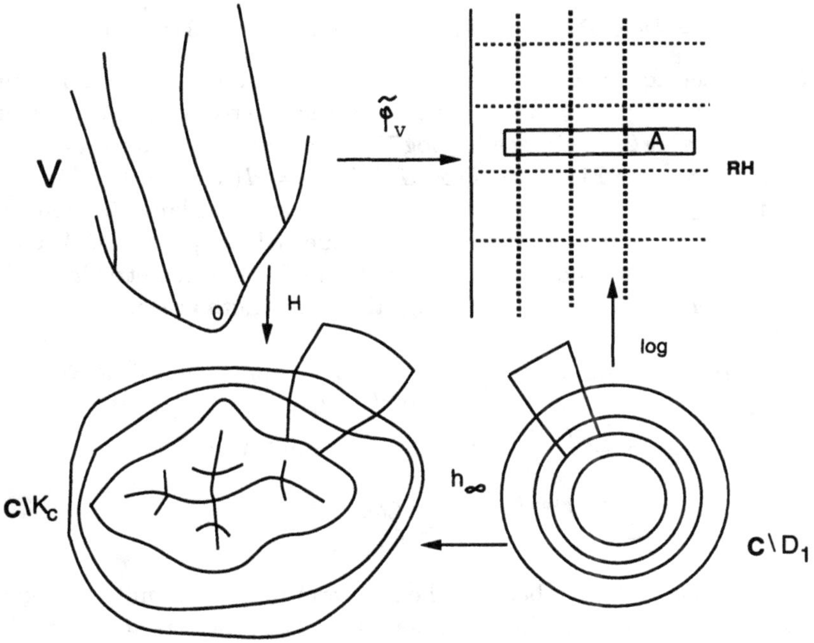

Fig. 5.2

Lemma 5.3. *Each connected component V of $\mathbf{C} \setminus \tilde{K}_c$ is periodic under the mapping $z \mapsto \lambda z$.*

Proof. Let $\mathbf{T} = (\mathbf{C} \setminus \{0\})/\{z \sim \lambda z\}$ be the torus obtained as the quotient space $\mathbf{C} \setminus \{0\}$ modulo the multiplication group generated by λ. Let $\pi : \mathbf{C} \setminus \{0\} \to \mathbf{T}$ be the canonical projection. If $\lambda^n V$ are all distinct, then π is injective on V. Let us show that it is impossible.

Let s_R be an equipotential curve of P_c and let S be the inverse image of s_R under H. Consider W_R the inverse image of $\mathbf{C} \setminus U_R$ under H where U_R is the domain bounded by s_R. Let $V' = \cup_{n \in \mathbf{Z}} \lambda^n V$ and let $a = \log R$. Then $S = \tilde{G}^{-1}(a)$. For any w in $\mathbf{RH}$ with the real part $\Re(w) \geq a2^n$,

$$\tilde{G}(\lambda^{-n}\tilde{\phi}_V^{-1}(w)) = 2^{-n}\tilde{G}(\tilde{\phi}_V^{-1}(w)) = 2^{-n}\Re(w) \geq a.$$

This implies that $\tilde{\phi}_V^{-1}(w)$ is in $\lambda^n W_R$.

We choose a point $w_0 = m + iv_0$ in $\mathbf{RH}$, where $m > 0$ and where $0 < v_0 < 2\pi$, we then select a rectangle (see Fig. 5.2)

$$A = \{w \in \mathbf{RH} \mid m \le \Re(w) \le m + L,\ |\Im w - \Im w_0| \le \delta\}$$

for $L > 0$, and for small $\delta > 0$, where $\Re$ and $\Im$ are the real and imaginary parts. We consider $\eta_V = \pi \circ \tilde{\phi}_V^{-1}$ and also the inverse image A' of A under $\tilde{\phi}_V$. For $L = n \log 2$, the image $h_\infty(\log^{-1}(A))$ crosses n fundamental domains of P_c in $\mathbf{C} \setminus K_c$ (see Fig. 5.2). From $P_c(H(z)) = H(\lambda z)$ and from the above argument, any connected component of A' crosses n fundamental domains of the map $z \mapsto \lambda z$. Thus the image of any connected component of A' under π makes n full turns around $\mathbf{T}$. Endow $\mathbf{T}$ with the Euclidean metric $|dz|/|z|$; the image of a horizontal line in A under η_V then has length at least n. Therefore,

$$\begin{aligned}\mathrm{Area}(\mathbf{T}) \ge \mathrm{Area}(\eta_V(A)) &= \int_{v_0-\delta}^{v_0+\delta}\int_m^{m+L} |\eta_V'(u+iv)|^2\, dudv \\ &\ge \frac{1}{L}\int_{v_0-\delta}^{v_0+\delta}\Big(\int_m^{m+L} |\eta_V'(u+iv)| du\Big)^2\, dv \\ &\ge \frac{1}{L}\int_{v_0-\delta}^{v_0+\delta} n^2\, dv = \frac{2\delta n}{\log 2}.\end{aligned}$$

This gives a contradiction because the last term tends to infinity as n goes to infinity, but $\mathrm{Area}(\mathbf{T}) = 2\pi \log|\lambda|$ is bounded. Thus there is an integer $k > 0$ such that $\lambda^k V = V$. ∎

Proof of Theorem 5.19. For any connected component V of $\mathbf{C}\setminus\tilde{K}_c$, the quotient $\pi(V)$ is an annulus in $\mathbf{T}$. Let γ be a unique closed geodesic of $\pi(V)$ with respect to the hyperbolic distance $d_{H,\pi(V)}$ and let $\tilde{\gamma}$ be its lift in V. Then $\tilde{\gamma} = \cup_{n\in\mathbf{Z}}\lambda^n\gamma$ is a geodesic in V with respect to the hyperbolic distance $d_{H,V}$. Therefore, $e_\theta = H(\tilde{\gamma})$ is an external ray (since it is a geodesic of $\mathbf{C}\setminus K_c$ with respect to the hyperbolic distance $d_{H,\mathbf{C}\setminus K_c}$). Since $\tilde{\gamma}$ tends to 0 at one end point and since H is the conjugacy on a neighborhood about 0, e_θ converges to z_0 at one end point (and to infinity at another end point). Since there is an integer $k > 0$ such that $\lambda^k V = V$, we have $\lambda^k\tilde{\gamma} = \tilde{\gamma}$. This implies that $P_c^{\circ k}(e_\theta) = e_\theta$. Hence e_θ is a periodic external ray landing at z_0. This proves that for each connected component V of $\mathbf{C}\setminus\tilde{K}_c$, there corresponds exactly one periodic external ray landing at z_0. Since $\mathbf{C}\setminus K_c$ is non-vacuous, there is at least one V. Hence there is at least one periodic external ray landing at z_0. This and Lemma 5.1 give the proof of the theorem. ∎

Let $P_c(z) = z^2 + c$ be a quadratic polynomial. Let $c(n) = P_c^{\circ n}(0)$ denote the nth critical value of P_c, let $CO = CO(c) = \{c(n)\}_{n=0}^{\infty}$ denote the critical orbit of P_c, and let $PCO = PCO(c) = \{c_n\}_{n=1}^{\infty}$ denote the post-critical orbit of P_c.

If 0 is not in the filled-in Julia set K_c of P_c, the Julia set $J_c = K_c$ is a Cantor set (see Theorem 5.1) and P_c on J_c is expanding. The polynomial P_c is called hyperbolic if its Julia set J_c is connected and contains no point in the closure of CO. For a hyperbolic quadratic polynomial P_c, there is an open neighborhood $U \supset J_c$ such that $U \subset V = P_c(U)$, such that V contains no point in $\overline{CO}$, and such that CO is in the immediate basin of the unique attractive or super-attractive periodic orbit in the complex plane $\mathbf{C}$ (see [MI2]). Conversely, if P_c has an attractive or super-attractive periodic orbit in $\mathbf{C}$, it is hyperbolic (see Theorem 5.5). For a hyperbolic polynomial P_c, one can prove that its Julia set J_c is locally connected (refer to [MI2]).

For a non-hyperbolic polynomial P_c with connected Julia set J_c, we can further classify it as neutral, non-recurrent, or recurrent as follows. It is neutral if it has a neutral periodic point; it is non-recurrent if it has no neutral periodic point and if 0 is not a limit point of PCO; it is recurrent if it has no neutral periodic point and if 0 is a limit point of PCO. A non-recurrent quadratic polynomial P_c is called sub-hyperbolic (or Misiurewicz or preperiodic) if PCO is finite and contains a repelling periodic point.

For a non-recurrent or recurrent quadratic polynomial P_c, its filled-in Julia set equals its Julia sets (see Theorem 5.5). The Julia set of a non-recurrent quadratic polynomial is locally connected (see [HUB]). In the following sections, we will pay attention to recurrent, in particular, infinitely renormalizable quadratic polynomials P_c. We would like to note that for a quadratic polynomial P_c with connected Julia set J_c, its filled-in Julia set K_c is locally connected if and only if its Julia set J_c is locally connected (see [MI2]). For the convenience, we will use, in the following sections, the term filled-in Julia set even in the case that the Julia set of a quadratic polynomial equals its filled-in Julia set.

5.5. Renormalizable Quadratic-Like Maps

Let (U_0, V_0, f_0) be a quadratic-like map and suppose its filled-in Julia set K_{f_0} is connected. It is renormalizable if there are an integer $n \geq 2$ and a subdomain U_1 of U_0 such that $0 \in U_1$ and such that $(U_1, V_1, f_0^{\circ n})$ is a quadratic-like map whose filled-in Julia set is connected, where $V_1 = f^{\circ n}(U_1)$. Let $f_1 = f_0^{\circ n}|U_1$ and let $K_{f_1} = K_{f_1}(n, U_1)$ be its filled-in Julia set. The domain U_1 is called a renormalization and (U_0, V_0, f_0) is called renormalizable about n. Otherwise, (U_0, V_0, f_0) is called non-renormalizable.

The quadratic-like map (U_0, V_0, f_0) is infinitely renormalizable if there is a strictly increasing sequence $\{m_k\}_{k=1}^{\infty}$ of integers $n \geq 2$ such that (U_0, V_0, f_0) is renormalizable about m_k for $k = 1, 2, \ldots$. Otherwise, (U_0, V_0, f_0) is called finitely renormalizable.

Let $\mathbf{R}$ be the real line. A quadratic-like map (U_0, V_0, f_0) is called real if $f_0(U_0 \cap \mathbf{R}) \subset V_0 \cap \mathbf{R}$ and $g = f_0|U_0 \cap \mathbf{R}$ is a real folding map. Let (U_0, V_0, f_0) be a real quadratic-like map and suppose its filled-in Julia set K_{f_0} is connected. Suppose $g = f_0|U_0 \cap \mathbf{R}$ has a fixed point $p \in \mathbf{R}$ with positive multiplier $f_0'(p)$. Let $p' \neq p$ be another inverse image of p under g, that is, $g(p') = p$. Conjugating by a Möbius transformation, we may assume that $p = -1$ and $p' = 1$ and that

$$[-1, 1] = \cap_{n=0}^{\infty} g^{-n}(V_0 \cap \mathbf{R}) = K_f \cap \mathbf{R}.$$

Hence g is a folding map with unique quadratic critical point 0.

Let g be a folding map of $[-1, 1]$ such that $g(-1) = g(1) = -1$ and such that 0 is a unique quadratic critical point. We say g is renormalizable about $n > 1$ if there is a subinterval I of $[-1, 1]$ such that $0 \in \mathring{I}$, such that $g^{\circ i}(I) \cap \mathring{I} = \emptyset$ for all $0 < i < n$, and such that $g^{\circ n}(I) \subseteq I$. Otherwise, g is non-renormalizable. We say g is infinitely renormalizable if there is a strictly increasing sequence $\{m_k\}_{k=1}^{\infty}$ of integers $n \geq 2$ such that g is renormalizable about m_k for all $k > 0$ (see §4.1). Otherwise, g is called finitely renormalizable. The next theorem shows that for a real quadratic-like map, both definitions of renormalization are essentially equivalent.

Theorem 5.20. *Let (U_0, V_0, f_0) be a real quadratic-like map and suppose its filled-in Julia set K_{f_0} is connected. Suppose f_0 has neither neutral, nor attractive, nor super-attractive periodic point. Then f_0 is renormalizable if and only if the folding map $g = f_0|[-1, 1]$ is renormalizable.*

Proof. Suppose the folding map $g = f_0|[-1, 1]$ is renormalizable. This means that there is a maximal closed subinterval I of $[-1, 1]$ and an integer

$n > 1$ such that 0 is in $\mathring{I}$, such that $g^{\circ i}(I) \cap \mathring{I} = \emptyset$ for all $0 < i < n$, and such that $g^{\circ n}(I) \subseteq I$. One of the endpoints of I is fixed by $g^{\circ n}$. It is a repelling fixed point. Take a neighborhood T' of I such that $f^{\circ n}|(L \cup R)$ is expanding, where $L \cup R = T' \setminus I$. Let $I_i = g^{\circ i}(I)$ for $0 \leq i \leq n$. The inverse h_n of $g^{\circ(n-1)} : I_1 \to I_n$ is a diffeomorphism and can be extended to a diffeomorphism on an open interval $T \supset I$. Take $T \subset T'$. Let $M' = h_n(T)$ and $M = g^{-1}(M')$. Then $\overline{M}$ is a subset of T. Because the critical orbit $CO = \{f_0^{\circ k}(0)\}_{k=0}^{\infty}$ of f_0 is in the real line $\mathbf{R}$, h_n can be extended to $V_1 = (V_0 \setminus \mathbf{R}) \cup T$ analytically. The image V_1' of V_1 under this extension is contained in $U' = (U_0 \setminus \mathbf{R}) \cup M'$. Let $U_1 = f_0^{-1}(V_1')$. Then $\overline{U}_1 \subset V_1$ and (U_1, V_1, f_1) for $f_1 = f_0^{\circ n}|U_1$ is a quadratic-like map. Since $g^{\circ n}(I) \subseteq I$, the filled-in Julia set K_{f_1} is connected. Therefore, (U_0, V_0, f_0) is renormalizable about n and U_1 is a renormalization.

Suppose (U_0, V_0, f_0) is renormalizable about $n > 1$. Let U_1 be a renormalization and set $V_1 = f^{\circ n}(U_1)$. Then (U_1, V_1, f_1) for $f_1 = f_0^{\circ n}|U_1$ is a quadratic-like map with the connected filled-in Julia set K_{f_1}. Let $I = K_{f_1} \cap \mathbf{R}$. Then $I = \cap_{i=0}^{\infty} f_1^{-i}(V_1 \cap \mathbf{R})$. For every $1 \leq i < n$, $f^{\circ i}(I) \cap \mathring{I} = \emptyset$ else $f^{\circ n}$ would have at least three fixed points in $I \cup f^{\circ i}(I)$ with one of them attractive, super-attractive, or parabolic. Since $f_1(0)$ is in I, $f_0^{\circ n}(I) \subseteq I$. Therefore, $g = f_0|[-1, 1]$ is a renormalizable folding map. ■

5.6. Two-Dimensional Yoccoz Puzzles and Renormalizability

In this section, we discuss a technique in the study of non-renormalizable quadratic polynomials and some of its applications to the renormalization theory.

Let $P_c(z) = z^2+c$ be a quadratic polynomial with connected filled-in Julia set K_c. The external ray e_0 of P_c is the only one fixed by P_c (see Fig. 5.3). It lands either at a repelling or parabolic fixed point β of P_c (see [MI2]). Suppose β is repelling. Applying Theorem 5.19, we see that e_0 is the only external ray landing at β. Thus $K_c \setminus \{\beta\}$ is connected. We call β the non-separate fixed point of P_c. Let $\alpha \neq \beta$ be the other fixed point of P_c. If α is either an attractive or a super-attractive fixed point, then $J_c = K_c \setminus (\cup_{n=0}^{\infty} P_c^{-n}(D(\alpha))$ for a small disk centered at α. The Julia set J_c is a Jordan curve; every external ray lands at a unique point in J_c (see Remark 5.6). If α is a repelling fixed point, there are at least two periodic external rays landing at α. We use $R_0(\alpha)$ to denote the union of a cycle of periodic external rays of period q landing at α (see Fig. 5.3). The set $R_0(\alpha)$ cuts $\mathbf{C}$ into finitely many simply connected domains Ω_0, Ω_1, ..., Ω_{q-1}. Each domain contains points in the Julia set J_c. Thus $K_c \setminus \{\alpha\}$ is disconnected. We call α the separate fixed point of P_c.

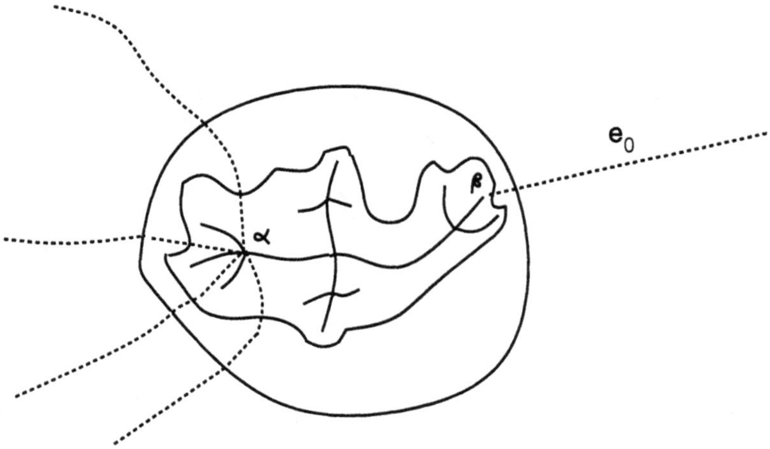

Fig. 5.3

Henceforth, we assume that the fixed points β and α are both repelling. Let s_r be a fixed equipotential curve of P_c and let U_r be the domain bounded by s_r. Then $(U_{\sqrt{r}}, U_r, P_c)$ is a quadratic-like map. The set $R_0(\alpha)$ cuts U_r into finitely many simply connected domains. Let C_0 be the closure of the domain containing 0, and let $B_{0,i}$ be the closure of the domain containing $P_c^{\circ i}(0)$ for $1 \leq i < q$. Since $R_0(\alpha)$ is forward invariant under P_c, the image under P_c of $C_0 \cap K_c$ or $B_{0,i} \cap K_c$, for every $1 \leq i < q$, is the union of some of $C_0 \cap K_c$, $B_{0,1} \cap K_c$, ..., $B_{0,q-1} \cap K_c$. The set

$$\eta_0 = \{C_0, B_{0,1}, \ldots, B_{0,q-1}\}$$

is called the original partition (with respect to $R_0(\alpha)$). We note that it is not a Markov partition because $P_c|C_0$ is a proper, holomorphic map of degree two. (But $P_c|B_{0,i}$ is a holomorphic diffeomorphism for every $1 \leq i < q$.)

Let $\Gamma_n = P_c^{-n}(\alpha)$ and let $R_n(\alpha) = P_c^{-n}(R_0(\alpha))$ for $n \geq 0$. The set $R_n(\alpha)$ is the union of some external rays landing at points in Γ_n; it cuts the domain $U_{r^{\frac{1}{2^n}}}$ into a finite number of simply connected domains. Let C_n be the closure of the domain containing 0 and let $B_{n,1}$, ..., B_{n,k_n} be the closures of others. Since $P_c(\Gamma_n) = \Gamma_{n-1}$, the image of C_n or $B_{n,i}$ under P_c is one of C_{n-1}, $B_{n-1,1}$, ..., $B_{n-1,k_{n-1}}$, for $1 \leq i \leq k_n$. Then $P_c|C_n$ is holomorphic, proper, branch covering map of degree two; all $P_c|B_{n,i}$ are holomorphic diffeomorphisms. The set

$$\eta_n = \{C_n, B_{n,1}, \ldots, B_{n,k_n}\}$$

is called the nth-partition (with respect to $R_0(\alpha)$). The sequence $\eta = \{\eta_n\}_{n=0}^{\infty}$ is called the two-dimensional Yoccoz puzzle for P_c (with respect to $R_0(\alpha)$). A similar puzzle for certain cubic polynomial was constructed by Branner and Hubbard [BRH]. Yoccoz used this puzzle while studying the local connectivity of a non-renormalizable quadratic polynomial.

Let $\Gamma_\infty = \cup_{n=0}^{\infty}\Gamma_n$. For any x in $K_c \setminus \Gamma_\infty$, there is one and only one sequence $\{D_n(x)\}_{n=0}^{\infty}$ such that $x \in D_n(x) \in \eta_n$. For any x in Γ_∞, there are q such sequences. We call such a sequence

$$x \in \cdots \subseteq D_n(x) \subseteq D_{n-1}(x) \subseteq \cdots \subseteq D_1(x) \subseteq D_0(x)$$

an x-end. In particular,

$$0 \in \cdots \subseteq C_n \subseteq C_{n-1} \subseteq \cdots \subseteq C_2 \subseteq C_1 \subseteq C_0$$

is called the critical end.

Suppose $D_{n+1} \subseteq D_n$ are domains in η_{n+1} and η_n for $n \geq 0$. Define $A_n = D_n \setminus \mathring{D}_{n+1}$. If $D_{n+1} \subset \mathring{D}_n$, then $\mathring{A}_n$ is a non-degenerate annulus and its modulus (denoted mod(A_n)) is greater than 0. Otherwise, A_n is a degenerate annulus and its modulus mod(A_n) is zero. The pair $D_{n+1} \subseteq D_n$ is critical if

$D_n = C_n$ and $D_{n+1} = C_{n+1}$; it is semi-critical if $D_n = C_n$ but $D_{n+1} \neq C_{n+1}$; it is non-critical if $D_n \neq C_n$. If it is non-critical, then

$$\text{mod}(A_n) = \text{mod}(P_c(A_n))$$

since $P_c : A_n \to P_c(A_n)$ is a conformal mapping. If it is critical, then

$$\text{mod}(A_n) = \frac{\text{mod}(P_c(A_n))}{2}$$

since $P_c : A_n \to P_c(A_n)$ is a proper, holomorphic, unramified covering map of degree two. If it is semi-critical, then

$$\frac{\text{mod}(P_c(A_n))}{2} \leq \text{mod}(A_n).$$

For a point x in K_c and an x-end

$$x \in \cdots \subseteq D_n(x) \subseteq D_{n-1}(x) \subseteq \cdots \subseteq D_1(x) \subseteq D_0(x),$$

we define

$$D_{nm}(x) = P_c^{\circ m}(D_{n+m}(x))(= D_n(P_c^{\circ m}(x)))$$

and

$$A_{nm}(x) = D_{nm}(x) \setminus \mathring{D}_{(n+1)m}(x)$$

for $n, m \geq 0$. The tableau

$$T(x) = (a_{nm})_{n\geq 0, m\geq 0}$$

is an $\infty \times \infty$-matrix defined as follows: $a_{nm} = 1$ if $D_{nm}(x) = C_n$, and $a_{nm} = 0$ if $D_{nm}(x) \neq C_n$. The tableau

$$T(0) = \left(a_{nm}^0\right)_{n\geq 0, m\geq 0}$$

is called the critical tableau. Note that

$$P_c^{\circ m}(x) \in \cdots \subseteq D_{nm}(x) \subseteq D_{(n-1)m}(x) \subseteq \cdots \subseteq D_{1m}(x) \subseteq D_{0m}(x)$$

is a $P_c^{\circ m}(x)$-end. In $T(x) = (a_{nm})_{n\geq 0, m\geq 0}$, $\{a_{nm}\}_{m=0}^{\infty}$ is called the nth-row and $\{a_{nm}\}_{n=0}^{\infty}$ is called the mth-column.

Lemma 5.4. *The tableau $T(x)$ satisfies the following rules:*

(T1) *if $a_{nm} = 1$ for some $n, m \geq 0$, then $a_{im} = 1$ for all $0 \leq i \leq n$,*

(T2) *if $a_{nm} = 1$ for some $n, m \geq 0$, then $a_{(n-i)(m+j)} = a^0_{(n-i)j}$ for all $0 \leq j \leq n$ and $0 \leq n-i \leq n-j$,*

(T3) *if*

i) *$a_{nm} = 1$ and $a_{(n+1)m} = 0$ for some $n, m \geq 0$ and if*

ii) *$a_{(n-i)(m+i)} = 1$ for some $1 \leq i \leq n$ and $a_{(n-j)(m+j)} = 0$ for all $0 < j < i$,*

then $a^0_{(n-i+1)i} = 1$ implies $a_{(n-i+1)(m+i)} = 0$.

$$
(\mathbf{T1}):\left(\begin{array}{cccc}
 & & m & \\
 & \cdots & \mathbf{1} & \cdots \\
 & \cdots & \vdots & \cdots \\
n & \cdots & \mathbf{1} & \cdots \\
 & \cdots & * & \cdots \\
 & \cdots & \vdots & \cdots
\end{array}\right),
$$

$$
(\mathbf{T2}):\left(\begin{array}{ccccccccccc}
 & & m & & & & & & & n+m & \\
 & \cdots & \mathbf{1} & \# & \# & \cdots & \# & \# & \# & \# & \cdots \\
 & \cdots & \mathbf{1} & \# & \# & \cdots & \# & \# & \# & * & \cdots \\
 & \cdots & \mathbf{1} & \# & \# & \cdots & \# & \# & * & * & \cdots \\
 & \vdots & \vdots & \textit{same} & \textit{as} & \vdots & \vdots & \vdots & \vdots & \vdots & \vdots \\
 & \vdots & \vdots & \textit{in} & \textit{the} & \vdots & \vdots & \vdots & \vdots & \vdots & \vdots \\
 & \vdots & \vdots & \textit{critical} & \textit{tableau} & \vdots & \vdots & \vdots & \vdots & \vdots & \vdots \\
 & \cdots & \mathbf{1} & \# & \# & \cdots & * & * & * & * & \cdots \\
 & \cdots & \mathbf{1} & \# & \# & \cdots & * & * & * & * & \cdots \\
 & \cdots & \mathbf{1} & \# & * & \cdots & * & * & * & * & \cdots \\
n & \cdots & \mathbf{1} & * & * & \cdots & * & * & * & * & \cdots \\
 & \cdots & * & * & * & \cdots & * & * & * & * & \cdots \\
 & \vdots & \vdots & \vdots & \vdots & \vdots & \vdots & \vdots & \vdots & \vdots & \vdots
\end{array}\right),
$$

$$
(\mathbf{T3}):\quad T(x) = \left(\begin{array}{r|ccccccccc}
 & & m & & & & & & m+i & \\
 & \cdots & \mathbf{1} & * & * & * & \cdots & * & \mathbf{1} & \cdots \\
 & \vdots & \vdots & \vdots & \vdots & \vdots & \vdots & \vdots & \vdots & \vdots \\
 & \cdots & \mathbf{1} & * & * & * & \cdots & * & \mathbf{1} & \cdots \\
n-i & \cdots & \mathbf{1} & * & * & * & \cdots & * & \mathbf{1} & \cdots \\
n-i+1 & \cdots & \mathbf{1} & * & * & * & \cdots & \mathbf{0} & \mathbf{0} & \cdots \\
 & \cdots & \mathbf{1} & * & * & * & \cdots & \mathbf{0} & \mathbf{0} & \cdots \\
 & \vdots & \vdots & \vdots & \vdots & \vdots & \vdots & \vdots & \vdots & \vdots \\
 & \cdots & \mathbf{1} & * & * & \mathbf{0} & \cdots & \mathbf{0} & \mathbf{0} & \cdots \\
 & \cdots & \mathbf{1} & * & \mathbf{0} & \mathbf{0} & \cdots & \mathbf{0} & \mathbf{0} & \cdots \\
 & \cdots & \mathbf{1} & \mathbf{0} & \mathbf{0} & \mathbf{0} & \cdots & \mathbf{0} & \mathbf{0} & \cdots \\
n & \cdots & \mathbf{1} & \mathbf{0} & \mathbf{0} & \mathbf{0} & \cdots & \mathbf{0} & \mathbf{0} & \cdots \\
n+1 & \cdots & \mathbf{0} & \mathbf{0} & \mathbf{0} & \mathbf{0} & \cdots & \mathbf{0} & \mathbf{0} & \cdots \\
 & \cdots & \mathbf{0} & \mathbf{0} & \mathbf{0} & \mathbf{0} & \cdots & \mathbf{0} & \mathbf{0} & \cdots \\
 & \vdots & \vdots & \vdots & \vdots & \vdots & \vdots & \vdots & \vdots & \vdots
\end{array}\right);
$$

$$
T(0) = \left(\begin{array}{r|cccccccc}
 & 0 & & & & & & i & \\
 & \mathbf{1} & * & * & * & \cdots & * & \mathbf{1} & \cdots \\
 & \vdots & \vdots & \vdots & \vdots & \vdots & \vdots & \vdots & \vdots \\
 & \mathbf{1} & * & * & * & \cdots & * & \mathbf{1} & \cdots \\
n-i & \mathbf{1} & * & * & * & \cdots & * & \mathbf{1} & \cdots \\
n-i+1 & \mathbf{1} & * & * & * & \cdots & \mathbf{0} & \mathbf{1} & \cdots \\
 & \mathbf{1} & * & * & * & \cdots & \mathbf{0} & * & \cdots \\
 & \vdots & \vdots & \vdots & \vdots & \vdots & \vdots & \vdots & \vdots \\
 & \mathbf{1} & * & * & \mathbf{0} & \cdots & \mathbf{0} & * & \cdots \\
 & \mathbf{1} & * & \mathbf{0} & \mathbf{0} & \cdots & \mathbf{0} & * & \cdots \\
 & \mathbf{1} & \mathbf{0} & \mathbf{0} & \mathbf{0} & \cdots & \mathbf{0} & * & \cdots \\
n & \mathbf{1} & \mathbf{0} & \mathbf{0} & \mathbf{0} & \cdots & \mathbf{0} & * & \cdots \\
n+1 & \mathbf{1} & \mathbf{0} & \mathbf{0} & \mathbf{0} & \cdots & \mathbf{0} & * & \cdots \\
 & \mathbf{1} & \mathbf{0} & \mathbf{0} & \mathbf{0} & \cdots & \mathbf{0} & * & \cdots \\
 & \vdots & \vdots & \vdots & \vdots & \vdots & \vdots & \vdots & \vdots
\end{array}\right).
$$

Proof. Rule **(T1)** is valid because

$$0 \in \cdots \subseteq C_n \subseteq C_{n-1} \subseteq \cdots \subseteq C_1 \subseteq C_0.$$

Rule **(T2)** is valid because if $a_{nm} = 1$, then $D_{im}(x) = C_i$ for all $0 \leq i \leq n$. Thus $D_{(n-i)(m+j)}(x) = f^{\circ j}(C_{n-i})$ for all $0 \leq j \leq n$ and all $0 \leq n-i \leq n-j$.

We prove Rule **(T3)**. Conditions **i)** and **ii)** imply that $P_c^{\circ i} : C_n \to C_{n-i}$ is a degree two holomorphic proper branch covering map. Condition **i)** and Rule **(T2)** imply $a_{k(m+j)} = a^0_{kj}$ for all $0 \leq j \leq n$ and all $0 \leq k \leq$

$n-j$. Now $a^0_{(n-i+1)i} = 1$ and $a^0_{j0} = 1$ for all $0 \le j < \infty$ imply that $P_c^{\circ i} : C_{n+1} \to C_{n-i+1}$ is a degree two holomorphic proper branch covering map. Thus $C_{n+1} = P_c^{-i}(C_{n-i+1}) \cap C_n$. Assume $a_{(n-i+1)(m+i)} = 1$. Then $D_{(n+1)m}(x) = P_c^{-i}(C_{n-i+1}) \cap C_n = C_{n+1}$. This contradicts $a_{(n+1)m} = 0$. ∎

Lemma 5.5. *For any domain D in η_n, for $n \ge 0$, $D \cap K_c$ is connected.*

Proof. Since the domain D is bounded by finitely many external rays $\Pi = \{e_{\theta_i}\}_{i=1}^m$ and by some equipotential curve, then $\partial D \cap K_c$ consists of a finite number of points $\{p_i\}_{i=1}^n$. Every p_i is a landing point of two external rays in Π. Suppose $D \cap K_c$ is not connected for some D in η_n. Then there are two disjoint open sets U and V such that $D \cap K_c = (D \cap K_c \cap U) \cup (D \cap K_c \cap V)$. Suppose that $p_1, \ldots, p_k$ are in U and that $p_{k+1}, \ldots, p_n$ are in V. The two external rays in Π landing at p_i cut $\mathbf{C}$ into two domains. Let W_i be the one which is disjoint from D. Then $U' = U \cup \cup_{i=1}^k W_i$ and $V' = V \cup \cup_{i=k+1}^n W_i$ are two disjoint open sets and $K_c = (U' \cap K_c) \cup (V' \cap K_c)$. This contradicts the fact that K_c is connected. ∎

Lemma 5.5 implies that for any x-end

$$x \in \cdots \subseteq D_n(x) \subseteq D_{n-1}(x) \subseteq \cdots \subseteq D_1(x) \subseteq D_0(x),$$

the intersection

$$L_x = \cap_{n=0}^{\infty} D_n(x)$$

is a compact connected non-empty set containing x. Let $T(x) = (a_{nm})_{n \ge 0, m \ge 0}$ be the tableau of an x-end. It is non-recurrent if there is an integer $N \ge 0$ such that $a_{nm} = 0$ for all $n \ge N$ and all $m \ge 1$. Otherwise, $T(x)$ is recurrent.

Lemma 5.6. *If $T(x)$ is non-recurrent, then $L_x = \{x\}$.*

Proof. Suppose $N \ge 0$ is an integer such that $a_{nm} = 0$ for all $n \ge N$ and all $m \ge 1$. Then, for $n > N$, every

$$P_c^{\circ(n-N-1)} : D_{(n-1)1}(x) \to D_{N(n-N)}(x)$$

is a holomorphic diffeomorphism. Thus for every $n > N$, $\mathrm{mod}(A_{n0}(x))$ is greater than or equal to $\mathrm{mod}(A_{N(n-N)}(x))/2$. There are only finitely many different annuli in $\{A_{Nm}(x)\}_{m=0}^{\infty}$ because η_N has only finitely many domains. If there are infinitely many non-degenerate annuli in $\{A_{Nm}(x)\}_{m=0}^{\infty}$, then there are infinitely many non-degenerate annuli in $\{A_{n0}(x)\}_{n=0}^{\infty}$ whose moduli are the same. This would imply that

$$\mathrm{mod}(D_0(x) \setminus L_x) \ge \sum_{n=0}^{\infty} \mathrm{mod}(A_{n0}(x)) = \infty.$$

Therefore $L_x = \{x\}$.

If there are only finitely many non-degenerate annuli in $\{A_{Nm}(x)\}_{m=0}^{\infty}$, The proof uses results from hyperbolic geometry. Let $B_{N,1}$ be the domain in η_N containing the critical value $P_c(0)$. Since $P_c^{\circ i}(x)$ do not enter C_{N+1} for all $0 < i < \infty$, then $P_c^{\circ i}(x)$ does not enter $B_{N,1}$ for all $2 \leq i < \infty$. Let us thicken $B_{N,i}$ to an open simply connected domain $\tilde{B}_{N,i}$ such that $B_{N,i} \subset \tilde{B}_{N,i}$ and such that $P_c(0)$ is not in $\tilde{B}_{N,i}$ for $1 < i \leq k_N$ where k_N is the number of elements in η_N. The map P_c has two inverse branches g_{i1} and g_{i2} defined on $\tilde{B}_{N,i}$ for every $1 < i \leq k_N$. We consider $\tilde{B}_{N,i}$ to be a hyperbolic Riemann surface with the hyperbolic distance $d_{H,i}$ for every $1 < i \leq k_N$. Then if g_{ik}, for $1 < i \leq k_N$ and $k = 1$ or 2, sends $B_{N,i}$ into $B_{N,j}$ for some $1 < j \leq k_N$, then it strictly contracts these hyperbolic distances; more precisely, there is a constant $0 < \lambda < 1$ such that $d_{H,j}(g_{ik}(x), g_{ik}(y)) < \lambda d_{H,i}(x, y)$ for x and y in $B_{N,i}$ and for $k = 0$ and 1. Therefore, there is a constant $C > 0$ such that for any $n > N$ and for any $D_{n1}(x)$,

$$d(D_{n1}(x)) = \max_{y,z \in D_{n1}(x)} |y - z| \leq C\lambda^{n-N-1}$$

since $D_{(n-i)i}(x)$ is in one of $B_{N,2}, \ldots, B_{N,k_N}$ for every $2 \leq i \leq n - N$. Thus $d(D_{n0}(x))$ tends to zero as n goes to infinity and $L_x = \{x\}$. ∎

The critical end

$$0 \in \cdots \subseteq C_n \subseteq C_{n-1} \subseteq \cdots \subseteq C_2 \subseteq C_1 \subseteq C_0$$

is important. Let $A_{n0}(0) = C_n \setminus \mathring{C}_{n+1}$.

Lemma 5.7. *If $\sum_{n=0}^{\infty} mod(A_{n0}(0)) = \infty$, then for any x in K_c and any x-end*

$$x \in \cdots \subseteq D_n(x) \subseteq D_{n-1}(x) \subseteq \cdots \subseteq D_1(x) \subseteq D_0(x),$$

$L_x = \{x\}$.

Proof. Consider the tableau $T(x) = (a_{nm})_{n\geq 0, m\geq 0}$. If $T(x)$ is non-recurrent, the lemma follows from Lemma 5.6.

Suppose $T(x)$ is recurrent. If there is a column which is entirely 1's, then there are integers $M \geq 0$ and $N \geq 0$ such that $a_{iM} = 1$ for all $i \geq 0$ and $a_{nm} = 0$ for all $n \geq N$ and $0 \leq m < M$. Thus $P_c^{\circ M} : D_{n0}(x) \to D_{(n-M)M} = C_{n-M}$ is a holomorphic diffeomorphism for every $n \geq N$. This implies

$$m(D_0(x) \setminus L_x) \geq \sum_{n=0}^{\infty} \text{mod}(A_{n0}(x)) \geq \sum_{n=N}^{\infty} \text{mod}(A_{n0}(x))$$

$$= \sum_{n=N-M}^{\infty} \text{mod}(A_{n0}(0)) = \infty.$$

So $L_x = \{x\}$.

Suppose that there is no column which is entirely 1's. Let $N > 0$ be an integer such that $a_{n0} = 0$ for all $n \geq N$. For any $n \geq N$, let $m_n > 0$ be the integer such that $a_{nm_n} = 1$ and $a_{ni} = 0$ for all $0 \leq i < m_n$. Then $P_c^{\circ m_n}$: $D_{(n+m_n-1)0}(x) \to D_{(n-1)m_n}(x) = C_{n-1}$ is a holomorphic diffeomorphism. Remember that $A_{(n-1)0}(x) = D_{(n-1)0}(x) \backslash \mathring{D}_{n0}(x)$ and $A_{(n-1)0}(0) = C_{n-1} \backslash \mathring{C}_n$. We have

$$\operatorname{mod}\big(A_{(n+m_n-1)0}(x)\big) = \operatorname{mod}\big(A_{(n-1)0}(0)\big).$$

Let $q_n = n + m_n - 1$. Then $q_N < q_{N+1} < \cdots < q_n < q_{n+1} < \cdots$. Thus

$$\begin{aligned}\operatorname{mod}(D_0(x) \setminus L_x) &\geq \sum_{n=N}^{\infty} \operatorname{mod}((A_{n0}(x)) \geq \sum_{n=N}^{\infty} \operatorname{mod}\big(A_{q_n 0}(x)\big) \\ &= \sum_{n=N}^{\infty} \operatorname{mod}\big(A_{(n-1)0}(0)\big) = \infty.\end{aligned}$$

This implies that $L_x = \{x\}$. ∎

The first column of the critical tableau $T(0) = (a_{nm}^0)_{n\geq 0, m\geq 0}$ is entirely 1's. If $T(0)$ has another column which is entirely 1's, that is, if there is an integer $m > 0$ such that $a_{im} = 1$ for all $i \geq 0$, then we call $T(0)$ a periodic critical tableau; the smallest such integer $m > 0$ is called its period.

Theorem 5.21 [YO4]. *The critical tableau $T(0)$ is periodic if and only if P_c is renormalizable.*

Proof. Suppose $T(0)$ is periodic. Let $n_1 > 0$ be the smallest integer such that $a_{in_1} = 1$ for all $i \geq 0$. Let $N \geq 0$ be the smallest integer such that $a_{ij} = 0$ for all $i \geq N$ and $0 < j < n_1$. For any $n \geq n_1 + N$, $P_c^{\circ n_1}$: $C_n \to C_{n-n_1}$ is a degree two proper holomorphic branch covering map. Thus $\{P^{\circ kn_1}(0)\}_{k=0}^{\infty}$ is contained in C_{n_1+N}. If $C_{n_1+N} \subset \mathring{C}_N$, then $P_c^{\circ n_1} : \mathring{C}_{n_1+N} \to \mathring{C}_N$ is a quadratic-like map with connected filled-in Julia set and $\mathring{C}_{n_1+N}$ is a renormalization about n_1. Thus P_c is renormalizable.

In general, let us consider a small open disk $D(\alpha)$ centered at the separate fixed point α of P_c such that

$$\overline{D}(\alpha) \subset D'(\alpha) = P_c^{\circ n_1}(D(\alpha))$$

and such that

$$D'(\alpha) \cap \{P_c^{\circ i}(0)\}_{i=0}^{n_1+N} = \emptyset.$$

Thicken C_0 and B_{0i} for $1 \le i \le q-1$ as follows. Suppose C_0 (respectively, B_{0i}) is bounded by two external rays R_{θ_1} and R_{θ_2} of angles θ_1 and θ_2. Let $\epsilon > 0$ be a small number such that

$$W_1 = \cup_{\theta_1-\epsilon<\theta<\theta_1+\epsilon}(R_\theta \setminus D_\alpha) \text{ and } W_2 = \cup_{\theta_2-\epsilon<\theta<\theta_2+\epsilon}(R_\theta \setminus D_\alpha)$$

are disjoint from $\{P_c^{\circ i}(0)\}_{i=0}^{n_1+N}$. Let

$$\tilde{C}_0 = (W_1 \cup C_0 \cup W_2 \cup D(\alpha)) \cap U_r \ \Big(\text{respectively, } \tilde{B}_{0i} = (W_1 \cup B_{0i} \cup W_2 \cup D(\alpha)) \cap U_r\Big)$$

where U_r is the domain bounded by the equipotential curve s_r. Let

$$\tilde{\eta}_0 = \{\tilde{C}_0, \tilde{B}_{01}, \ldots, \tilde{B}_{0(q-1)}\}$$

and let

$$\tilde{\eta}_n = P^{-n}(\tilde{\eta}_n) = \{\tilde{C}_n, \tilde{B}_{n1}, \ldots, \tilde{B}_{nk_n}\}$$

for $1 \le n \le n_1 + N$. The diffeomorphism $g = P_c^{-(n_1-1)} : C_N \to P_c(C_{n_1+N})$ can be extended to $\tilde{C}_N$. Let B' be the image of $\tilde{C}_N$ under g. Then $\tilde{C}_{n_1+N} = P_c^{-1}(B')$. Let $\mathring{\tilde{C}}_n$ denote the interior of $\tilde{C}_n$ for $0 \le n \le n_1 + N$. Then $\tilde{C}_{n_1+N} \subset \mathring{\tilde{C}}_N$. Thus

$$P_c^{\circ n_1} : \mathring{\tilde{C}}_{n_1+N} \to \mathring{\tilde{C}}_N$$

is a quadratic-like map and $\mathring{\tilde{C}}_{n_1+N}$ is a renormalization about n_1. This proves the "only if" part.

Now suppose P_c is renormalizable. Let U_1 be a renormalization about n_1, that is, (U_1, V_1, f_1) is a quadratic-like map with connected filled-in Julia set K_{f_1} where $f_1 = P_c^{\circ n_1}|U_1$ and $V_1 = f_1(U_1)$. The map f_1 has two fixed points β_1 and α_1 in U_1. Let

$$\alpha_1 \in \cdots \subseteq D_n(\alpha_1) \subseteq D_{n-1}(\alpha_1) \subseteq \cdots \subseteq D_1(\alpha_1) \subseteq D_0(\alpha_1)$$

be an α_1-end. There is a $D_k(\alpha_1)$ such that $K_{f_1} \subset D_k(\alpha_1)$ and $D_k(\alpha_1) = C_k$. Since

$$K_{f_1} \subseteq f_1^{-1}(U_1 \cap C_k) \subseteq U_1 \cap C_k,$$

then $P_c^{\circ n_1}$ sends C_{k+in_1} to $C_{k+(i-1)n_1}$ for all $i > 0$. Thus $T(0)$ is periodic. It is the "if" part. ■

We define a function τ on the set $\mathbf{N}$ of natural numbers by using the critical tableau $T(0) = (a^0_{st})_{s\ge0,t\ge0}$ as follows: $\tau(n) = m$ if $a^0_{(n-i)i} = 0$ for all $0 < i < n-m$ but $a^0_{m(n-m)} = 1$; if there is no such integer $m \geq 0$, then $\tau(n) = -1$.

If the critical tableau $T(0)$ is periodic, then there are integers $n_1 > 0$ and $N \geq 0$ such that $a_{in_1} = 1$ for all $i \geq 0$ and such that $a_{ij} = 0$ for all $i \geq N$ and all $0 < j < n_1$. Thus $\tau(n) = n - n_1$ for all $n \geq N + n_1$.

If the critical tableau $T(0)$ is non-recurrent, then there is the smallest integer $N \geq 0$ such that $a_{nm} = 0$ for all $n \geq N$ and all $m > 0$. Thus the image $\tau(\mathbf{N})$ is contained in the finite set $\{-1, 0, 1, \ldots, N-1\}$.

If the critical tableau $T(0)$ is not periodic and is recurrent, then every row $\{a^0_{st}\}_{t=0}^{\infty}$ of $T(0)$ has infinitely many 1's and every column $\{a^0_{st}\}_{s=0}^{\infty}$ except for the 0th-column has 0. An integer $n \geq 0$ is noble if for every entry a^0_{nk} such that $a^0_{nk} = 1$, we have $a^0_{(n+1)k} = 1$.

Lemma 5.8. *If the critical tableau $T(0)$ is not periodic and is recurrent, then the function τ satisfies the following properties:*

(i) *For any integer $m \geq 0$, $\tau^{-1}(m)$ is not empty.*

(ii) *If $m \geq 0$ is noble, then $\tau^{-1}(m)$ contains at least two different integers.*

(iii) *If $\tau(n) = m$ and if m is noble, then n is also noble.*

(iv) *If $\tau^{-1}(m)$ contains only one integer n, then n is noble.*

Proof. We prove **(i)** first. Consider any m^{th}-row in $T(0)$ for $m \geq 0$. Let $k > 0$ be the integer such that $a^0_{mi} = 0$ for all $0 < i < k$ and such that $a^0_{mk} = 1$. From **(T1)**, $a^0_{(m+k-i)i} = 0$ for all $0 < i < k$. Thus $\tau(m+k) = m$.

To prove **(ii)**, suppose $m \geq 0$ is noble. Let k be the same integer as that in the proof of **(i)**. Let $m_1 > m$ be the integer such that $a^0_{m_1 k} = 1$ and such that $a^0_{ik} = 0$ for all $i > m_1$. Consider $a^0_{(m_1-k)(2k)}$, $a^0_{(m_1-2k)(3k)}$, $\ldots$, and $a^0_{(m_1-(i-1)k)(ik)}$ where $m_1 - ik \leq m < m_1 - (i-1)k$. From the tableau rules **(T1)** and **(T3)**, $a^0_{(m_1-k+1)(2k)} = 0$, $a^0_{(m_1-2k+1)(3k)} = 0$, $\ldots$, $a^0_{(m_1-(i-1)k+1)(ik)} = 0$. If $m = m_1 - ik$, from the tableau rules **(T1)** and **(T2)**, $a^0_{m(ik)} = 1$. Since m is noble, $a^0_{(m+1)(ik)} = 1$. But from the tableau rules **(T1)** and **(T3)**, $a^0_{(m+1)(ik)} = 0$. The contradiction implies that $m > m_1 - ik$.

Now from the tableau rule **(T2)**, $a^0_{m(k+m_1-m)} = 0$. Let $k_1 > k + m_1 - m$ be the integer such that $a^0_{mi} = 0$ for all $k + m_1 - m < i < k_1$ and such that $a^0_{mk_1} = 1$. Then $a^0_{(m+k_1-i)i} = 0$ for all $0 < i < k_1$ but $a^0_{mk_1} = 1$. This says that $\tau(m + k_1) = m$.

To prove **(iii)**, suppose $\tau(n) = m$ where m is noble. For any $a^0_{nk} = 1$, since $a^0_{(n-i)i} = 0$ for all $0 < i < n-m$ but $a^0_{m(n-m)} = 1$ and since the tableau rules **(T1)** and **(T2)**, we have $a^0_{(n-i)(k+i)} = 0$ for all $0 < i < n-m$ but $a^0_{m(k+n-m)} = 1$. Since m is noble, $a^0_{(m+1)(n-m)} = 1$. Assume $a^0_{(n+1)k} = 0$.

From the tableau rule **(T3)**, $a^0_{(m+1)(k+n-m)} = 0$. This contradicts to that m is noble. Thus $a^0_{(n+1)k} = 1$. This means that n is noble.

Now we prove **(iv)**. Suppose $n > 0$ is the only integer such that $\tau(n) = m$. We first consider $a^0_{(m+1)(n-m)}$. If $a^0_{(m+1)(n-m)} = 0$, then we would have an integer $k > n - m$ such that $a^0_{mi} = 0$ for all $n - m < i < k$ and such that $a^0_{mk} = 1$. From the tableau rule **(T1)**, $a^0_{(m+k-i)i} = 0$ for all $0 < i < k$. This would imply that $\tau(m + k) = m$, which contradicts the assumption. Thus, $a^0_{(m+1)(n-m)} = 1$. If there is an entry $a^0_{nk_1} = 1$ with $a^0_{(n+1)k_1} = 0$ (where $k_1 \geq n - m$), from the tableau rules **(T1)** and **(T3)**, $a^0_{m(n-m+k_1)} = 1$ and $a^0_{(m+1)(n-m+k_1)} = 0$. Consider the smallest integer $k_2 > n - m + k_1$ such that $a^0_{mi} = 0$ for all $k_1 + n - m < i < k_2$ and such that $a^0_{mk_2} = 1$. From the tableau rule **(T1)**, $a^0_{(m+k_2-i)i} = 0$ for $k_1 + n - m < i < k_2$. So we can find another integer $n_0 \geq k_2 - k_1 + m > n$ such that $\tau(n_0) = m$. This would contradict the assumption. Therefore, n must be noble. ∎

Theorem 5.22 **[YO4]**. *Suppose $P_c(z) = z^2 + c$ is a recurrent quadratic polynomial. The critical tableau $T(0)$ is periodic if and only if L_0 contains more than one point.*

Proof. We use the same notation as in the proof of Theorem 5.21 and the proof of Lemma 5.8. Suppose $T(0)$ is periodic. Then for $n > N + n_1$

$$P_c^{\circ n_1} : C_{n+1} \to C_{n-n_1+1}$$

is a degree two proper holomorphic branch covering map. Replacing C_{n+1} by $\tilde{C}_{n+1}$ if it is necessary, we may assume that this map is a quadratic-like map. Since L_0 is the filled-in Julia set of this map, it contains more than one point. This is the "only if" part.

To prove the "if" part, suppose $T(0)$ is not periodic. We prove that L_0 contains only one point. Since P_c is recurrent, there are infinitely many 1's in every row of $T(0)$, that is, $T(0)$ is recurrent. Consider the first partition

$$\eta_1 = \{C_1, B_{11}, \ldots, B_{1(q-1)}, B_{01}, \ldots B_{0(q-1)}\},$$

where $B_{0i} = B_{0,i}$ for all $1 \leq i < q$ and where $B_{1i} \subseteq C_0$ and $P_c(B_{1i}) = B_{0,i}$ for all $1 \leq i < q$. (Remember that $\eta_0 = \{C_0, B_{0,1}, \ldots, B_{0,q-1}\}$ is the original partition.) Let $c(n) = P_c^{\circ n}(0)$. If the critical orbit $CO = \{c(n)\}_{n=0}^{\infty}$ is contained in the union

$$C_1 \cup B_{01} \cup \ldots \cup B_{0(q-1)},$$

then $T(0)$ is periodic of period q. Hence there must be one critical value $c(n)$ in $B_{11} \cup \cdots \cup B_{1(q-1)}$. Let $c(n)$ be in B_{1i}. The annulus $A_{0n}(0) = C_0 \setminus \mathring{B}_{1i}$ is non-degenerate. Pull back $A_{0n}(0)$ by P_c along $A_{i(n-i)}(0)$ for all $0 \leq i \leq n$; we get a non-degenerate annulus $A_{n0}(0)$.

Now consider $\tau^{-k}(n)$. For each m in $\tau^{-k}(n)$,

$$\operatorname{mod}(A_{m0}(0)) \geq \frac{\operatorname{mod}(A_{n0}(0))}{2^k}.$$

If the number of $\tau^{-k}(n)$ is greater than or equal to 2^k for every $k > 0$, then

$$\begin{aligned}\operatorname{mod}(C_0 \setminus L_0) &\geq \sum_{m=1}^{\infty} \operatorname{mod}(A_{m0}(0)) \geq \sum_{k=1}^{\infty} \sum_{m \in \tau^{-k}(n)} \operatorname{mod}(A_{m0}(0)) \\ &\geq \sum_{k=1}^{\infty} \operatorname{mod}(A_{n0}(0)) = \Big(\operatorname{mod}(A_{n0}(0))\Big) \sum_{k=1}^{\infty} 1 = \infty.\end{aligned}$$

So $L_0 = \{0\}$.

If there is an integer $k > 0$ such that the number of $\tau^{-k}(n)$ is less than 2^k, then there are pre-images $m > q$ of n under iterates of τ such that m is the only pre-image of q under τ. From **(iii)** in Lemma 5.8, m is noble. Hence $\tau^{-k}(m)$ are noble and contain at least 2^k different integers for all $k \geq 1$. Moreover

$$\operatorname{mod}(A_{p0}(0)) = \frac{\operatorname{mod}(A_{m0}(0))}{2^k}$$

for every p in $\tau^{-k}(m)$. Therefore,

$$\begin{aligned}\operatorname{mod}(C_0 \setminus L_0) &\geq \sum_{k=1}^{\infty} \operatorname{mod}(A_{k0}(0)) \geq \sum_{k=1}^{\infty} \sum_{p \in \tau^{-k}(m)} \operatorname{mod} \\ &\qquad (A_{p0}(0)) \geq \sum_{k=1}^{\infty} \operatorname{mod}(A_{m0}(0)) = \Big(\operatorname{mod}(A_{m0}(0))\Big) \sum_{k=1}^{\infty} 1 = \infty.\end{aligned}$$

Again we have $L_0 = \{0\}$. This completes the "if" part. ■

Theorem 5.23 [YO4]. *If $P_c(z) = z^2 + c$ is a recurrent non-renormalizable quadratic polynomial, then its filled-in Julia set K_c $(= J_c)$ is locally connected.*

Proof. Let α be the separate fixed point of P_c. Construct the two-dimensional Yoccoz puzzle for P_c (with respect to $R_0(\alpha)$). For any x in K_c, let

$$x \in \cdots \subseteq D_n(x) \subseteq D_{n-1}(x) \subseteq \cdots \subseteq D_1(x) \subseteq D_0(x)$$

be an x-end. Since P_c is recurrent and non-renormalizable $T(0)$ is recurrent and is not periodic. Lemma 5.7 and the proof of Theorem 5.22 imply that the diameter $d(D_n(x))$ tends to zero as n goes to infinity.

If x is not a preimage of α under any iterate of P_c, then x is an interior point of $D_n(x)$ for all $n \geq 0$. From Lemma 5.5, $\{D_n(x)\}$ is a basis of connected neighborhoods at x. If x is a preimage of α under some iterate of P_c, then there are q different x-ends,

$$x \in \cdots \subseteq D_{i,n}(x) \subseteq D_{i,(n-1)}(x) \subseteq \cdots \subseteq D_{i,1}(x) \subseteq D_{i,0}(x)$$

where q is the period of the external rays landing at α. Let $\tilde{D}_n(x) = \cup_{i=1}^{q} D_{i,n}(x)$. Then x is an interior point of $\tilde{D}_n$. Since $K_c \cap D_{1,n}(x)$, ..., $K_c \cap D_{q,n}(x)$ have a common point x, from Lemma 5.5, $K_c \cap \tilde{D}_n(x)$ is connected. So $\{\tilde{D}_n(x)\}_{n=0}^{\infty}$ is a basis of connected neighborhoods at x. ∎

From Theorem 5.15, all arguments in this section apply to a quadratic-like map. Suppose that (U, V, f) is a quadratic-like map and that its filled-in Julia set K_f is connected. Suppose two fixed points β and α of f are repelling. Let β be the non-separate fixed point of f, that is, $K_f \setminus \{\beta\}$ is connected, and let α be the separate fixed point of f, that is, $K_f \setminus \{\alpha\}$ is disconnected. Since (U, V, f) is hybrid equivalent to a quadratic polynomial P_c, there is a quasiconformal mapping H defined on V such that

$$H \circ f = P_c \circ H$$

on U. We call $e_{\theta,f} = H^{-1}(e_\theta \cap H(U))$ the external ray of angle θ of f where e_θ is the external ray of P_c of angle θ. (The construction of $e_{\theta,f}$ depends on H but we may think it is fixed.)

Two points $H(\beta)$ and $H(\alpha)$ are non-separate and separate fixed points of P_c, respectively. Suppose Γ is the union of a cycle of periodic external rays landing at $H(\alpha)$. Let $\Gamma' = H^{-1}(\Gamma \cap H(U))$. The set Γ' cuts the domain U into q domains. Each of them contains points in the filled-in Julia set K_c. Let C_0 be the domain containing 0 and let $B_{0,i}$ be the domain containing $f^{\circ i}(0)$ for $1 \leq i < q$. The partition

$$\eta_0 = \{C_0, B_{0,1}, \ldots, B_{0,q-1}\}$$

is called the original partition for f (with respect to Γ'). Let $\Gamma'_n = f^{-n}(\Gamma')$ and $U_n = f^{-n}(U)$. Then Γ'_n cuts U_n into finitely many domains. Let C_n be the domain containing 0 and $B_{n,i}$ for $1 \leq i \leq k_n$ be others. Then

$$\eta_n = \{C_n, B_{n,1}, \ldots, B_{n,k_n}\}$$

is called the nth-partition for f (with respect to Γ'). We use $f^{-n}(\eta_0)$ to denote η_n, i.e., $\eta_n = f^{-n}(\eta_0)$, for $1 \leq n < \infty$. We have that $f(C_n)$ and $f(B_{n,i})$ for $1 \leq i \leq k_n$ are in η_{n-1} for $n > 0$ (set $k_0 = q-1$). We call $\eta = \{\eta_n\}_{n=0}^{\infty}$ the two-dimensional Yoccoz puzzle of f (with respect to Γ'). Let

$$\Lambda = \cup_{n=0}^{\infty} f^{-n}(\alpha).$$

Let

$$L_0 = \cap_{n=0}^{\infty} C_n$$

be the connected component of $K_f \setminus \Lambda$ containing 0. (The sets Λ and L_0 are independent of H and Γ'.) We state Theorems 5.22 and 5.23 in the following form.

Theorem 5.24 [YO4]. *Suppose (U, V, f) is a recurrent quadratic-like map. Then (U, V, f) is renormalizable if and only if L_0 contains more than one point. Moreover, if (U, V, f) is non-renormalizable, then any connected component of $K_f \setminus \Lambda$ consists of only one point and K_f is locally connected.*

Proof. All properties in the theorem are invariant under a homeomorphism. Thus the theorem follows from Theorems 5.22 and 5.23. ∎

5.7. Infinitely Renormalizable Quadratic Julia Sets

Suppose (U, V, f) is a renormalizable quadratic-like map with connected filled-in Julia set K_f. Let $\eta = \{\eta_n\}_{n=0}^{\infty}$ be the two-dimensional Yoccoz puzzle for f. From the previous section, the critical tableau $T(0) = (a_{nm}^0)_{n\geq 0, m\geq 0}$ is periodic of period n_1. Let

$$0 \in \cdots \subseteq C_n \subseteq C_{n-1} \subseteq \cdots \subseteq C_1 \subseteq C_0$$

be the critical end. There is an integer $N > 0$ such that $a_{ij}^0 = 0$ for all $i \geq N$ and all $0 < j < n_1$. Let $f_0 = f^{\circ n_1}|\mathring{C}_{N+n_1}$. Then $f_0 : \mathring{C}_{N+n_1} \to \mathring{C}_N$ is a proper, holomorphic, branch cover of degree two. Assume $C_{N+n_1} \subset \mathring{C}_N$. (Otherwise, we can replace C_n with $\tilde{C}_n$ (see Theorem 5.21).) Then $(\mathring{C}_{N+n_1}, \mathring{C}_N, f_0)$ is a quadratic-like map and its filled-in Julia set is $L_0 = \cap_{n=0}^{\infty} C_n$.

Theorem 5.25 [SU6,MC1,JI16]. *Suppose (U, V, f) is a renormalizable quadratic-like map with connected filled-in Julia set K_f. For any renormalization U_1 about n_1, let $f_1 = f^{\circ n_1}|U_1$ and let $V_1 = f_1(U_1)$. Then the filled-in Julia set K_{f_1} (or the Julia set J_{f_1}) of (U_1, V_1, f_1) is always L_0 (or ∂L_0).*

Proof. Let $U' = \mathring{C}_{N+n_1} \cap U_1$ and let U'' be the connected component of U' containing 0. Let $f_2 = f^{\circ n_1}|U''$ and $V'' = f_2(U'')$. Then $f_2 : U'' \to V''$ is a degree two branch covering. It is also proper because

$$f_2^{-1}(K) = f_0^{-1}(K) \cap f_1^{-1}(K)$$

for any compact set K of V''. Since $C_{N+n_1} \subset \mathring{C}_N$ and $\overline{U}_1 \subset V_1$, then

$$C_{N+n_1} \cap \overline{U}_1 \subset \mathring{C}_N \cap V_1.$$

Thus $\overline{U''} \subset V''$ because they are the connected components of U' and $\mathring{C}_N \cap V_1$ containing 0. Both U'' and V'' are simply connected and isomorphic to a disc because they are intersections of simply connected domains each of which is isomorphic to a disc. Therefore, (U'', V'', f_2) is a quadratic-like map. Let K_{f_2} be the filled-in Julia set of (U'', V'', f_2). Since a filled-in Julia set is completely invariant and since 0 is in U'', the two inverse images of 0 under f_2 are in K_{f_2}. But these two points are also inverse images of 0 under f_0 and under f_1. Therefore, they are both in L_0 and in K_{f_1}. Using this argument, the set Ξ of all inverse images of 0 under iterates of f_2 is contained in K_{f_2} and is also contained in L_0 and in K_{f_1}. Therefore,

$$K_{f_1} = K_{f_2} = L_0 \qquad (\text{ or } J_{f_1} = J_{f_2} = \partial L_0 \ )$$

because each of ∂L_0, $J_{f_1} = \partial K_{f_1}$, and $J_{f_2} = \partial K_{f_2}$ is the limit set of Ξ. ∎

As we saw in §5.5, the definition of the filled-in Julia set of a renormalization about n_1 *a priorily* depends on choices of domains in renormalization. But the renormalized filled-in Julia set is actually canonical; it is independent of choices of domains in renormalization and is the intersection of all the critical pieces in the two-dimensional Yoccoz puzzle from Theorem 5.25.

Suppose (U_1, V_1, f_1) is a quadratic-like map with connected filled-in Julia set K_1. We call K_1 (or J_1) a quadratic filled-in Julia set (or quadratic Julia set). It is renormalizable if the corresponding quadratic-like map is renormalizable. Suppose both fixed points of f_1 are repelling. Let β_1 and α_1 be the non-separate and separate fixed points of f_1, i.e., $K_1 \setminus \{\beta_1\}$ is still connected and $K_1 \setminus \{\alpha_1\}$ is disconnected. Let $\Lambda_1 = \cup_{n=0}^{\infty} f_1^{-n}(\alpha_1)$. Let $K_2 = L_0$ be the connected component of $K_1 \setminus \Lambda_1$ containing 0. From Theorem 5.24, K_1 is renormalizable if and only if K_2 contains more than one point. The quadratic filled-in Julia set K_2 is called the renormalization of K_1.

Inductively, let K_i be the renormalization of K_{i-1}. Let $f_i = f_{i-1}^{\circ n_{i-1}}$ for $i \geq 2$, where n_{i-1} is the period of the critical tableau of the two-dimensional Yoccoz puzzle for $(U_{i-1}, V_{i-1}, f_{i-1})$. Let β_i and α_i be the non-separate and the separate fixed points of f_i, i.e., $K_i \setminus \{\beta_i\}$ is still connected and $K_i \setminus \{\alpha_i\}$ is disconnected. Let $\Lambda_i = \cup_{n=0}^{\infty} f_i^{-n}(\alpha_i)$ and let K_{i+1} be the connected component of $K_i \setminus \Lambda_i$ containing 0. Then K_i is renormalizable if and only if K_{i+1} contains more than one point. Here K_i, for $i > 1$, is called the ith-renormalization of K_1. From Theorem 5.24 and Theorem 5.25, we have an equivalent definition of an infinitely renormalizable quadratic-like map.

Definition 5.5. *Suppose that (U_1, V_1, f_1) is a recurrent quadratic-like map and that K_1 is its filled-in Julia set. The quadratic filled-in Julia set K_1 is finitely renormalizable if there is an integer $m \geq 1$ such that K_1, ..., K_m contains more than one point and such that K_{m+1} contains only the point 0. In other words, K_1 is infinitely renormalizable if all K_i contain more than one point.*

Theorem 5.24 is generalized as

Theorem 5.26 [YO4]. *Suppose that (U_1, V_1, f_1) is a recurrent quadratic-like map and that K_1 is its filled-in Julia set. If K_1 is finitely renormalizable, then it is locally connected.*

Proof. Let m be the integer in Definition 5.5. Let α_m be the separate fixed point of f_m. Let Γ_m be a cycle of periodic external rays of f_1 landing at α_m (refer to the end of §5.6). Using Γ_m, we can construct the two-dimensional

Yoccoz puzzle for f_1 (with respect to Γ_m): let η_0^m be the set consisting of the closures of the connected components of $V_1 \setminus \Gamma_m$. Let $\eta_n^m = f_1^{-n}(\eta_0^{m_i})$ for $n \geq 1$. Let C_n^m be the member of η_n^m containing 0. Since f_m is non-renormalizable, we use a proof similar to that of Theorem 5.22 to show that $\sum_{n=0}^{\infty} \mathrm{mod}(A_{n0}^m(0)) = \infty$, where $A_{n0}^m(0) = C_n^m \setminus \mathring{C}_{n+1}^m$. Applying Lemma 5.7, for every x-end

$$x \in \cdots \subseteq D_n^m(x) \subseteq D_{n-1}^m(x) \subseteq \cdots \subseteq D_1^m(x) \subseteq D_0^m(x),$$

$L_x = \cap_{n=0}^{\infty} D_n^m(x)$ contains only x. By using a similar argument to the proof of Theorem 5.23, we can now show that K_1 is locally connected. ∎

Now let us consider an infinitely renormalizable quadratic-like map

$$(U_1, V_1, f_1).$$

Let K_1 be the filled-in Julia set of f_1. Let K_i be the ith-renormalization of K_1. (Note that in this case, K_1 equals the Julia set J_1 of f_1.) Then $\mathcal{K} = \{K_i\}_{i=1}^{\infty}$ is a sequence of renormalizations of K_1. Let $\{(U_i, V_i, f_i)\}_{i=1}^{\infty}$ be a sequence of renormalizations with filled-in Julia set K_i where $f_i = f_{i-1}^{\circ n_{i-1}}$ and where n_{i-1} is the period of the critical tableau of the two-dimensional Yoccoz puzzle for $(U_{i-1}, V_{i-1}, f_{i-1})$, for $i \geq 2$. Here U_i is a renormalization of $(U_{i-1}, V_{i-1}, f_{i-1})$. We describe (U_1, V_1, f_1) as $(n_1, n_2, \ldots)$-infinitely renormalizable. Let $c(n) = f_1^{\circ n}(0)$. The critical orbit of f_1 is $CO = \{c(n)\}_{n=0}^{\infty}$. Let $GCO = \cup_{k=0}^{\infty} \cup_{n=0}^{\infty} f_1^{-k}(c(n))$ be the grand critical orbit of f_1.

Definition 5.6. *An infinitely renormalizable quadratic-like map (U_1, V_1, f_1) has a priori complex bounds if there are a constant $\lambda > 0$ and a sequence of renormalizations $\{(U_{i_k}, V_{i_k}, f_{i_k})\}_{k=1}^{\infty}$ of f_1 such that*

$$\mathrm{mod}(V_{i_k} \setminus \overline{U}_{i_k}) \geq \lambda$$

for all $k \geq 1$.

Theorem 5.27 [JI16]. *Suppose (U_1, V_1, f_1) is an infinitely renormalizable quadratic-like map having the a priori complex bounds. Its filled-in Julia set K_1 is locally connected at every point in GCO.*

Proof. Suppose, without loss of generality, that $\{U_i, V_i, f_i\}_{i=1}^{\infty}$ is the sequence of renormalizations in Definition 5.6. Let $\lambda > 0$ be the constant in Definition 5.6. Then $\{U_i\}_{i=1}^{\infty}$ is a sequence of nested domains containing 0. Consider the annulus $A_i = U_i \setminus \overline{U}_{i+1}$ for $i \geq 1$. For each $i \geq 1$, let $cv_i = f_i(0)$.

Since $f_{i+1} = f_i^{\circ n_i} : U_{i+1} \to V_{i+1}$ is quadratic-like, $cv_i \notin f_i^{\circ 2}(U_{i+1})$. Let γ_i be a curve in $V_i \setminus f_i^{\circ 2}(U_{i+1})$ connecting cv_i and a point on the boundary of V_i such that $\gamma_i \cap f_i^{-1}(\gamma_i) = \emptyset$. Let $0 \in U_i' \subset U_i$ be the connected component of the pre-image of $V_i \setminus \gamma_i$ under $f_i^{\circ 2}$. Then $f_i^{\circ 2} : U_i' \to V_i \setminus \gamma_i$ is a degree two branch covering. Moreover, $f_i : U_i' \to f_i(U_i') \subset U_i$ is also a degree two branch covering map. Thus $f_i : U_i \setminus \overline{U'}_i \to V_i \setminus \overline{f(U_i')}$ is a degree two covering map. This implies that

$$\mathrm{mod}(U_i \setminus \overline{U'}_i) = \frac{1}{2}\mathrm{mod}(V_i \setminus \overline{f_i(U_i')}).$$

But $V_i \setminus \overline{U}_i$ is a sub-annulus of $V_i \setminus \overline{f_i(U_i')}$. So

$$\mathrm{mod}(U_i \setminus \overline{U'}_i) \geq \frac{1}{2}\mathrm{mod}(V_i \setminus \overline{U}_i) > \frac{\lambda}{2}.$$

Remember that U_{i+1} is the domain of the renormalization

$$f_{i+1} = f_i^{n_i} : U_{i+1} \to V_{i+1}$$

where $n_i \geq 2$. We have $U_{i+1} \subset U_i'$. Hence $U_i \setminus \overline{U'}_i$ is a sub-annulus A_i. So $\mathrm{mod}(A_i) > \lambda/2$. Let $A_\infty = \cap_{i=1}^{\infty} U_i$. Since $U_1 \setminus A_\infty = \cup_{i=1}^{\infty} A_i$,

$$\mathrm{mod}(U_1 \setminus A_\infty) \geq \sum_{i=1}^{\infty} \mathrm{mod}(A_i) = \infty.$$

Thus, $A_\infty = \{0\}$. This implies that the diameter $d(U_i)$ tends to 0 as i goes to infinity.

Let α_i be the separate fixed point of f_i. Let Γ_i be a cycle of periodic external rays of f_1 landing at α_i (refer to the end of §5.6). Let η_0^i be the set consisting of the closures of the connected components of $U_1 \setminus \Gamma_i$ and let $\eta_n^i = f_1^{-n}(\eta_1^i)$ for $n \geq 1$. Then $\eta^i = \{\eta_n^i\}_{n=0}^{\infty}$ is a puzzle for f_1. Let C_n^i be the member of η_n^i containing 0. Consider the corresponding critical end

$$0 \in \cdots \subseteq C_n^i \subseteq C_{n-1}^i \subseteq \cdots \subseteq C_1^i \subseteq C_0^i$$

and the corresponding critical tableau $T^i(0) = (a_{nm}^0(i))_{n\geq 0, m\geq 0}$. Since (U_1, V_1, f_1) is infinitely $(n_1, n_2, \ldots)$-renormalizable, $T^i(0)$ is periodic of period $m_i = \prod_{j=1}^{i} n_j$. There is an integer $N > 0$ such that $a_{nm}^0(i) = 0$ for $n \geq N$ and $0 < i < m_i$. Thus

$$f_1^{\circ m_i} = f_i^{\circ n_i} : \mathring{C}_{m_i+N}^i \to \mathring{C}_N^i$$

is a degree two proper holomorphic map. We may assume that $C^i_{m_i+N} \subset \mathring{C}^i_N$ (otherwise, we can modify C^i_n as the proof of Theorem 5.21). Therefore,

$$f_{i+1} = f_1^{\circ m_i} = f_i^{\circ n_i} : \mathring{C}^i_{m_i+N} \to \mathring{C}^i_N$$

is a quadratic-like map. Since

$$K_{i+1} = \cap_{j=0}^{\infty} f_1^{-jm_i}(C^i_N) = \cap_{j=0}^{\infty} C^i_{jm_i+N},$$

there is a $C^i_{k(i)}$ contained in U_i. The diameter $d(C^i_{k(i)})$ of $C^i_{k(i)}$ tends to zero as i goes to infinity. From Lemma 5.5, $C^i_{k(i)} \cap K_1$ is connected. So $\{C^i_{k(i)}\}_{i=1}^{\infty}$ is a basis of connected neighborhoods at 0.

For any $x = f_1^{\circ n}(0)$, consider $\{D_{i,n}(x) = f_1^{\circ n}(C^i_{k(i)})\}_{i=1}^{\infty}$. It is a basis of connected neighborhoods at x. For any y in $f_1^{-m}(x)$ (where $x = f_1^{\circ n}(0)$ and where 0 is not in $f^{-i}(x)$ for any $0 < i \leq m$), there is an open neighborhood W of y such that $f_1^{\circ m} : W \to f^{\circ m}(W)$ is a homeomorphism. Let g be its inverse. Then $\{g(D_{i,n}(x))\}_{i=1}^{\infty}$ is a basis of connected neighborhoods at y. ∎

Suppose (U_1, V_1, f_1) is an infinitely $(n_1, n_2, \ldots)$-renormalizable quadratic-like map. Let $\mathcal{K} = \{K_i\}_{i=1}^{\infty}$ be the sequence of the renormalizations constructed in the paragraph before Definition 5.5. We call the puzzle and the grid

$$\{\{\eta^i_n\}_{n=0}^{\infty}\}_{i=1}^{\infty} \qquad \text{and} \qquad \{T^i(0)\}_{i=1}^{\infty}$$

constructed in Theorem 5.27 the three-dimensional Yoccoz puzzle (with respect to $\{\Gamma_i\}_{i=1}^{\infty}$) and the corresponding three-dimensional critical tableau for (U_1, V_1, f_1).

Definition 5.7. *An infinitely $(n_1, n_2, \ldots)$-renormalizable quadratic-like map (U_1, V_1, f_1) is unbranched if there are a constant $\lambda > 0$ and a sequence of domains $\{W_k\}_{k=1}^{\infty}$ such that $W_k \supset K_{i_k}$, such that*

$$mod(W_k \setminus K_{i_k}) \geq \lambda,$$

and such that $W_k \setminus K_{i_k}$ contains no critical values of f_1.

Theorem 5.28 [JI16]. *Suppose (U_1, V_1, f_1) is an infinitely renormalizable unbranched quadratic-like map having the a priori complex bounds. Then its filled-in Julia set K_1 is locally connected.*

Proof. Suppose, without loss of generality, that $i_k = i$ in Definition 5.7, and that $\{U_i, V_i, f_i\}_{i=1}^{\infty}$ is a sequence of renormalizations in Definition 5.6. Let $\lambda > 0$ be a constant satisfying Definitions 5.6 and 5.7.

Let $\{\{\eta_n^i\}_{n=0}^{\infty}\}_{i=1}^{\infty}$ be the three-dimensional Yoccoz puzzle for (U_1, V_1, f_1). Let $\{C_{k(i)}^i\}_{i=1}^{\infty}$ be the basis of connected neighborhoods constructed in Theorem 5.27. By choosing $k(i)$ large enough and by modifying W_i, we can have

$$\frac{\lambda}{2} \leq \mathrm{mod}(W_i \setminus C_{k(i)}^i) \leq 2\lambda$$

for all $i \geq 1$.

If $x = 0$, Theorem 5.27 says that K_1 is locally connected at x. For each $x \neq 0$ in K_1, there are two cases: either (1) x is non-recurrent, which means that there is an integer $i \geq 1$ such that $\{f_1^{\circ n}(x)\}_{n=0}^{\infty} \cap C_{k(i)}^i = \emptyset$; or else (2) x is recurrent.

In case (1), we prove it by applying the results in hyperbolic geometry. Let x be a non-recurrent point. Then there is a $C_{k(i)}^i$ such that the orbit $O(x) = \{f_1^{\circ n}(x)\}_{n=0}^{\infty}$ is disjoint from the interior of $C_{k(i)}^i$. Set $N = k(i)$. Let

$$\eta_N^i = \{C_N^i, B_{N,1}, B_{N,2}, \ldots, B_{N,q}\}$$

be the Nth-partition in the ith-puzzle $\{\eta_n^i\}_{n=0}^{\infty}$, here q is the number of elements in η_N^i. Assume that $B_{N,1}$ contains $f_1(0)$. The orbit $f_1(O(x))$ is disjoint with $\mathring{B}_{N,1}$ since $O(x)$ is disjoint with $\mathring{C}_{N+1}^i$. Let us thicken $B_{N,j}$ to an open simply connected domain $\tilde{B}_{N,j}$ such that $B_{N,j} \subset \tilde{B}_{N,j}$ and such that $f_1(0)$ is not in $\tilde{B}_{N,j}$ for $2 \leq j \leq q$. Every $f_1|\tilde{B}_{N,j}$, $2 \leq j \leq q$, has two inverse branches g_{j0} and g_{j1}. Consider $\tilde{B}_{N,j}$ as a hyperbolic Riemann surface with the hyperbolic distance $d_{H,j}$ for every $2 \leq j \leq q$. Then if g_{js}, for $2 \leq j \leq q$ and $s = 0$ or 1, sends $B_{N,j}$ into $B_{N,j'}$ for some $2 \leq j' \leq q$, then it strictly contracts these hyperbolic distances; more precisely, there is a constant $0 < \lambda < 1$ such that $d_{H,j'}(g_{js}(x), g_{js}(y)) < \lambda d_{H,j}(x,y)$ for x and y in $B_{N,j}$ and for $s = 0$ and 1. Let

$$x \in \cdots \subseteq D_n^i(x) \subseteq D_{n-1}^i(x) \subseteq \cdots \subseteq D_1^i(x) \subseteq D_0^i(x)$$

be any x-end in the ith-puzzle. Then $f_1^{\circ m}(D_n^i(x))$ for $n - m > N$ is in $B_{N,j}$ for some $2 \leq j \leq q$. Therefore, there is a constant $C > 0$ such that for any $D_n^i(x)$ and for any $n > N$,

$$d(D_n^i(x)) = \max_{y,z \in D_n^i(x)} |y - z| \leq C\lambda^{n-N}.$$

Thus $d(D_n^i(x))$ tends to zero as n goes to infinity. Following the proof of Theorem 5.23, we can find a basis of connected neighborhoods at x. Therefore, K_1 is locally connected at x.

In case (2), $f_1^{\circ n}(x)$ enters $C^i_{k(i)}$ infinitely many times. For each $i \geq 1$ such that $x \notin C^i_{k(i)}$, consider the ith puzzle $\eta^i = \{\eta^i_n\}_{n=0}^{\infty}$. For the corresponding x-end,

$$x \in \cdots \subseteq D^i_n(x) \subseteq D^i_{n-1}(x) \subseteq \cdots \subseteq D^i_1(x) \subseteq D^i_0(x),$$

let $T^i(x) = (a_{nm}(i))_{n\geq 0, m\geq 0}$ be the corresponding tableau. Let q_i be the integer such that $a_{k(i)t} = 0$ for all $0 \leq t < q_i$ and such that $a_{k(i)q_i} = 1$. Let $p_i = k(i) + q_i$. Then $f_1^{\circ q_i} : D^i_{p_i}(x) \to C^i_{k(i)}$ is a proper holomorphic diffeomorphism. Let

$$g_{i,x} : C^i_{k(i)} \to D^i_{p_i}(x)$$

be its inverse. Since there is no critical value of f_1 in $W_i \setminus C^i_{k(i)}$, $g_{i,x}$ can be extended to a proper holomorphic diffeomorphism on W_i which we still denote as $g_{i,x}$.

From the proof of the previous theorem, the diameter $d(C^j_{k(j)})$ tends to zero as j goes to infinity. So does the diameter $d(W_j)$. For each $i \geq 1$ such that $x \notin C^i_{k(i)}$, we can find an integer $j = j(i) > i$ such that $W_j \subset C^i_{k(i)}$. Let $x_i = f_1^{\circ q_i}(x)$ and $x_j = f_1^{\circ q_j}(x) = f_1^{\circ(q_j - q_i)}(x_i)$. Consider the jth-puzzle $\eta^j = \{\eta^j_n\}_{j=1}^{\infty}$. Let

$$x_i \in \cdots \subseteq D^j_n(x_i) \subseteq D^j_{n-1}(x_i) \subseteq \cdots \subseteq D^j_1(x_i) \subseteq D^j_0(x_i)$$

be the corresponding x_i-end and let $T^j(x_i) = (b_{nm}(j))_{n\geq 0, m\geq 0}$ be the corresponding tableau. Then one can check that $f_1^{\circ(q_j - q_i)}$ is a proper holomorphic diffeomorphism from $D^j_{k(j)+q_j-q_i}(x_i)$ to $C^j_{k(j)}$. In other words, $b_{k(j)t} = 0$ for $0 \leq t < q_j - q_i$, but $b_{k(j)(q_j - q_i)} = 1$. Let

$$g_{ij} : C^j_{k(j)} \to D^j_{k(j)+q_j-q_i}(x_i)$$

be the inverse of $f_1^{\circ(q_j - q_i)} : D^j_{k(j)+q_j-q_i}(x_i) \to C^j_{k(j)}$. Then g_{ij} can be extended to W_j since $W_j \setminus C^j_{k(j)}$ contains no critical values of f_1. Since $C^i_{k(i)}$ is bounded by external rays landing at some pre-images of α_i under iterations of f_1 and is a part of an invariant net under f_1, $W_{ij} = g_{ij}(W_j)$ is contained in $C^i_{k(i)}$. Thus

$$\operatorname{mod}(W_i \setminus W_{ij}) \geq \frac{\lambda}{2}.$$

Consider $X_i = g_{i,x}(W_i)$ and $X_j = g_{j,x}(W_j) = g_{i,x}(W_{ij})$. Then

$$\operatorname{mod}(X_i \setminus X_j) \geq \frac{\lambda}{2}.$$

Therefore, inductively, we can find a sequence of domains $\{X_{i_s}\}_{s=1}^{\infty}$ such that

$$\operatorname{mod}(X_{i_s} \setminus X_{i_{s+1}}) \geq \frac{\lambda}{2}$$

for $s \geq 1$. Thus the diameter of X_{i_s} tends to zero as s goes to infinity. For each puzzle $\eta^{i_s} = \{\eta_n^{i_s}\}_{n=0}^{\infty}$, consider both the x-end,

$$x \in \cdots \subseteq D_n^{i_s}(x) \subseteq D_{n-1}^{i_s}(x) \subseteq \cdots \subseteq D_1^{i_s}(x) \subseteq D_0^{i_s}(x)$$

and the corresponding tableau $T^{i_s}(x) = (a_{nm}(i_s))_{n\geq 0, m\geq 0}$. For each $k(i_s)$, there is an integer q_{i_s} such that $a^i_{k(i_s)t} = 0$ for $0 \leq t < q_{i_s}$ and such that $a^i_{k(i_s)q_{i_s}} = 1$. Then for $p_{i_s} = k(i_s) + q_{i_s}$, $f_1^{\circ q_{i_s}} : D_{p_{i_s}}(x) \to C_{k(i_s)}$ is a proper holomorphic diffeomorphism. This implies that

$$D_{p_{i_s}}^{i_s}(x) \subseteq X_{i_s}.$$

So the diameter $d(D_{p_{i_s}}^{i_s}(x))$ tends to zero as s goes to infinity. From Lemma 5.5, $\{D_{p_{i_s}}^{i_s}(x)\}_{s=1}^{\infty}$ forms a basis of connected neighborhoods at x. ∎

Remark 5.7. Theorems 5.25, 5.27, and 5.28 are first proved in [JI13]. The self-contained discussion of these theorems is given in [JI16].

5.8. On Sullivan's Sector Theorem

Let $I = [0,1]$ be the closed unit interval. Let $\mathcal{E}_0$ be the set of all functions G such that

(1) $G : I_G \supseteq I \to G(I_G)$ is a homeomorphism and $G(0) = 0$, $G(1) = 1$, and

(2) G can be extended to be a schlicht function g on $\mathbf{C}_G = (\mathbf{C} \setminus \mathbf{R}) \cup \mathring{I}_G$ preserving upper- and lower-half planes.

We assume I_G is the maximum interval satisfies **(1)** and **(2)** for each G in $\mathcal{E}_0$ and call it the definition interval of G. We do not distinguish g and G anymore. Take

$$S_\gamma(z) = r^{\frac{1}{\gamma}} e^{\frac{\theta}{\gamma} i} : \mathbf{C} \setminus \{x < 0\} \to \mathbf{C}$$

as the standard γ-root where $z = re^{\theta i}$ for $r > 0$, $-\pi < \theta < \pi$, and $\gamma > 1$. For every $a \leq 0$, we call $L_a(z) = ES_\gamma(z - a) + F$ a γ-root at a where $E = 1/\left((1-a)^{\frac{1}{\gamma}} - (-a)^{\frac{1}{\gamma}}\right)$ and $F = -(-a)^{\frac{1}{\gamma}} E$ are determined by $L_a(0) = 0$ and $L_a(1) = 1$. Then L_a is an element in $\mathcal{E}_0$ whose definition interval is $[a, \infty)$.

Suppose that L_a is a γ-root at a and that G is an element in $\mathcal{E}_0$ whose definition interval is I_G. We say that L_a and G are compatible if $[a, 1] \subset G(I_G)$. For a compatible pair L_a and G, let $a' = G^{-1}(a)$, $J = [a', 1]$, $L \cup R = I_G \setminus J$, and $b = \min\{|L|, |R|\}$.

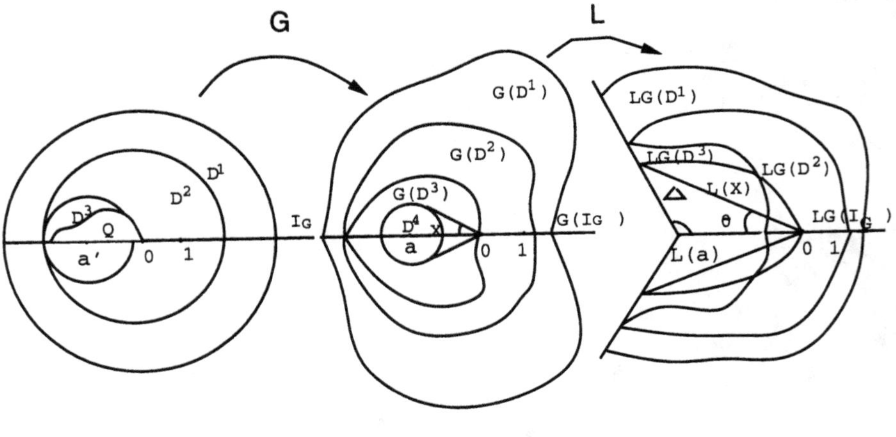

Fig. 5.4

We consider several disks related to a compatible pair L_a and G (see Fig. 5.4). Set D^1 to be the closed disk centered at the middle point $(1 + a')/2$ of $[a', 1]$ with diameter $1 + |a'| + 2b$. Set D^2 to be the closed disk centered at $(1 + a')/2$ with diameter $k = \min\{1 + |a'| + b, 2(1 + |a'|)\}$. Set D^3 to be the maximum closed disk centered at a' and contained in D^2. Then the radius of D^3 is $d = \min\{b/2, (1+|a'|)/2\}$. We note that

$$D^3 \subset D^2 \subset D^1.$$

The map G is a schlicht function on D^1. Let $\mu = (1+|a'|)/b$ and $\nu = (2\mu+3)^4$. From Koebe's distortion theorem (Theorem 2.5), for any ξ and η in D^2,

$$\nu^{-1} \leq \frac{|G'(\xi)|}{|G'(\eta)|} \leq \nu.$$

Using the fact that $G(I) = I$, there is at least one η in D^2 such that $|G'(\eta)| = 1$. Hence

$$\nu^{-1} \leq |G'(\xi)| \leq \nu$$

for all ξ in D^2. This implies that

$$\frac{|a'|}{\nu} \leq |a| \leq |a'|\nu.$$

Let $\mathbf{UH} = \{z = x + yi \in \mathbf{C} \mid y > 0\}$ be the upper-half plane. For any z in $\mathbf{UH}$, let $\theta(z) = \arg(z)$.

Theorem 5.29. *Suppose that $L = L_a$ and G are a pair of compatible elements in $\mathcal{E}_0$. There is a constant $0 < \theta < \pi$ depending only on μ such that the image $L(G(\mathbf{UH}))$ of the upper-half plane under $L \circ G$ contains an open triangle Δ based on $[L(a), 0]$ whose angle at $L(a)$ is π/γ and whose angle at 0 is θ.*

Proof. The image $G(D^3)$ of D^3 under G contains the closed disk D^4 centered at a with radius d/ν. Similarly, for any $a' \leq x \leq 1$, consider the closed disk $D(x)$ centered at x with radius d. Then $D(x) \subset D^2$ and $G(D(x))$ contains the closed disk centered at $G(x)$ with radius d/ν. This implies that the convex-hull X of $\{0\} \cup D^4$ is contained in $G(D^2)$. Either $X = D^4$ or $X \cap \mathbf{UH}$ has an angle at 0. In the later case, the angle φ of $X \cap \mathbf{UH}$ at 0 has $\sin\varphi = d/(|a|\nu)$.

Since L is a γ-root for $\gamma > 1$, the convex set $L(X)$ contains a triangle Λ based on $[L(a), 0]$ whose angle at $L(a)$ is π/γ and whose angle ω at 0 can be calculated from

$$\frac{\sin\omega}{\sin(\frac{\pi}{\gamma}+\omega)} = \left(\frac{d}{|a|\nu}\right)^{\frac{1}{\gamma}}$$

through the law of sines. Because $d = \min\{b/2, (1+|a'|)/2\}$,

$$\frac{d}{|a|} \geq \frac{d}{|a'|\nu} \geq \min\{\frac{1}{2\mu\nu}, \frac{1}{2\nu}\}.$$

Hence Λ contains a triangle Δ based on $[L(a), 0]$ whose angle at $L(a)$ is still π/γ and whose angle θ at 0 is calculated from

$$\frac{\sin\theta}{\sin(\frac{\pi}{\gamma}+\theta)} = \left(\min\{\frac{1}{2\mu\nu}, \frac{1}{2\nu}\}\right)^{\frac{1}{\gamma}}.$$

■

Suppose $Q = (L \circ G)^{-1}(\overline{\Delta})$, where Δ is the triangle obtained in Theorem 5.29. Then

$$Q \subset D^2 \subset D^1.$$

Suppose $\{(L_i, G_i)\}_{i=0}^{n}$ is a sequence of compatible pairs in $\mathcal{E}_0$ where L_i is a γ-root at a_i. Let a_i', b_i, k_i, d_i, μ_i, ν_i, and θ_i be the numbers, and Δ_i and Q_i be the sets corresponding to each compatible pair L_i and G_i. Let

$$\mathcal{L} = L_n \circ G_n \circ \cdots \circ L_i \circ G_i \circ \cdots \circ L_0 \circ G_0.$$

Then $\mathcal{L}$ is a schlicht function defined on $\mathbf{C}_I = (\mathbf{C} \setminus \mathbf{R}) \cup \mathring{I}$.

Definition 5.8. *We call $\mathcal{L}$ a root-like map if there are constants $C > 0$ and $\lambda > 1$ such that*

(i) $a_0 = 0$ *and* $a_1 \geq 1/C$,

(ii) $|a_j| \geq \max\{(\lambda^{j-i}/C)|a_i|, \left(1+(\lambda-1)/C\right)|a_i|\}$ *for all* $1 \leq i < j \leq n$, *and*

(iii) $\mu_i < C$ *for all* $1 \leq i \leq n$.

Theorem 5.30 (Sullivan's Sector Theorem). *Suppose $\mathcal{L}$ is a root-like map. There is a constant $\theta > 0$ depending only on λ and C such that the image of the upper-half plane under $\mathcal{L}$ is contained in the sector*

$$\mathbf{Sec}_\theta = \{z \in \mathbf{C} \mid 0 \leq \arg(z) \leq \pi - \theta\}.$$

Before we prove this theorem, we introduce some basic results in hyperbolic geometry. Let $\mathbf{C}_I = (\mathbf{C} \setminus \mathbf{R}) \cup \mathring{I}$ be a planar domain. Then $q(z) = -z^2/(1-z^2)$ is a diffeomorphism from the upper-half plane $\mathbf{UH}$ onto $\mathbf{C}_I$. Consider $\mathbf{UH}$ to be a hyperbolic plane with the Poincaré metric $d_{H,\mathbf{UH}}s = |dz|/y$ for $z = x + yi$. This metric induces a hyperbolic metric $d_{H,\mathbf{C}_I}s = q_*(|dz|/y)$ on $\mathbf{C}_I$. The planar domain $\mathbf{C}_I$ under this metric is a hyperbolic Riemann surface. Let $d = d_{H,\mathbf{C}_I}$ be the induced hyperbolic distance. We note that q maps the positive imaginary line in $\mathbf{UH}$ onto the interval I and maps the real line, which is the boundary of $\mathbf{UH}$, onto the set $\mathbf{R} \setminus \mathring{I}$.

Lemma 5.9. *A hyperbolic neighborhood $\Phi(r) = \{z \in \mathbf{C}_I \mid d(z, I) < r\}$ is the union of two Euclidean disks D^+ and D^-, symmetric to each other with respect to I, centered at c^+ and at $c^- = -c^+$, with the same radius $R^+ = R^-$. Moreover,*

$$R^+ = \frac{1}{2\sin\beta} \qquad \text{and} \qquad c^+ = \frac{1}{2} + \frac{\cot\beta}{2} i$$

where

$$\beta = 4\cot^{-1}(e^r)$$

is the angle at 0 between ∂D^+ and the negative real line (and the angle at 1 between ∂D^+ and the ray $[1, \infty)$).

Proof. Consider the pre-image $\Phi' = q^{-1}(\Phi(r))$. It is a hyperbolic neighborhood in **UH** and consists of all points in **UH** whose hyperbolic distances to the half-line $l_+ = \{z = yi \mid y > 0\}$ are less than r. The boundary $\partial\Phi'$ consists of two rays starting from 0. Thus Φ' is a sector, symmetric with respect to the half-line l_+. Suppose $\beta/2$ is the outer angle of this sector (with respect to the real line). Since a geodesic in **UH** is a semi-circle or half-line perpendicular to the real line, it is easy to check that

$$\log\left(\cot\frac{\beta}{4}\right) = r.$$

Therefore, $\Phi(r)$ is the union of two disks D^+ and D^- symmetric with respect to I. The angle between ∂D^+ (or ∂D^-) and the negative real line is β. Moreover, every point z in ∂D^+ (or ∂D^-) views I under the same angle β, that is, every triangle $\Delta(0z1)$ has the angle β at z. Now consider the point u such that the segment $\overline{1u}$ is a diameter of D^+. The triangle $\Delta(u01)$ is a right triangle. We can calculate the length $2R^+$ of the segment $1u$ and length $|u|$ of the segment $\overline{0u}$ as follows:

$$2R^+ = \frac{1}{\sin\beta}, \qquad |u| = \cot\beta.$$

Therefore,

$$R^+ = \frac{1}{2\sin\beta} \qquad \text{and} \qquad c^+ = \frac{1}{2} + \frac{\cot\beta}{2} i.$$

■

Lemma 5.10. *Let z be a point in $\mathbf{C}_I$ and let $\Phi(r) = D^+ \cup D^-$ be the smallest hyperbolic neighborhood containing z. The Euclidean radius R^+ of D^+ (and D^-) is*

$$\frac{|z-1|}{2\sin(\arg(z))}.$$

Proof. Let u be the point in ∂D^+ such that the segment $\overline{1u}$ is a diameter of D^+. The angle of the triangle $\Delta(0z1)$ at z and the angle of the triangle $\Delta(0u1)$ at u are both β (see the previous lemma). Now applying the law of sines,

$$\frac{\sin(\arg(z))}{|1-z|} = \sin\beta = \frac{1}{|1-u|}.$$

Therefore,

$$2R^+ = \frac{|z-1|}{\sin(\arg(z))}.$$

∎

Suppose $\mathcal{L}$ is a root-like map. Then there is a constant angle σ and a constant $C_0 > 0$ depending only on C such that $\theta_i \geq \sigma$ and $\nu_i \leq C_0$ for all $0 \leq i \leq n$.

For each $1 \leq i \leq n-1$, let $A_{i+1} = Q_{i+1} \setminus \Delta_i$. Let $A_1 = Q_1 \setminus \{z \in \mathbf{C} \mid 0 \leq \arg(z) < \pi - \pi/\gamma\}$. Suppose $\Phi(r_i) = D_i^+ \cup D_i^-$ is the smallest hyperbolic neighborhood in $\mathbf{C}_I$ containing A_i if $A_i \neq \emptyset$ and let D_i^0 be the smallest disk centered at $1/2$ containing $\Phi(r_i)$. Let R_i^+ be the Euclidean radius of D_i^+ and let R_i be the Euclidean radius of D_i^0. From Lemmas 5.9 and 5.10, since z and 1 are both in D_i^2, we have that

$$R_i \leq 2R_i^+ \leq \frac{k_i}{\sin\theta_i} \leq \frac{2(1+|a_i'|)}{\sin\theta_i} \leq \frac{2(1+C)|a_i'|}{\sin\sigma} = C_1|a_i'|$$

where $k_i = \min\{1+|a_i'|+b_i, 2(1+|a_i'|)\}$ and where $C_1 = 2(1+C)/\sin\sigma$ is a positive constant depending only on C (see Fig. 5.5).

For every $1 \leq i \leq n$, the definition interval of $\phi_i = L_i \circ G_i$ is $[a_i', a_i''] = [a_i', \infty) \cap I_{G_i}$. According to **(iii)** of Definition 5.8, the right-endpoint a_i'' satisfies

$$a_i'' \geq \frac{1+|a_i'|}{C} + 1 \geq 1 + \frac{|a_i'|}{C}.$$

For every $1 \leq i < j \leq n$,

$$|a_j'| \geq \frac{\lambda^{j-i}}{CC_0^2}|a_i'|.$$

Therefore, if $\tau = \min\{1, 1/(CC_0)^2\}$, we have

$$[a_j', a_j''] \supset I_{i,\tau} = [\lambda^{j-i}\tau a_i', \ \ \lambda^{j-i}\tau|a_i'| + 1].$$

In other words, $\phi_j = L_j \circ G_j$ are schlicht functions on $\mathbf{C}_{I_{i,\tau}} = (\mathbf{C} \setminus \mathbf{R}) \cup I_{i,\tau}$ for all $1 \leq i < j \leq n$.

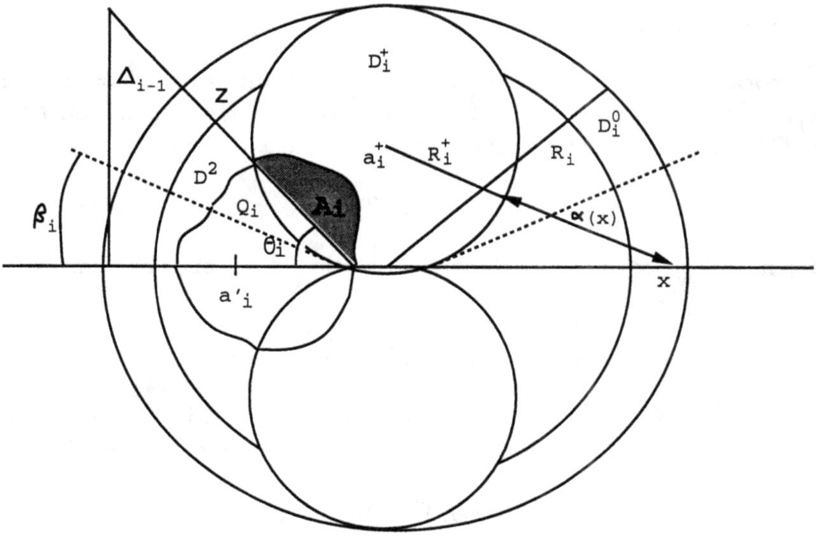

Fig. 5.5

■

Lemma 5.11. *There is a fixed integer $n_0 > 0$ depending only on C and on λ such that for any $0 \leq i < n$ and any $j \geq i + n_0$,*

$$|a'_j| + \frac{1}{2} > \lambda^{j-i-n_0} R_i \quad \text{and} \quad a''_j - \frac{1}{2} > \lambda^{j-i-n_0} R_i.$$

Proof. Take m_0 as the biggest integer such that

$$m \leq \frac{\log(\frac{C_1}{\tau})}{\log \lambda}.$$

From the above estimates, one can see that $n_0 = \max\{m_0 + 1, 0\}$ is the integer satisfying the lemma. ■

Lemma 5.12. *There is a constant $C_2 > 0$ depending only on λ such that for any $0 \leq i < n - n_0$, let $\Pi_i = L_n \circ G_n \circ \cdots \circ L_j \circ G_j \circ \cdots \circ L_{i+n_0+1} \circ G_{i+n_0+1}$, the distortion of Π_i on $\Phi(r_i)$ is bounded by C_2, more precisely,*

$$\frac{|\Pi'_i(\xi)|}{|\Pi'_i(\eta)|} \leq C_2$$

for all ξ *and* η *in* $\Phi(r_i)$.

Proof. Consider the disk D_j^5 centered at $1/2$ with radius $t_j = \min\{|a'_j| + \frac{1}{2}, |a''_j| - \frac{1}{2}\}$. The composition $\phi_j = L_j \circ G_j$ is a schlicht function on D_j^5. Let $r_{ij} = CR_i/|t'_j|$ be the ratio of the radii of D_i^0 and D_j^5. Then

$$r_{ij} \leq \lambda^{i+n_0-j}$$

for $i + n_0 + 1 \leq j \leq n$. From Koebe's distortion theorem (Theorem 2.5),

$$\frac{|\phi'_j(\xi)|}{|\phi'_j(\eta)|} \leq \Big(\frac{1+r_{ij}}{1-r_{ij}}\Big)^4 \leq \Big(\frac{1+\lambda^{-k}}{1-\lambda^{-k}}\Big)^4$$

for $0 < k = j - i - n_0 \leq n - i - n_0$ and for all ξ and η in D_i^0. Since ϕ_j is a schlicht function on $\mathbf{C}_I$, it contracts the hyperbolic distance d on $\mathbf{C}_I$. Thus

$$\phi_j(\Phi(r_i)) \subset \Phi(r_i) \subset D_i^0.$$

Therefore, by the chain rule,

$$\frac{|\Pi'_i(\xi)|}{|\Pi'_i(\eta)|} = \prod_{j=n_0+1}^{n} \frac{|\phi'_j(\xi)|}{|\phi'_j(\eta)|} \leq C_2 = \Big(\prod_{k=1}^{\infty} \frac{1+\lambda^{-k}}{1-\lambda^{-k}}\Big)^4$$

for all ξ and η in $\Phi(r_i)$. ∎

Lemma 5.13. *There is a constant* $C_3 > 0$ *depending only on* C *such that for* $0 \leq i \leq n$, *for* $i \leq j \leq i + n_0$, *for* $\phi_j = L_j \circ G_j$, *and for all* ξ *and* η *in* D_i^+ *(or* D_i^-*),*

$$\frac{|\phi'_j(\xi)|}{|\phi'_j(\eta)|} \leq C_3.$$

Proof. Suppose x is a real number. Let $\alpha = \alpha(x) = \min\{|x - z| \mid z \in D_i^+ \cap \mathbf{UH}\}$. Suppose that $c = c_i^+$ and that $R = R_i^+$ are the center and the radius of D_i^+. Suppose $h = h_i^+$ is the length of the segment $\overline{c\frac{1}{2}}$ (the straight line connecting c and $1/2$). Since the two triangles $\Delta(x\frac{1}{2}c)$ and $\Delta(0\frac{1}{2}c)$ are both right triangles, then

$$(\alpha + R)^2 = (\frac{1}{2} - x)^2 + h^2$$

and

$$R^2 = (\frac{1}{2})^2 + h^2.$$

Therefore,

$$\frac{\alpha + R}{R} = \sqrt{1 + \frac{x^2 - x}{R^2}}.$$

This implies that there is a constant $0 < C_4 < 1$ depending only on C such that for $x = a'_j$ or $x = a''_j$,

$$\frac{R}{\alpha + R} \leq C_4.$$

Now consider the largest disk D_j^6 centered at c such that ϕ_j is a schlicht function on it. Then the radius of D_j^6 is greater than or equal to $\min\{R + \alpha(a'_j), R + \alpha(a''_j)\}$. From Koebe's distortion theorem (Theorem 2.5),

$$\frac{|\phi'_j(\xi)|}{|\phi'_j(\eta)|} \leq C_3 = \Big(\frac{1 + C_4}{1 - C_4}\Big)^4$$

for any ξ and η in D_i^+. ∎

Combining Lemmas 5.12 and 5.13, we obtained the following estimate:

Lemma 5.14. *There is a constant $C_5 > 0$ depending on λ and on C such that for any $0 \leq i < n$, let $\Sigma_i = L_n \circ G_n \circ \cdots \circ L_j \circ G_j \circ \cdots \circ L_i \circ G_i$, the distortion of Σ_i on D_i^+ (or D_i^-) is bounded by C_5, more precisely,*

$$\frac{|\Sigma'_i(\xi)|}{|\Sigma'_i(\eta)|} \leq C_5$$

for all ξ and η in D_i^+ (or D_i^-).

Proof. Since each ϕ_j is a schlicht function on $\mathbf{C}_I$, it contracts the hyperbolic distance d on $\mathbf{C}_I$. So

$$\phi_j(D_i^+) \subset D_i^+$$

for all $i \leq j \leq n$. If $n - i \leq n_0$, then from Lemma 5.13 and the chain rule,

$$\frac{|\Sigma'_i(\xi)|}{|\Sigma'_i(\eta)|} \leq C_3^{n_0}$$

for all ξ and η in D_i^+.

Now we consider $n - i > n_0$ and write $\Sigma_i = \Pi_i \circ \Theta_i$ where $\Theta_i = L_{i+n_0} \circ G_{i+n_0} \circ \cdots \circ L_i \circ G_i$ and where $\Pi_i = L_n \circ G_n \circ \cdots \circ L_{i+n_0+1} \circ G_{i+n_0+1}$. From Lemma 5.13,

$$\frac{|\Theta'_i(\xi)|}{|\Theta'_i(\eta)|} \leq C_3^{n_0}$$

for all ξ and η in D_i^+, and from Lemma 5.12,

$$\frac{|\Pi_i'(\xi)|}{|\Pi_i'(\eta)|} \leq C_2$$

for all ξ and η in D_i^+. Again, because Θ_i is a schlicht function on $\mathbf{C}_I$ and contracts the hyperbolic distance d on $\mathbf{C}_I$, we have $\Theta_i(D_i^+) \subset D_i^+$. Therefore, from the chain rule,

$$\frac{|\Sigma_i'(\xi)|}{|\Sigma_i'(\eta)|} \leq C_5 = C_2 C_3^{n_0}$$

for all ξ and η in D_i^+. ∎

Lemma 5.15. *Suppose G is in $\mathcal{E}_0$ and D is a closed simply connected convex domain with $I \subset D \subset \mathbf{C}_G$. Then for all $z = x + yi$ with $y > 0$ in D*

$$\sin\Big(\arg(G(z))\Big) \geq \frac{\sin(\arg(z))}{N_0}$$

where $N_0 = \sup_{\xi,\eta \in D} |G'(\xi)/G'(\eta)|$ measures the distortion of G on D.

Proof. Since G maps $\mathbf{UH}$ into itself, it contracts the hyperbolic metric dz/y on $\mathbf{UH}$. Suppose $z = x + yi$ with $y > 0$ and $G(z) = X + Yi$. Then $|G'(z)|y \leq Y$. Therefore,

$$\sin(\arg(z)) = \frac{y}{|z|} \leq \frac{Y}{|z||G'(z)|} = \sin\Big(\arg(G(z))\Big)\frac{|G(z)|}{|z||G'(z)|}.$$

So

$$\frac{\sin(\arg(z))}{N_0} \leq \sin\Big(\arg(G(z))\Big)$$

for all $z = x + yi$ with $y > 0$ in D. ∎

Now we complete the proof of Theorem 5.30 as follows.

Proof of Theorem 5.30. For any z_0 in $\mathbf{UH}$, let $z_{i+1} = L_i(G_i(z_i))$ for $0 \leq i \leq n$. Since $0 \leq \arg(z_1) \leq \pi/\gamma$, the smallest positive integer i such that z_i lies in Δ_i must either be bigger than zero or not exist. If such a positive integer does not exist, then $0 \leq \arg(z_{n+1}) \leq \pi - \theta_n \leq \pi - \sigma$. Now let us suppose that this smallest number exists and is $i_0 + 1$. Then z_{i_0} is in $A_{i_0} \neq \emptyset$ which is a subset of $D_{i_0}^+$. Since $0 < \arg(z_{i_0}) \leq \pi - \theta_i$, Lemmas 5.14 and 5.15 assure us there is a constant angle $0 < \theta \leq \sigma$ depending only on λ and C such that $0 < \arg(z_{n+1}) \leq \pi - \theta$. ∎

Sullivan's Sector Theorem is first proved by Sullivan in [SU4]. The formulation and the discussion given in this section is from [JI16]. Other formulations and discussions can be found in [MV2], in [FAR], and in [FLE].

5.9. The Feigenbaum Quadratic Polynomial

Consider the real quadratic family $\{P_t(x) = t - (t+1)x^2\}_{0\leq t\leq 1}$. From the kneading theory (see [COE,MIT,MV2]), there is a Feigenbaum-like map in this family, which is infinitely $(2, 2, \ldots)$-renormalizable. Sullivan [SU4] (see also [MV2]) proved that it is unique in the family. The proof of Sullivan can be outlined as follows. Suppose f_1 and f_2 are two Feigenbaum-like maps in the family. Let h be a homeomorphism such that $f_2 \circ h = h \circ f_1$. The attractors Λ_1 and Λ_2 of f_1 and f_2 have bounded geometry (see §4.6). According to this, $h|\Lambda_1$ can be extended to a quasisymmetric homeomorphism $\tilde{h}$. Furthermore, $\tilde{h}$ can be extended to a quasiconformal mapping $\tilde{H}$ of the complex plane $\mathbf{C}$. The map $(U_{r,1}, U_{r^2,1}, f_1)$ (or $(U_{r,2}, U_{r^2,2}, f_2)$ is a quadratic-like map where $U_{r,1}$ (or $U_{r,2}$) is the domain bounded by an equipotential curve $s_{r,1}$ of f_1 (or $s_{r,2}$ of f_2). One can pull back $\tilde{H}$ from the corresponding fundamental domains $U_{r^2,1} \setminus U_{r,1}$ and $U_{r^2,2} \setminus U_{r,2}$ and get another quasiconformal mapping H such that $H \circ f_1 = f_2 \circ H$ on $U_{r,1}$ (compare to Theorem 4.4). Further, the filled-in Julia set K_{f_1} of f_1 supports no invariant line field. This implies that H is conformal when restricted to K_{f_1}. So f_1 and f_2 are hybrid equivalent. From Douady and Hubbard's theorem (see Theorem 5.15), $f_1 = f_2$. The unique Feigenbaum-like map in the real quadratic family is called the Feigenbaum polynomial.

Let $f(z) = t_\infty - (1 + t_\infty)z^2$ be the Feigenbaum polynomial and let K be its filled-in Julia set. Let $U = U_r$ and $V = U_{r^2}$ be two domains bounded by equipotential curves s_r and s_{r^2}. Then (U, V, f) is a quadratic-like map. Let $m_k = 2^k$ for $k \geq 0$. From Chapter Four, there is a sequence of nested intervals $\{I_k = [-a_k, a_k]\}_{k=1}^{\infty}$ such that

(a) $f^{\circ m_k}$ is monotone when restricted to $[-a_k, 0]$ and to $[0, a_k]$,

(b) $f^{\circ m_k}(I_k) \subset I_k$,

(c) I_k, $f(I_k)$, ..., $f^{\circ(m_k-1)}(I_k)$ have pairwise disjoint interiors.

Let $c(i) = f^{\circ i}(0)$ be the ith critical value of f and let $J_k(i)$ be the interval bounded by $c(i)$ and $c(m_k + i)$ for $k \geq 0$ and $1 \leq i \leq m_k$. We note that $J_k(0) = J_k(m_k)$. Then $f : J_k(0) \to J_k(1)$ is folding for all $k \geq 1$ and $f : J_k(i) \to J_k(i+1)$ is a homeomorphism for every $k \geq 1$ and $1 \leq i < m_k$. Let $\zeta_k = \{J_k(i)\}_{0\leq i<m_k}$ for $k \geq 0$. Let $I_k(i) = f^{\circ i}(I_k)$, $0 \leq i < m_k$, and let $\xi_k = \{I_k(i)\}_{0\leq i<m_k}$. Note that $J_k(i) \subseteq I_k(i)$. Let $LI_k(i)$ and $RI_k(i)$ be the intervals in ξ_k adjacent to $I_k(i)$ and on the left and right sides of $I_k(i)$, respectively (there is only $LI_k(1)$ or $RI_k(2)$ in ξ_k). Let $LI_k^+(i)$ be the smallest interval containing $LI_k(i)$ and the left end-point of $I_k(i)$ and let $RI_k^+(i)$ be the smallest interval containing $RI_k(i)$ and the right end-point of $I_k(i)$, for $i = 0$

or $3 \le i < m_k$ (see Fig. 4.4). Let $LI_k^+(2) = [-1, c(2)]$ and $RI_k^+(1) = [c(1), 1]$. Similarly, we can define $LJ_k(i)$ and $RJ_k(i)$ (and $LJ_k^+(i)$ and $RJ_k^+(i)$) for $0 \le i < m_k$. From Lemma 4.2, there is a constant $C > 0$ such that

$$\min\{|LI_k^+(i)|, |RI_k^+(i)|\} \ge C|I_k(i)| \text{ and } \min\{|LJ_k^+(i)|, |RJ_k^+(i)|\} \ge C|J_k(i)|.$$

for all $k \ge 1$ and $0 \le i < m_k$.

Consider the slit domain $V_0 = V \setminus [c(1), \infty)$. The map $f|V_0$ has two inverse branches (see Fig. 5.6)

$$g_0 : V_0 \to U_{0,0} = U \cap \mathbf{LH} \quad \text{and} \quad g_1 : V_0 \to U_{0,1} = U \cap \mathbf{RH},$$

where $\mathbf{LH} = \{z = x + yi \in \mathbf{C} \mid x < 0\}$ and $\mathbf{RH} = \{z = x + yi \in \mathbf{C} \mid x > 0\}$ are the left and right half planes.

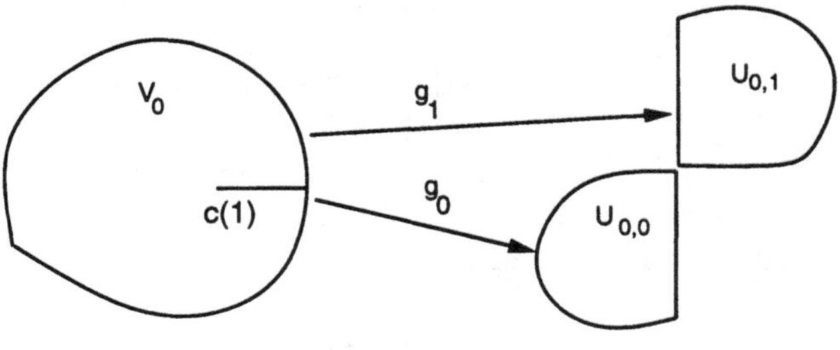

Fig. 5.6

For each $k \ge 1$, let $g_{s(i)} : J_k(i+1) \to J_k(i)$ for $1 \le i < m_k$ where $s(i) = 0$ if $J_k(i) \subset \mathbf{LH}$, and otherwise $s(1) = 1$. Then

$$\mathcal{A}_k = g_{s(1)} \circ g_{s(2)} \circ \cdots \circ g_{s(m_k-1)} : J_k(m_k) = J_k(0) \to J_k(1)$$

is a homeomorphism and can be extended homeomorphically to the maximum closed interval

$$T_k(m_k) \supseteq LJ_k^+(m_k) \cup J_k(m_k) \cup RJ_k^+(m_k).$$

Furthermore, $\mathcal{A}_k$ can be extended analytically to $V_k = (V \setminus \mathbf{R}) \cup \mathring{T}_k(m_k)$. Let us continue to use $\mathcal{A}_k$ to denote this extension. Let $U_k' = \mathcal{A}_k(V_k)$, and let $U_k = f^{-1}(U_k')$ be the pre-image of U_k' under f. Since f has no attractive and no parabolic periodic point, then $g_0(\mathcal{A}_k(T_k(m_k)))$ and $g_1(\mathcal{A}_k(T_k(m_k)))$ are contained strictly in $T_k(m_k)$. Thus, $\overline{U}_k \subset V_k$ and

$$f^{\circ m_k} : U_k \to V_k$$

is a quadratic-like map (see Fig. 5.7). Let K_k be its filled-in Julia set. Then K_k is the kth-renormalization of K.

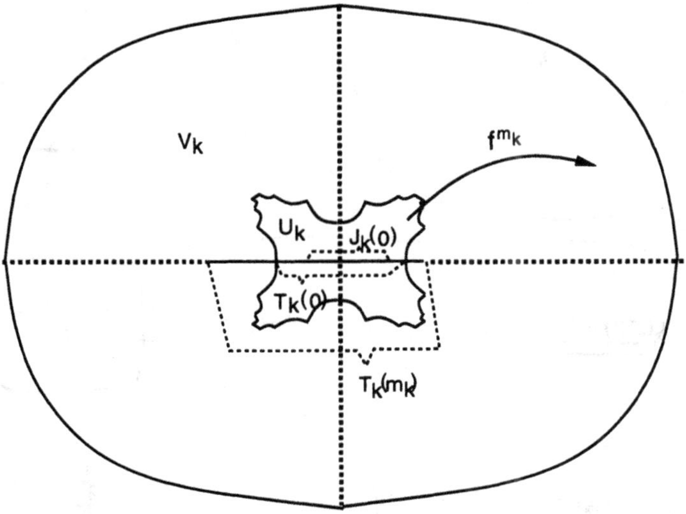

Fig. 5.7

Let $T_k(i) = g_{s(i)}(T_k(i+1))$ for $i = m_k - 1,\ m_k - 2,\ \ldots,\ 1$. Let $T_k(0) = f_0^{-1}(T_k(1))$. Then

$$J_k(0) = J_k(m_k) \subseteq I_k \subseteq T_k(0) \subset T_k(m_k).$$

The interval $T_k(m_k)$ is bounded by two critical values $c(q(k))$ and $c(r(k))$ of f; one of them is a maximum value and the other is a minimum value of $f^{\circ m_k}$. Suppose $T_k(1) = [d_1, e_1]$ where $d_1 < c(1) < e_1$. Let $d_i = f^{\circ i}(d_1)$ and let $e_i = f^{\circ i}(e_1)$ for $1 \leq i \leq m_k$. Note that d_{m_k} and e_{m_k} are $c(r(k))$ and $c(q(k))$. We normalize $T_k(i)$ into the unit interval $[0,1]$: let $l_i : T_k(i) \to [0,1]$ be the linear map such that $l_i(d_i) = 0$ and $l_i(e_i) = 1$ and let

$$\tilde{g}_{s(i)} = l_i \circ g_{s(i)} \circ l_{i+1}^{-1}$$

for $i = m_k - 1, \ldots, 2, 1$. The map $\tilde{g}_{s(i)}$ fixes 0 and 1 and is a univalent function defined on

$$\mathbf{C}_{[0,1],V} = \Big((\mathbf{C} \setminus \mathbf{R}) \cup (0,1)\Big) \cap V.$$

The restriction $\tilde{g}_{s(i)}|[0,1]$ is a homeomorphism of $[0,1]$. For $k \geq 1$, let

$$\mathcal{L}_k = \tilde{g}_{s(1)} \circ \tilde{g}_{s(2)} \circ \cdots \circ \tilde{g}_{s(m_k-1)}.$$

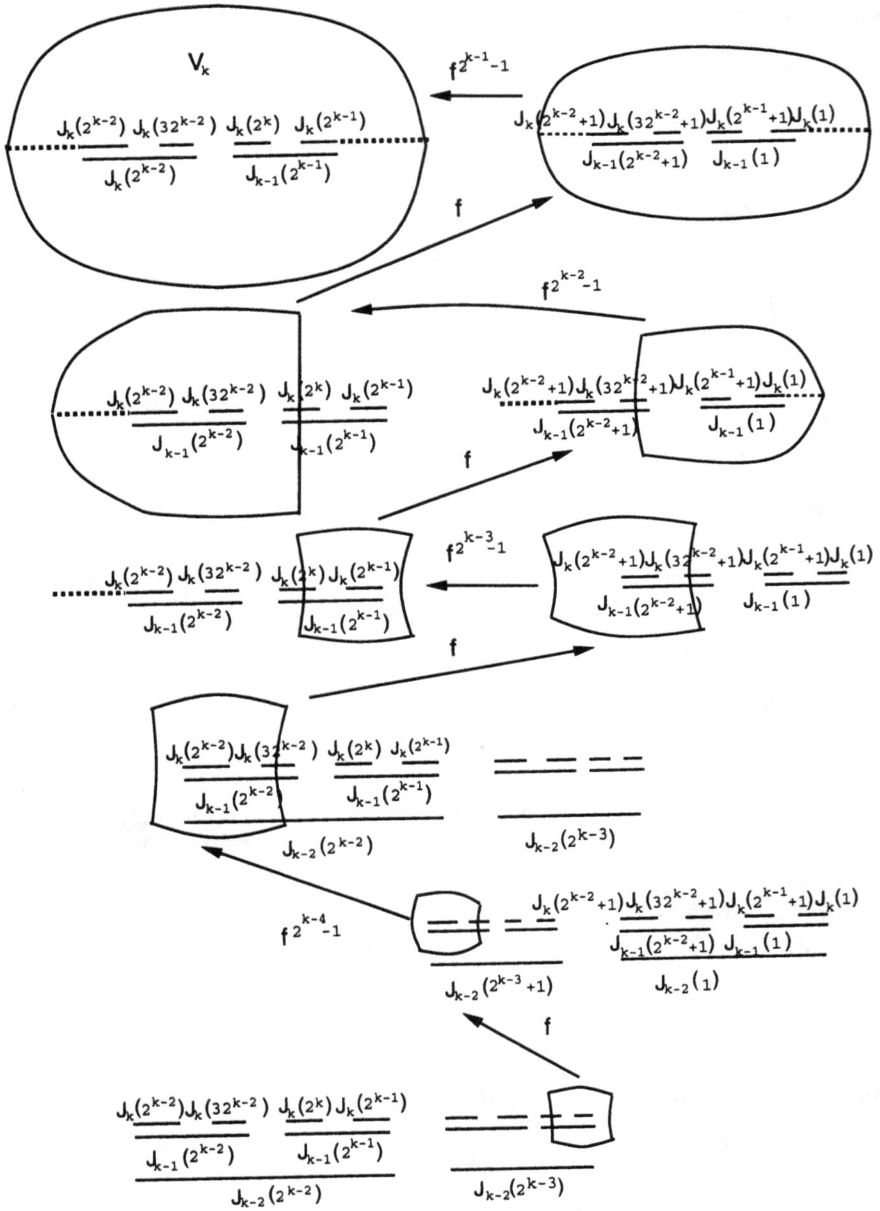
V_k
f^{2^{k-1}-1}
J_k(2^{k-2}) J_k(32^{k-2}) J_k(2^k) J_k(2^{k-1})
J_k(2^{k-2})
J_{k-1}(2^{k-1})
J_k(2^{k-2}+1) J_k(32^{k-2}+1) J_k(2^{k-1}+1) J_k(1)
J_{k-1}(2^{k-2}+1) J_{k-1}(1)
f
f^{2^{k-2}-1}
J_{k-1}(2^{k-2})
f^{2^{k-3}-1}
J_{k-2}(2^{k-2})
J_{k-2}(2^{k-3})
f^{2^{k-4}-1}
J_{k-2}(2^{k-3}+1)
J_{k-2}(1)

Fig. 5.8

Lemma 5.16. *The mappings* $\{\mathcal{L}_k\}_{k=3}^{\infty}$ *are uniformly root-like maps.*

Proof. Remember that $m_k = 2^k$. For any $k \geq 3$, consider the homeomorphism $f^{\circ(2^{k-1}-1)} : J_k(2^{k-1}+1) \to J_k(2^k) = J_k(0)$. Its inverse $g_{s(2^{k-1}+1)} \circ \cdots \circ g_{s(2^k-1)} : J_k(2^k) \to J_k(2^{k-1}+1)$ can be extended to $T_{k-1}(2^{k-1})$, that is, we can consider

$$g_{s(2^{k-1}+1)} \circ \cdots \circ g_{s(2^k-1)} : T_{k-1}(2^{k-1}) \to T_{k-1}(1).$$

Let

$$G_0 = \tilde{g}_{s(2^{k-1}+1)} \circ \cdots \circ \tilde{g}_{s(2^k-1)}$$

and let

$$L_{a_0} = \tilde{g}_{s(2^{k-1})}.$$

Then G_0 is a univalent map of $\mathbf{C}_{[0,1],V}$ such that $G_0(0) = 0$ and $G_0(1) = 1$ and such that $G_0|[-1,1]$ is a homeomorphism of $[0,1]$. The map L_{a_0} is a square root at 0 (see Fig. 5.8).

Consider the homeomorphism $f^{\circ(2^{k-2}-1)} : J_k(2^{k-2}+1) \to J_k(2^{k-1})$. Its inverse $g_{s(2^{k-2}+1)} \circ \cdots \circ g_{s(2^{k-1}-1)} : J_k(2^{k-1}) \to J_k(2^{k-2}+1)$ can be extended to $T_{k-2}(2^{k-2})$, that is, we can consider

$$G_1' = g_{s(2^{k-2}+1)} \circ \cdots \circ g_{s(2^{k-1}-1)} : T_{k-2}(2^{k-2}) \to T_{k-2}(1).$$

Let

$$G_1 = \tilde{g}_{s(2^{k-2}+1)} \circ \cdots \circ \tilde{g}_{s(2^{k-1}-1)}$$

and let

$$L_{a_1} = \tilde{g}_{s(2^{k-2})}.$$

The map G_1 is a univalent map of $\mathbf{C}_{[0,1],V}$ and $G_1|[-1,1]$ is a homeomorphism of $[0,1]$. The map L_{a_1} is a square root at a_1. The pre-image of $c(1)$ under G_1' is $c(2^{k-2})$. One of the end-points of $T_k(2^{k-1})$ is 0; the other one is between $J_{k-1}(0)$ and one of $LJ_{k-1}(0)$ or $RJ_{k-1}(0)$. From Lemma 4.2, two components of $T_{k-2}(2^{k-2}) \setminus T_k(2^{k-1})$ have lengths greater than a constant C (obtained from Lemma 4.2) times the length of $T_k(2^{k-1})$. Thus (L_{a_1}, G_1) is a compatible pair and satisfies **(i)** and **(iii)** of Definition 5.8 (see Fig. 5.8).

Next we consider the homeomorphism $f^{\circ(2^{k-3}-1)} : J_k(2^{k-3}+1) \to J_k(2^{k-2})$. Its inverse $g_{s(2^{k-3}+1)} \circ \cdots \circ g_{s(2^{k-2}-1)} : J_k(2^{k-2}) \to J_k(2^{k-3}+1)$ can be extended to $T_{k-3}(2^{k-3})$; that is, we can consider

$$G_2' = g_{s(2^{k-3}+1)} \circ \cdots \circ g_{s(2^{k-2}-1)} : T_{k-3}(2^{k-3}) \to T_{k-3}(1).$$

Let

$$G_2 = \tilde{g}_{s(2^{k-3}+1)} \circ \cdots \circ \tilde{g}_{s(2^{k-2}-1)}$$

and let

$$L_{a_2} = \tilde{g}_{s(2^{k-3})}.$$

The map G_2 is a univalent map of $\mathbf{C}_{[0,1],V}$ and $G_2|[-1,1]$ is a homeomorphism of $[0,1]$. The map L_{a_2} is a square root at a_2. The preimage of $c(1)$ under G_2' is $c(2^{k-3})$. The interval $T_k(2^{k-2})$ is contained in $T_{k-1}(2^{k-1})$. From Lemma 4.2, two components of $T_{k-3}(2^{k-3})\setminus T_k(2^{k-2})$ have lengths greater than a constant C (obtained from Lemma 4.2) times the length of $T_k(2^{k-2})$. Thus (L_{a_2}, G_2) is a compatible pair and satisfies **(iii)** of Definition 5.8 (see Fig. 5.8).

In general, for $3 < i \leq k-1$, consider the homeomorphism $f^{\circ(2^{k-i}-1)}$: $J_k(2^{k-i}+1) \to J_k(2^{k-i+1})$. Its inverse $g_{s(2^{k-i}+1)}\circ\cdots\circ g_{s(2^{k-i+1}-1)} : J_k(2^{k-i+1}) \to J_k(2^{k-i}+1)$ can be extended to $T_{k-i}(2^{k-i})$; that is, we can consider

$$G'_{i+1} = g_{s(2^{k-i}+1)} \circ \cdots \circ g_{s(2^{k-i+1}-1)} : T_{k-i}(2^{k-i}) \to T_{k-i}(1).$$

Let

$$G_{i+1} = \tilde{g}_{s(2^{k-i}+1)} \circ \cdots \circ \tilde{g}_{s(2^{k-i+1}-1)}$$

and let

$$L_{a_{i+1}} = \tilde{g}_{s(2^{k-i})}.$$

The map G_{i+1} is a univalent map of $\mathbf{C}_{[0,1],V}$ and $G_{i+1}|[0,1]$ is a homeomorphism. The map $L_{a_{i+1}}$ is a square root at a_{i+1}. The preimage of $c(1)$ under G'_{i+1} is $c(2^{k-i})$. The interval $T_k(2^{k-i+1})$ is contained in $T_{k-i+2}(2^{k-i+2})$. From Lemma 4.2, two components $T_{k-i}(2^{k-i})\setminus T_k(2^{k-i+1})$ have lengths greater than a constant C (obtained from Lemma 4.2) times the length of $T_k(2^{k-i+1})$. Thus $(L_{a_{i+1}}, G_{i+1})$ is a compatible pair and satisfies **(iii)** of Definition 5.8.

One of the endpoints of $T_k(2^{k-i}+1)$ is to the left of $c(2^{k-i}+1)$; the other is in the interval in ζ_k which is adjacent to $J_k(2^{k-i}+1)$. The branch point of $g_{s(2^{k-i})}$ is always $c(1)$. From Lemma 4.2, we have a constant $\lambda > 1$ such that $\{a_i\}_{i=1}^{k}$ satisfies **(ii)** of Definition 5.8 for $C = 1$.

For all $k \geq 3$, we therefore decompose

$$\mathcal{L}_k = L_{a_k} \circ G_k \circ L_{a_{k-1}} \circ G_{k-1} \circ \cdots \circ L_{a_1} \circ G_1 \circ L_{a_0} \circ G_0.$$

From the construction above, $\mathcal{L}_k$, $3 \leq k < \infty$, are uniformly root-like map. ∎

Consider the renormalizations

$$f^{\circ m_k} : U_k \to V_k$$

for $1 \leq k < \infty$ where $V_k = (V \setminus \mathbf{R}) \cup \mathring{T}_k(m_k)$. Let $U_k \cap \mathbf{R} = [-o_k, o_k]$. Let $w_{k,\theta}$ be the the ray starting at o_k with slop $\tan\theta$ for $0 < \theta < \pi/2$. Let $R_{k,\theta}$ be the domain containing 0 bounded by $w_{k,\theta}$, $-w_{k,\theta}$, $\overline{w}_{k,\theta}$, and $-\overline{w}_{k,\theta}$.

Lemma 5.17. *There is a constant angle* $\theta_0 > 0$ *such that*

$$U_k \subseteq R_{k,\theta_0}$$

for all $k \geq 3$ *(see Fig. 5.9).*

Proof. Consider $f^{\circ(m_k-1)} : J_k(1) \to J_k(0)$. Its inverse has the maximum extension

$$\mathcal{A}_k = g_{s(1)} \circ g_{s(2)} \circ \cdots \circ g_{s(m_k-1)} : V_k = (V \setminus \mathbf{R}) \cup \mathring{T}_k(m_k) \to U'_k.$$

Remember that

$$\mathcal{L}_k = l_1 \circ \mathcal{A}_k \circ l_{m_k}^{-1}$$

where l_{m_k} and l_1 are the linear maps normalizing T_{m_k} and $T_k(1)$, respectively, to $[0,1]$. Let

$$w'_{k,\theta} = \{z \in \mathbf{C} \mid \arg(z - d_1) = \theta, \Im(z) > 0\}$$

be the ray starting at d_1 with angle $0 \leq \theta \leq \pi$ where $U'_k \cap \mathbf{R} = T_k(1) = [d_1, e_1]$ with $d_1 < c(1) < e_1$. Let $R'_{k,\theta}$ be the sector containing $c(1)$ bounded by $w'_{k,\theta}$ and $\overline{w'}_{k,\theta}$. Applying Theorem 5.30, there is a constant angle $0 < \theta_1 \leq \pi$ such that U'_k is contained in a sector domain R'_{k,θ_1} (see Fig. 5.9). Since $f(z) = t_\infty - (1 + t_\infty)z^2$, there is a constant angle $0 < \theta_0 < \pi/2$ depending on θ_1 such that $U_k = f^{-1}(U'_k)$ is contained in R_{k,θ_0} (see Fig. 5.9). ∎

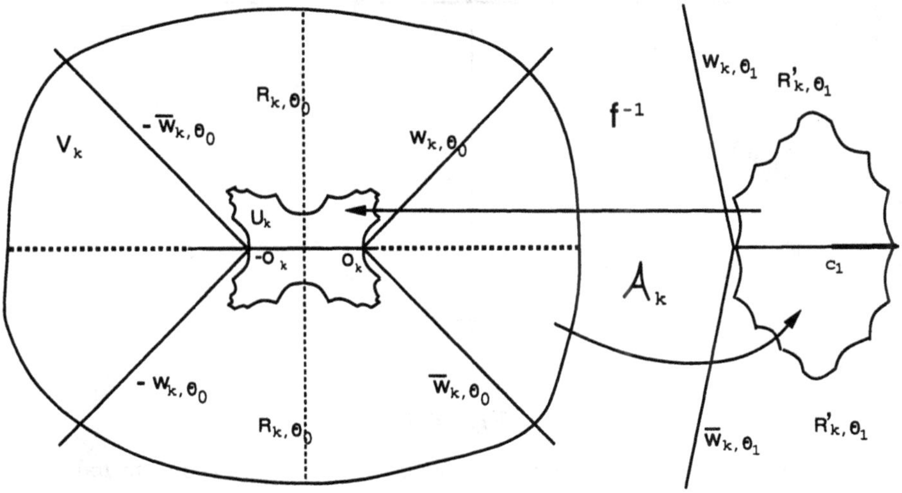

Fig. 5.9

Take $I_0 = (-1, 1)$. Let $\mathbf{C}_{I_0} = (\mathbf{C} \setminus \mathbf{R}) \cup I_0$. Let $d = d_{H,\mathbf{C}_{I_0}}$ be the hyperbolic distance on $\mathbf{C}_{I_0}$. Let $\Omega_r = \{z \in \mathbf{C}_{I_0} \mid d(z, I_0) < r\}$ be a hyperbolic neighborhood. From Lemma 5.9, Ω_r is the union of two disks D_β^+ and D_β^- centered at $c_\beta^+ = i \cot \beta$ and $c_\beta^- = -i \cot \beta$ with radii $R_\beta^+ = R_\beta^- = 1/\sin \beta$ where β is the angle between ∂D_β^+ and the line $[1, \infty)$ at 1 (see Fig. 5.10). Using the law of cosines for the triangle $\Delta(c0z)$, for any $z = re^{i\phi}$ in ∂D_β^+,

$$r = \cot \beta \sin \phi + \sqrt{\csc^2 \beta - \cot^2 \beta \cos^2 \phi}.$$

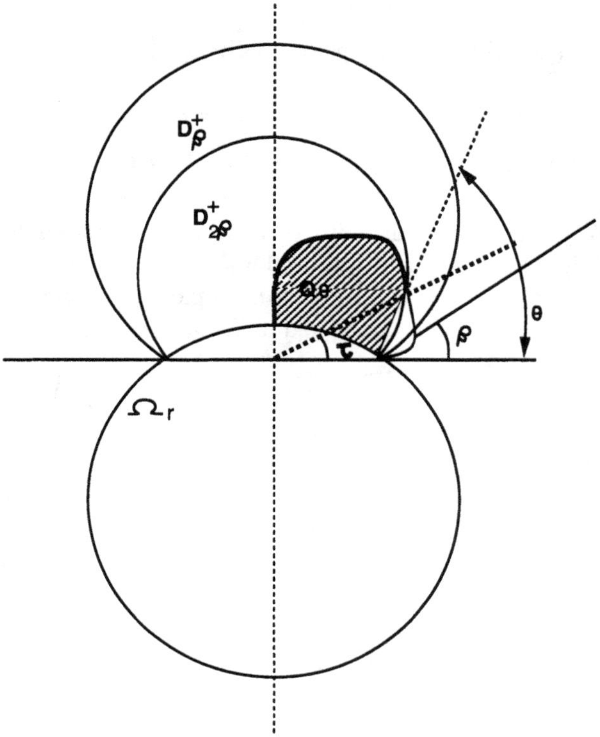

Fig. 5.10

Let $q(z) = \sqrt{z}$ be the square root from $\mathbf{C} \setminus \{x < 0\}$ to the right half-plane **RH**. Let $\Pi_\beta^+ = q(D_\beta^+)$. For $0 < \beta < \pi/4$, consider $\Omega_{r'} = D_{2\beta}^+ \cup D_{2\beta}^-$. Let $z_0 = re^{i\tau} \neq 1$ be the intersection point of $\partial D_{2\beta}^+$ and Π_β^+. Then τ is the unique non-zero solution of the equation

$$\sqrt{\cot\beta\sin(2\phi)+\sqrt{\csc^2\beta-\cot^2\beta\cos^2(2\phi)}}$$
$$=\cot(2\beta)\sin\phi+\sqrt{\csc^2(2\beta)-\cot^2(2\beta)\cos^2\phi}.$$

Thus $\tau = \tau(\beta)$ is a strictly increasing function and $\tau \to 0$ as $\beta \to 0$. Let $z_0 - 1 = \tilde{r}e^{i\theta}$. Then $\theta = \theta(\tau)$ is a strictly increasing function and $\theta \to 0$ as $\tau \to 0$. Let $\theta = \theta \circ \tau(\beta)$. It is a strictly decreasing function and $\theta \to 0$ as $\beta \to 0$. Let $\beta = \beta(\theta)$ be its inverse function.

For any $0 < \theta_0 < \pi/2$, let $0 < \beta_0 = \beta(\theta_0) < \pi/2$ and let

$$\mathcal{Q}_{\theta_0} = \{z \in \Pi_{\beta_0} \mid \theta_0 \le \arg(z-1) \le \frac{\pi}{2}\}.$$

Then $\mathcal{Q}_{\theta_0} \subseteq D^+_{2\beta_0}$. Let $\overline{\mathcal{Q}}_{\theta_0} = \{\overline{z} \mid z \in \mathcal{Q}_{\theta_0}\}$ and let $\mathcal{S}_{\theta_0} = \mathcal{Q}_{\theta_0} \cup (-\mathcal{Q}_{\theta_0}) \cup \overline{\mathcal{Q}}_{\theta_0} \cup (-\overline{\mathcal{Q}}_{\theta_0})$. For a number $0 < \nu_0 < 1$, let

$$\nu_0 \cdot \mathcal{S}_{\theta_0} = \{w = \nu_0 \cdot z \mid z \in \mathcal{S}_{\theta_0}\}.$$

Let $A_{\theta_0} = \Omega_r \setminus (\nu_0 \mathcal{S}_{\theta_0})$. From the above calculation, we have

Lemma 5.18. *There is a constant $C = C(\theta_0, \nu_0) > 0$ depending on θ_0 and on ν_0 such that the modulus mod(A_{θ_0}) of the annulus A_{θ_0} is greater than C.*

Proof. Let $a = \mathrm{diam}(\nu_0\mathcal{S}_{\theta_0})$ be the diameter of $\nu_0\mathcal{S}_{\theta_0}$ and let $b = d(\partial(\nu_0\mathcal{S}_{\theta_0}), \partial\Omega_r)$ be the distance between $\partial(\nu_0\mathcal{S}_{\theta_0})$ and $\partial\Omega_r$. Then a/b is bounded from above by a constant depending only on θ_0 and on ν_0. This implies the lemma (by using Grötzsch argument (refer to [AH1])). ∎

Suppose $0 < C_0 < 1$ is a constant. If we use $q_a(z) = \sqrt{z-a}/\sqrt{1-a}$ for $|a| < C_0$ to replace $q(z)$ in the above calculation, then Lemma 5.18 has a generalized version.

Lemma 5.18′. *There is a constant $C = C(\theta_0, \nu_0, C_0) > 0$ depending on θ_0, ν_0, and C_0 such that the modulus mod(A_{θ_0}) of the annulus A_{θ_0} is greater than C.*

Theorem 5.31 [SU4]. *The Feigenbaum quadratic polynomial has the a priori complex bounds and is unbranched.*

Proof. Consider the renormalizations

$$f^{\circ m_k} : U_k \to V_k$$

for $1 \le k < \infty$. From Lemma 5.17, there is a constant angle $0 < \theta_0 < \pi/2$ such that

$$U_k \subseteq R_{k,\theta_0}$$

for $k > 0$. Let $\beta_0 = \beta(\theta_0)$.

Consider $f^{\circ(m_k-1)} : J_k(1) \to J_k(0)$. Its inverse has the maximum extension

$$\mathcal{A}_k = g_{s(1)} \circ g_{s(2)} \circ \cdots \circ g_{s(m_k-1)} : V_k \to U'_k$$

where $U'_k \cap \mathbf{R} = T_k(1) = [d_1, e_1]$ with $d_1 < c(1) < e_1$. From Theorem 4.10, there is a constant $C_0 > 0$ such that $C_0^{-1} \le |c(1) - d_1|/|e_1 - c(1)| < C_0$ for all $k > 0$. We normalize $\mathring{T}_k(m_k) = (d_{m_k}, e_{m_k})$ to $(-1, 1)$ by the linear map s_1 such that $s_1(d_{m_k}) = 1$ and $s_1(e_{m_k}) = -1$. We normalize $(0, o_k)$ to $(0, 1)$ by the linear map s_2 such that $s_2(0) = 0$ and $s_2(o_k) = 1$. We normalize $\mathring{T}_k(1) = (d_1, e_1)$ to $(-1, 1)$ by the linear map s_3 such that $s_3(e_1) = -1$ and $s_3(d_1) = 1$. Let $a = s_3(c(1))$. Then there is a constant we still denote it as $0 < C_0 < 1$ such that $|a| < C_0$ for all $k > 0$. There is an integer $n_0 > 0$ such that for any $k > n_0$, $\Omega_r = D^+_{\beta_0} \cup D^-_{\beta_0}$ is contained in $s_1(V)$. Let $\mathcal{B}_k = s_3 \circ \mathcal{A}_k \circ s_1^{-1}$ and let $q_a(z) = s_2 \circ g_{s m_k} \circ s_3^{-1}$. Then q_a is comparable with $\sqrt{z-a}/\sqrt{1-a}$. Since $\mathcal{B}_k$ contracts the hyperbolic distance $d_{H, \mathbf{C}_{I_0}}$, then $\mathcal{B}_k(\Omega_r) \subseteq \Omega_r$. From Lemma 5.17,

$$X'_k = q_a(\mathcal{B}_k(\Omega_r)) \subseteq \mathcal{S}_{\theta_0} = \mathcal{Q}_{\theta_0} \cup (-\mathcal{Q}_{\theta_0}) \cup \overline{\mathcal{Q}}_{\theta_0} \cup (-\overline{\mathcal{Q}}_{\theta_0}).$$

Let $X''_k = s_1 \circ s_3^{-1}(X'_k)$. Then $X''_k \subseteq \nu_0 S_{\theta_0}$ for all $k > 0$ where $\nu_0 > 0$ is a constant obtained from Lemma 4.2. Let $Y''_k = \Omega_r$. Then from Lemma 5.18$'$,

$$\mathrm{mod}(Y''_k \setminus X''_k) > C$$

for all $k > n_0$ where $C > 0$ is a constant.

Now let $X_k = s_1^{-1}(X''_k)$ and let $Y_k = s_1^{-1}(Y''_k)$. Then

$$f^{\circ m_k} : X_k \to Y_k$$

is quadratic-like map and $\mathrm{mod}(Y_k \setminus X_k) > C$ for all $k > n_0$. This means that (U, V, f) has the *a priori* complex bounds.

Let $W_k = Y_k \setminus (LJ_k(0) \cup RJ_k(0))$. Applying Lemma 4.2 and the above argument, there is a constant $C' > 0$ such that $\mathrm{mod}(W_k \setminus K_k) > C'$ for all $k > n_0$ where K_k is the filled-in Julia set of $f^{\circ m_k} : X_k \to Y_k$. But $W_k \cap CO = \{c(jm_k)\}_{j=0}^{\infty}$. So (U, V, f) is unbranched. This completes the proof. ∎

Theorem 5.31 and Theorem 5.28 now give us that

Corollary 5.2 [JI16]. *The Julia set of the Feigenbaum quadratic polynomial is locally connected (see Fig. 5.11).*

Remark 5.8. Corollary 5.2 is first announced in [JIH] without giving the proof of Sullivan's result. The complete proof is given in [JI16]. The method here can be applied to a larger class of infinitely renormalizable quadratic polynomials containing all bounded type ones. It is conjectured that every infinitely renormalizable map in the real quadratic family has the *a priori* complex bounds and is unbranched. Following this conjecture and Theorem 5.28 would be that the filled-in Julia set of every real infinitely renormalizable quadratic polynomial is locally connected. See the recent work of Lyubich and Yampolsky [LYY], Levin and Van Strien [LES], and Graczyk and Świątek [GRS] which attacks the conjecture.

The Julia set of the Feigenbaum polynomial and some equipotential curve.

Fig. 5.11

5.10. The Mandelbrot Set at Certain Infinitely Renormalizable Points

Consider the complex quadratic family $\{P_c(z) = z^2 + c\}_{c\in\mathbf{C}}$. Let K_c be the filled-in Julia set of P_c, which is the set of points z such that the orbit $\{P_c^{\circ n}(z)\}_{n=0}^{\infty}$ is bounded. Let $J_c = \partial K_c$ be the Julia set of P_c. The Mandelbrot set $\mathcal{M}$ (see Fig. 5.12) is, by definition, the set of parameters c such that J_c is connected. Let $CO(c) = \{P_c^{\circ n}(0)\}_{n=0}^{\infty}$ be the critical orbit of P_c. From Corollary 5.1,

$$\mathcal{M} = \{c \in \mathbf{C} \mid CO(c) \text{ is bounded }\}.$$

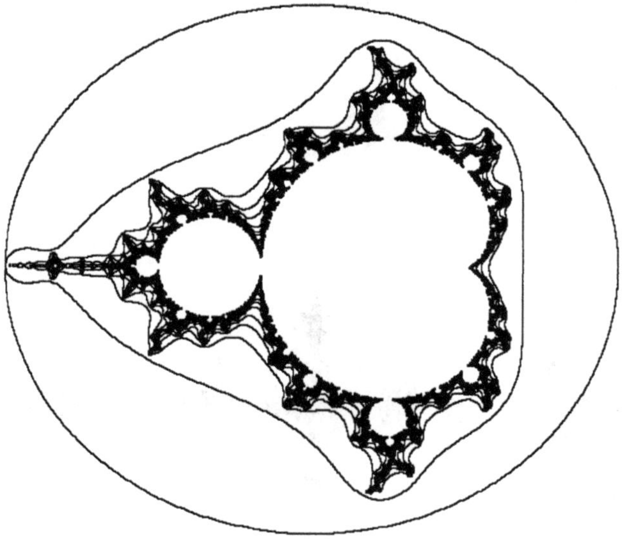

The boundary of the Mandelbrot set and some equipotential curves.

Fig. 5.12

Let p be a periodic point of P_c of period $k \geq 1$ and let $\lambda_p = (P_c^{\circ k})'(p)$ be the multiplier of P_c at p. The point p is attractive if $0 < |\lambda_p| < 1$; super-attractive if $\lambda_p = 0$; repelling if $|\lambda_p| > 1$; parabolic if $\lambda_p^n = 1$ for some integer $n \geq 1$; irrationally indifferent if $\lambda_p = \exp(2\pi\theta i)$ where θ is irrational. Let $\mathcal{HP}$ be the set of parameters c such that P_c is hyperbolic. Then

$$\mathcal{HP} = \{c \in \mathcal{M} \mid P_c \text{ has an attractive or super-attractive periodic point in } \mathbf{C} \}.$$

It is easy to check that $\mathcal{HP}$ is open. An important conjecture is that

Conjecture 5.1. *The set $\mathcal{HP}$ is dense in $\mathcal{M}$.*

Douady and Hubbard [DH2] proved that this conjecture follows from

Conjecture 5.2. *The Mandelbrot set $\mathcal{M}$ is locally connected.*

Let $PCO(c) = \{P_c^{\circ n}(0)\}_{n=1}^{\infty}$ be the post-critical orbit of P_c. We can further classify points in $\mathcal{M} \setminus \mathcal{HP}$ as

$$\mathcal{PA} = \{c \in \mathcal{M} \mid P_c \text{ has a parabolic periodic point }\};$$

$$\mathcal{II} = \{c \in \mathcal{M} \mid P_c \text{ has an irrationally indifferent periodic point }\};$$

$$\mathcal{NR} = \{c \in \mathcal{M} \mid c \notin \mathcal{PA} \cup \mathcal{II} \cup \mathcal{HP} \text{ and } 0 \notin \overline{PCO(c)}\};$$

$$\mathcal{FR} = \{c \in \mathcal{M} \mid c \notin \mathcal{PA} \cup \mathcal{II} \cup \mathcal{NR} \cup \mathcal{HP} \text{ and } P_c \text{ is finitely renormalizable }\};$$

and

$$\mathcal{IR} = \{c \in \mathcal{M} \mid c \notin \mathcal{PA} \cup \mathcal{II} \cup \mathcal{NR} \cup \mathcal{HP} \text{ and } P_c \text{ is infinitely renormalizable }\}.$$

A parameter c in $\mathcal{NR}$ is called a preperiodic (or Misiurewicz or sub-hyperbolic) point if $PCO(c)$ is finite; equivalently, if $p = P_c^{\circ m}(0)$ is a repelling periodic point of P_c for an integer $m \geq 1$ (actually, $m > 1$ because c has only one pre-image 0 under P_c). A parameter c in $\mathcal{IR}$ is called infinitely renormalizable. The proof of the following theorem can be found in [CAG]. The reader may also refer to the book of Carleson and Gamelin [CAG] for some results on the basic structure of $\mathcal{M}$.

Lemma 5.19. *The set of all preperiodic points c is dense in the boundary $\partial\mathcal{M}$ of $\mathcal{M}$.*

A consequence of the recent work of Yoccoz (see [HUB]) is that $\mathcal{M}$ is locally connected at all points which are not infinitely renormalizable. Thus, there remain only infinitely renormalizable points for which to complete the proof of Conjecture 5.2. In this section, we extend the work of Yoccoz and prove that

Theorem 5.32 [JI17]. *There is a subset Υ in $\mathcal{M}$ such that* **(1)** *Υ is dense in the boundary $\partial\mathcal{M}$ of the Mandelbrot set $\mathcal{M}$,* **(2)** *every c in Υ is infinitely renormalizable, and* **(3)** *$\mathcal{M}$ is locally connected at every point c in Υ.*

Let us start the construction of the subset Υ in Theorem 5.32 around the point $-2 \in \mathcal{M}$. Consider the polynomial $P(z) = z^2 - 2$. Its Julia set

J is the interval $[-2, 2]$. It has the non-separate fixed point $\beta = 2$ and the separate fixed point α, this means that $J \setminus \{\beta\}$ is still connected and $J \setminus \{\alpha\}$ is disconnected. Let γ be a fixed topological curve isomorphic to a circle such that $P : U \to V$ is a quadratic-like map where U is the domain bounded by γ and V is the domain bounded by $P(\gamma)$. For example, we can take a fixed equipotential curve (see §5.4) as the curve γ. Let Γ be a topological curve in $\overline{\mathbf{C}}$ isomorphic to a circle such that $\infty \in \Gamma$, such that $P(\Gamma) = \Gamma$, and such that $\Gamma \cap J = \{\alpha\}$. For example, we can take the union of two external rays (see §5.4) of P which land at α and ∞ as the topological curve Γ. Then $\mathbf{C} \setminus \Gamma$ consists of two disjoint domains. Let D denote the closure of the domain bounded by γ. Let $\tilde{\Gamma} = P^{-1}(\Gamma)$ and let $0 \in D_0$ be the closure of the domain bounded by γ and $\tilde{\Gamma}$ (see Fig. 5.13).

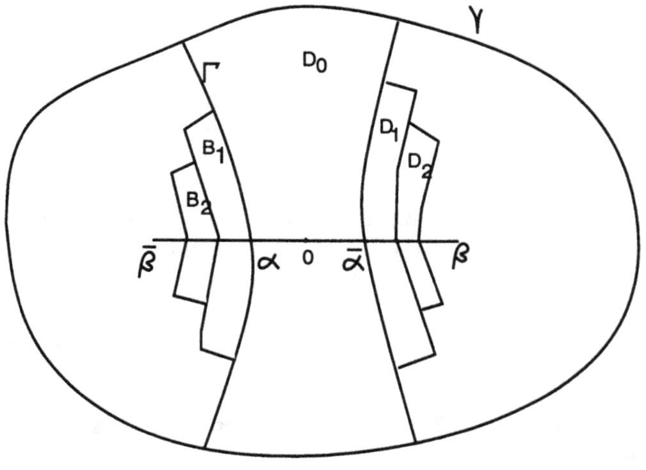

Fig. 5.13

The restriction $P|(\mathbf{C} \setminus (-\infty, -2])$ has two inverse branches

$$G_0 : \mathbf{C} \setminus (-\infty, -2] \to \mathbf{LH} = \{z = x + yi \in \mathbf{C} \mid x < 0\}$$

and

$$G_1 : \mathbf{C} \setminus (-\infty, -2] \to \mathbf{RH} = \{z = x + yi \in \mathbf{C} \mid x > 0\}.$$

Let $D_n = G_1^{\circ n}(D_0)$ and let $B_n = G_0(D_{n-1})$ for $n \geq 1$ (see Fig. 5.13). Since 2 is an expanding fixed point of P and $P(-2) = 2$, the diameter $\mathrm{diam}(B_n)$ tends to zero exponentially as n goes to infinity.

Consider the graph $\mathcal{G}_{-2}$ which consists of the boundaries of all B_n for $1 \leq n < \infty$, the boundaries of all D_n for $0 \leq n < \infty$, and $P^{-1}(\beta)$. For a small neighborhood Ω of $\mathcal{G}_{-2}$, $P|\Omega$ is a hyperbolic endomorphism. From the structural stability theorem (see [SHU,PRZ]), any C^1-perturbation f of $P|\Omega$ is topologically conjugate to $P|\Omega$, i.e., there is a homeomorphism $h : \Omega \to h(\Omega)$ (close to the identity) such that $f \circ h = h \circ P$ on $P^{-1}(\Omega)$. Therefore, the graph $\mathcal{G}_{-2}$ is preserved for P_c as long as c is close to -2. Let $0 \notin U_0$ be a small neighborhood about -2 such that $\mathrm{diam}(U_0) \leq 1$ and such that the corresponding graph $\mathcal{G}_c$ for P_c exists for c in U_0. For c in U_0, let $h_c : G_{-2} \to G_c$ be the homeomorphism such that $P_c \circ h_c = h_c \circ P$, let $\gamma_c = h_c(\gamma)$, let D_c be the closure of the domain bounded by γ_c, let $\Gamma_c = h_c(\Gamma \cap D)$, and let $\tilde{\Gamma}_c = h_c(\tilde{\Gamma} \cap D)$. Let $0 \in D_{c,0}$ be the closure of the domain bounded by γ_c and $\tilde{\Gamma}_c$. Let $B_{c,n}$ and $D_{c,n}$ be the connected components of the inverse image of $D_{c,n-1}$ under P_c and in the same sides as B_n and D_n, respectively, for $n = 1$, $2, \ldots$. Let $\beta(c)$ and $\alpha(c)$ be the non-separate and separate fixed points of P_c. Let $\tilde{\beta}(c)$ be another inverse image of $\beta(c)$ under P_c. One can check that $\Gamma_c \cap J_c = \{\alpha_c\}$ and that $B_{c,n}$ and $D_{c,n}$ tend to $\tilde{\beta}(c)$ and $\beta(c)$, respectively, as n goes to infinity. Since $\beta(c)$ is an expanding fixed point of P_c and since there is a constant $\mu > 1$ such that $|P_c'(\beta(c))| \geq \mu$ for all c in U_0, the diameter $\mathrm{diam}(B_n)$ tends to 0 uniformly on U_0 and the set $B_n(c)$ approaches to $\tilde{\beta}(c)$ uniformly on U_0 as n goes to infinity. Let

$$W_n = \{c \mid c \in B_n(c)\}$$

be a domain in the parameter space. Since the equation $P_c(0) - \tilde{\beta}(c) = 0$ has a unique solution -2 in U_0, the Rouché Theorem assures that $P_c(0) - x = 0$ has a unique solution for any x in $B_n(c)$ and any large n and the solution is close to -2. Therefore, there is an integer $N_0 > 0$ such that for $n \geq N_0$, $W_n \subset U_0$. That is, for $n \geq N_0$, $\mathrm{diam}(W_n) \leq 1$ (see Fig. 5.14).

Lemma 5.20. *The intersection $\tilde{\mathcal{M}}_n = W_n \cap \mathcal{M}$ for $n \geq N_0$ is connected.*

Proof. The boundary $B_n(c)$ consists of four topological curves $\gamma_n(c)$, $\gamma_{n+1}(c)$, $\Gamma_n(c)$, and $\overline{\Gamma}_n(c)$ where $P_c^{\circ n}(\gamma_n(c)) = \gamma(c)$, $P_c^{\circ(n+1)}(\gamma_{n+1}(c)) = \gamma(c)$, $P_c^{\circ n}(\Gamma_n(c) \cup \overline{\Gamma}_n(c)) \subset \Gamma(c)$. The domain W_n is bounded by four topological curves $\kappa_n = \{c \mid c \in \gamma_n(c)\}$, $\kappa_{n+1} = \{c \mid c \in \gamma_{n+1}(c)\}$, $\Psi_n = \{c \mid c \in \Gamma_n(c)\}$, and $\overline{\Psi}_n = \{c \mid c \in \overline{\Gamma}_n(c)\}$, where $\kappa_n \cap \mathcal{M} = \{a_n\}$ and $\kappa_{n+1} \cap \mathcal{M} = \{a_{n+1}\}$. Since $\overline{\mathbf{C}} \setminus \mathcal{M}$ is simple connected, κ_n (or κ_{n+1}) can be extended to a topological curve isomorphic to a circle in $\overline{\mathbf{C}} \setminus \mathcal{M}$ which passes ∞ and a_n (or a_{n+1}). Let us still denote these extended curves as κ_n and κ_{n+1}. Then κ_n (or κ_{n+1}) cuts $\mathbf{C}$ into two domains, one of which, denote as X_n (or X_{n+1}), is disjoint with

W_n. The sets X_n, X_{n+1}, and W_n divide the Mandelbrot set $\mathcal{M}$ into three connected components.

Assume $\tilde{\mathcal{M}}_n$ is disconnected. Let U and V be two non-empty domains such that $U \cap V = \emptyset$ and such that $(U \cap \tilde{\mathcal{M}}_n) \cup (V \cap \tilde{\mathcal{M}}_n) = \tilde{\mathcal{M}}_n$. Assume $a_n \in U$ and $a_{n+1} \in V$ (other cases can be proved similarly). Let $\tilde{U} = U \cup X_n$ and $\tilde{V} = V \cup X_{n+1}$. Then $\tilde{U}$ and $\tilde{V}$ are two non-empty domains such that $\tilde{U} \cap \tilde{V} = \emptyset$ and such that $(\tilde{U} \cap \mathcal{M}) \cup (\tilde{V} \cap \mathcal{M}) = \mathcal{M}$. This would say that $\mathcal{M}$ is disconnected. It is a contradiction. ■

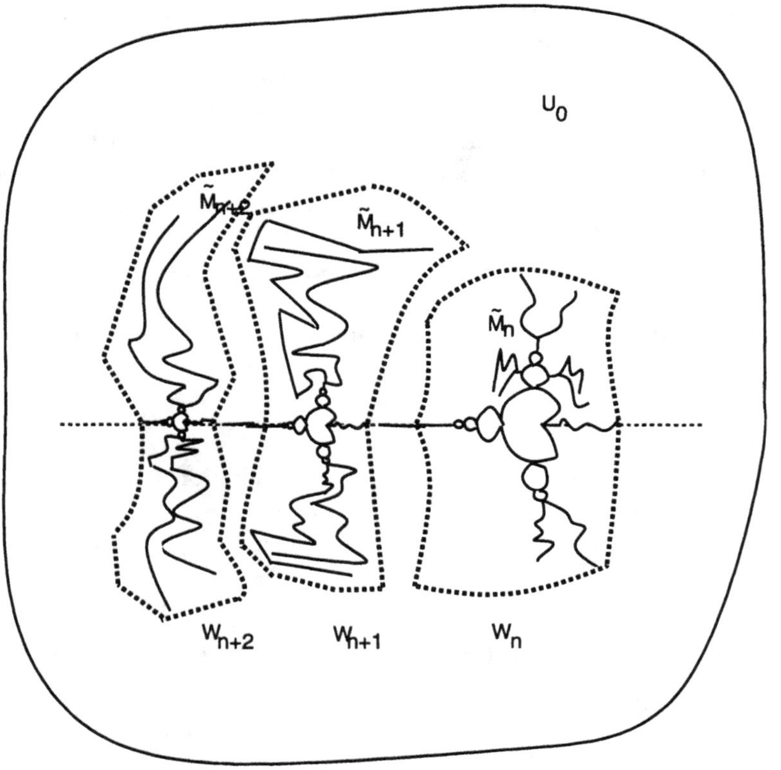

Fig. 5.14

For each W_n where $n \geq N_0$, since $c \in B_n(c)$, $C_n(c) = P_c^{-1}(B_n(c))$ is the closure of a connected and simply connected domain which contains 0 and which is a subdomain of $D_0(c)$. Let

$$F_{n,c} = P_c^{\circ(n+1)} : \mathring{C}_n(c) \to \mathring{D}_0(c).$$

Then it is a quadratic-like map (see Fig. 5.15). The family

$$\{F_{n,c} : \mathring{C}_n(c) \to \mathring{D}_0(c) \mid c \in W_n\}$$

is full (see [DH3]). So W_n contains a copy $\mathcal{M}_n$ of the Mandelbrot set $\mathcal{M}$ (see [DH3]). For $c \in \mathcal{M}_n$, the Julia set $J_{F_{n,c}}$ of $F_{n,c} : \mathring{C}_n(c) \to \mathring{D}_0(c)$ is connected. Therefore, for

$$c \in \Upsilon_1 = \cup_{n \geq N_0}^{\infty} \mathcal{M}_n,$$

P_c is once renormalizable.

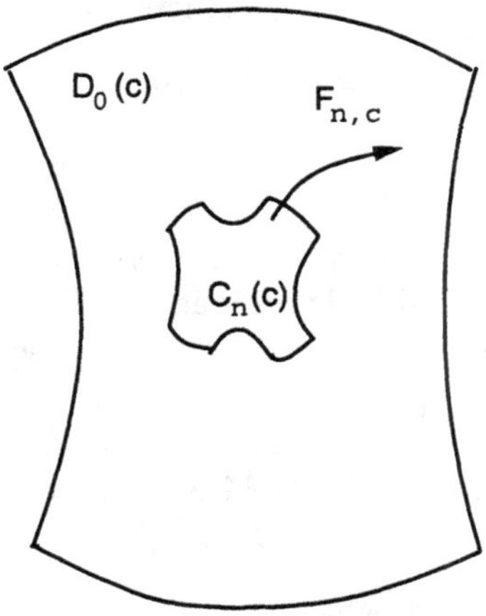

Fig. 5.15

For a fixed integer $i_0 \geq N_0$, consider W_{i_0} and $\mathcal{M}_{i_0}$; there is a parameter $c_{i_0} \in \mathcal{M}_{i_0}$ such that

$$F_{i_0} = F_{i_0, c_{i_0}} : \mathring{C}_{i_0} = \mathring{C}_{i_0}(c_{i_0}) \to \mathring{D}_{i_0} = \mathring{D}_0(c_{i_0})$$

is hybrid equivalent to $P(z) = z^2 - 2$. The quadratic-like map $F_{i_0} : \mathring{C}_{i_0} \to \mathring{D}_{i_0}$ has the non-separate fixed point β_{i_0} and the separate fixed point α_{i_0}. Let $\tilde{\beta}_{i_0}$ be another pre-image of β_{i_0} under F_{i_0}. Let Γ_{i_0} be a topological curve in $\overline{\mathbf{C}}$ isomorphic to a circle such that $\infty \in \Gamma_{i_0}$, such that $F_{i_0}(\Gamma_{i_0}) = \Gamma_{i_0}$, and such that $\Gamma_{i_0} \cap J_{c_{i_0}} = \{\alpha_{i_0}\}$. For example, we can take the union of two external rays (see §5.4) of $P_{c_{i_0}}$ which land at α_{i_0} and at ∞ as the topological curve Γ_{i_0}. Then $\mathbf{C} \setminus \Gamma_{i_0}$ consists of two disjoint domains. Let $D_{i_0 0}$ be the domain which contains 0 and which is bounded by ∂C_{i_0} and $\tilde{\Gamma}_{i_0} = F_{i_0}^{-1}(\Gamma_{i_0})$. Let $\tilde{\beta}_{i_0} \in E_{i_0 0}$

and $\beta_{i_0} \in E_{i_0 1}$ be the components of the closure of $C_{i_0} \setminus D_{i_0 0}$. Let $G_{i_0 0}$ and $G_{i_0 1}$ be the inverses of $F_{i_0}|E_{i_0 0}$ and $F_{i_0}|E_{i_0 1}$. Let

$$D_{i_0 n} = G_{i_0 1}^{\circ n}(D_{i_0 0})$$

and let

$$B_{i_0 n} = G_{i_0 0}(D_{i_0 (n-1)})$$

for $n \geq 1$. Again by the structural stability theorem, the points β_{i_0}, $\tilde{\beta}_{i_0}$, and α_{i_0} and the topological curves Γ_{i_0} and ∂C_{i_0} are preserved by a small perturbation c of c_{i_0}. Similar to the argument after Fig. 5.13, we can find a small neighborhood U_{i_0} about c_{i_0} with $\mathrm{diam}(U_{i_0}) \leq 1/2$ such that the corresponding domains $B_{i_0}(c)$ and $D_{i_0}(c)$ can be constructed for P_c for $c \in U_{i_0}$. Let

$$W_{i_0 n} = \{c \in \mathbf{C} \mid F_{i_0,c}(0) \in B_{i_0 n}(c)\}$$

be the closure of a domain in the parameter space.

The diameter $\mathrm{diam}(B_{i_0 n}(c))$ tends to zero uniformly on U_{i_0} and the set $B_{i_0 n}(c)$ approaches to $\tilde{\beta}_{i_0}(c)$ uniformly on U_{i_0} as n goes to infinity. Since the equation $F_{i_0,c}(0) - \tilde{\beta}_{i_0}(c) = 0$ has a unique solution c_{i_0}, the Rouché Theorem implies that $F_{i_0,c}(0) - x = 0$ has a unique solution near c_0 for any $x \in B_{i_0 n}(c)$ and any n large. Thus, there is an integer $N_{i_0} \geq 0$ such that for $n \geq N_{i_0}$, $W_{i_0 n} \subseteq U_{i_0}$. This implies that $\mathrm{diam}(W_{i_0 n}) \leq 1/2$. Similar to the proof of Lemma 5.20, we have $\tilde{\mathcal{M}}_{i_0 n} = W_{i_0} \cap \mathcal{M}$ is connected for $n \geq N_{i_0}$. For each c in $W_{i_0 n}$, $n \geq N_{i_0}$, let $C_{i_0 n}(c) = F_{i_0,c}^{-1}(B_{i_0 n}(c))$. Then

$$F_{i_0 n,c} = F_{i_0,c}^{\circ(n+1)} : \mathring{C}_{i_0 n}(c) \to \mathring{D}_{i_0 0}(c)$$

is a quadratic-like map. Moreover,

$$\{F_{i_0 n,c}\mathring{C}_{i_0 n}(c) \to \mathring{D}_{i_0 0}(c) \mid c \in W_{i_0 n}\}$$

is a full family. Thus $W_{i_0 n}$ contains a copy $\mathcal{M}_{i_0 n}$ of the Mandelbrot set $\mathcal{M}$. For

$$c \in \Upsilon_2 = \cup_{i_0 \geq N_0} \cup_{i_1 \geq N_{i_0}} \mathcal{M}_{i_0 i_1},$$

P_c is twice renormalizable.

We use the induction to complete the construction of the subset $\Upsilon(-2)$ around -2. Suppose we have constructed W_w where $w = i_0 i_1 \ldots i_{k-1}$ and $i_0 \geq N_0$, $i_1 \geq N_{i_1}$, ..., $i_{k-1} \geq N_{i_0 i_1 \ldots i_{k-2}}$. Let $v = i_0 \ldots i_{k-2}$. There is a parameter $c_w \in \mathcal{M}_w$ such that

$$F_w = F_{w,c_w} : \mathring{C}_w = \mathring{C}_w(c_w) \to \mathring{D}_w = \mathring{D}_{v0}(c_w)$$

is hybrid equivalent to $P(z) = z^2 - 2$. The quadratic-like map $F_w : \mathring{C}_w \to \mathring{D}_w$ has the non-separate fixed point β_w and the separate fixed point α_w. Let $\tilde{\beta}_w$ be another pre-image of β_w under F_w. Let Γ_w be a topological curve in $\overline{\mathbf{C}}$ isomorphic to a circle such that $\infty \in \Gamma_w$, such that $P(\Gamma_w) = \Gamma_w$, and such that $\Gamma_w \cap J_{c_w} = \{\alpha_w\}$. For example, we can take the union of two external rays (see §5.4) of P_{c_w} which land at α_w and at ∞ as the topological curve Γ_w. Then $\mathbf{C} \setminus \Gamma_w$ consists of two disjoint domains. Let D_{w0} be the domain which contains 0 and which is bounded by ∂C_w and $\tilde{\Gamma}_w = F_w^{-1}(\Gamma_w)$. Let $\tilde{\beta}_w \in E_{w0}$ and $\beta_w \in E_{w1}$ be the components of the closure of $C_w \setminus D_{w0}$. Let G_{w0} and G_{w1} be the inverses of $F_w|E_{w0}$ and $F_w|E_{w1}$. Let

$$D_{wn} = G_{w1}^{\circ n}(D_{w0})$$

and let

$$B_{wn} = G_{w0}(D_{w(n-1)})$$

for $n \geq 1$. By the structure stability theorem, the points β_w, $\tilde{\beta}_w$, and α_w and the topological curves Γ_w and ∂C_w are preserved by a small perturbation c of c_w. We can find a small neighborhood U_w about c_w with $\mathrm{diam}(U_w) \leq 1/2^k$ such that the corresponding domains $D_{wn}(c)$ and $B_{wn}(c)$ can be constructed for P_c, $c \in U_w$. Let

$$W_{wn} = \{c \in \mathbf{C} \mid F_{w,c}(0) \in B_{wn}(c)\}$$

be the closure of a domain in the parameter space.

The diameter $\mathrm{diam}(B_{wn}(c))$ tends to zero uniformly on U_w and the set $B_{wn}(c)$ approaches to $\tilde{\beta}_w(c)$ uniformly on U_w as n goes to infinity. Since the equation $F_{w,c}(0) - \tilde{\beta}_w(c) = 0$ has a unique solution c_w, the Rouché Theorem implies that $F_{w,c}(0) - x = 0$ has a unique solution near c_w for any $x \in B_{wn}(c)$ and any n large. Thus, there is an integer $N_w \geq 0$ such that for $n \geq N_w$, $W_{wn} \subseteq U_w$. This implies that $\mathrm{diam}(W_{wn}) \leq 1/2^k$. Similar to the proof of Lemma 5.20, we have $\tilde{\mathcal{M}}_{wn} = W_{wn} \cap \mathcal{M}$ for $n \geq N_w$ is connected.

For each c in W_{wn} where $n \geq N_w$, let $C_{wn}(c) = F_{w,c}^{-1}(B_{wn}(c))$. Then

$$F_{wn,c} = F_{w,c}^{\circ(n+1)} : \mathring{C}_{wn}(c) \to \mathring{D}_{w0}(c)$$

is a quadratic-like map. Moreover,

$$\{F_{wn,c} : \mathring{C}_{wn}(c) \to \mathring{D}_{w0}(c) \mid c \in W_{wn}\}$$

is a full family. Therefore, W_{wn} contains a copy $\mathcal{M}_{wn} \subset \tilde{\mathcal{M}}_{wn}$ of the Mandelbrot set $\mathcal{M}$. For

$$c \in \Upsilon_{k+1} = \cup_w \cup_{i_k \geq N_w} \mathcal{M}_{wi_k},$$

P_c is $(k+1)$-*times* renormalizable where $w = i_0 i_1 \dots i_{k-1}$ runs over all sequences of integers of length k in the induction.

We have thus constructed a subset $\Upsilon(-2) = \cap_{k=1}^{\infty} \Upsilon_k$ such that for each $c \in \Upsilon(-2)$, P_c is infinitely renormalizable and such that -2 is a limit point of $\Upsilon(-2)$. For each $c \in \Upsilon(-2)$, there is a corresponding sequence $w_\infty = i_0 i_1 \dots i_k \dots$ of integers such that

$$\{c\} = \cap_{k=0}^{\infty} W_{i_0 \dots i_k}.$$

Since

$$\tilde{\mathcal{M}}_{i_0 \dots i_k} = W_{i_0 \dots i_k} \cap \mathcal{M}$$

is connected,

$$\{W_{i_0 \dots i_k}\}_{k=0}^{\infty}$$

is a basis of connected neighborhoods of the Mandelbrot $\mathcal{M}$ at c. In other words, the Mandelbrot set $\mathcal{M}$ is locally connected at c.

Now we construct a subset $\Upsilon(c_0)$ for every preperiodic point c_0. Let c_0 be a preperiodic point in $\mathcal{M}$ and let J_{c_0} be its Julia set. Then there is the smallest integer $m \geq 1$ such that $p = P_{c_0}^{\circ m}(0)$ is a repelling periodic point of P_{c_0} of period $k \geq 1$. Let α be the separate fixed point of P_{c_0}. Without loss of generality, we assume that P_{c_0} is non-renormalizable. (If P_{c_0} is renormalizable, it must be finitely renormalizable. We would then take α as the separate fixed point of the last renormalization of P_{c_0} (see §5.7). Let Γ be the union of topological curves in $\overline{\mathbf{C}}$ isomorphic to a circle such that each of curves passes ∞ and α, such that $P_{c_0}(\Gamma) = \Gamma$, and such that $\Gamma \cap J_{c_0} = \{\alpha\}$. For example, we can take the union of a cycle of external rays (see §5.4) of P_{c_0} which land at α and at ∞ as the union Γ. Then $\mathbf{C} \setminus \Gamma$ consists of finitely many disjoint domains. Let γ be a topological curve isomorphic to a circle such that $P_{c_0} : U \to V$ is a quadratic-like map where U and V are the domains bounded by γ and $P_{c_0}(\gamma)$, respectively. We can construct the two-dimensional Yoccoz puzzle as follows (refer to §5.6 for the notation). Let C_{-1} be closure of the domain bounded by γ. Then Γ cuts $\mathring{C}_{-1}$ into a finite number of domains. Let η_0 denote the set of the closures of these domains. Let $\eta_n = P_{c_0}^{-n}(\eta_0)$. Let C_n be the member of η_n containing 0 for $n \geq 0$.

Let

$$p \in \dots \subseteq D_n(p) \subseteq D_{n-1}(p) \subseteq \dots \subseteq D_1(p) \subseteq D_0(p)$$

be a p-end, that means that $p \in D_n(p)$, $D_n(p) \subseteq D_{n-1}(p)$, and $D_n(p) \in \eta_n$. Let

$$c_0 \in \dots \subseteq E_n(c_0) \subseteq E_{n-1}(c_0) \subseteq \dots \subseteq E_1(c_0) \subseteq E_0(c_0)$$

be a c_0-end, that means that $c_0 \in E_n(p)$, $E_n(p) \subseteq E_{n-1}(p)$, and $E_n(p) \in \eta_n$. We have

$$P_{c_0}^{\circ(m-1)}(E_{n+m-1}(c_0)) = D_n(p).$$

Since the diameter $\operatorname{diam}(D_n(p))$ tends to zero as $n \to \infty$ and since p is a repelling periodic point, we can find an integer $l \geq m$ such that $|(P_{c_0}^{\circ k})'(x)| \geq \lambda > 1$ for all $x \in D_l(p)$ and such that

$$P_{c_0}^{\circ(m-1)} : E_{l+m-1}(c_0) \to D_l(p)$$

is a homeomorphism. Let $q \geq 0$ be the integer such that

$$f = P_{c_0}^{\circ q} : D_l(p) \to C_{r_0}$$

is a homeomorphism, where C_{r_0} is the domain containing 0 in η_{r_0}, $r_0 \geq 0$. There is an integer $r > r_0$ such that $r + q > l$ and such that $B_0 = f^{-q}(C_r)$ does not contain p. Thus

$$P_{c_0}^{\circ q} : B_0 \to C_r$$

is a homeomorphism. Define

$$B_n = \left(P_{c_0}^{\circ nk} | D_{l+nk}(p)\right)^{-1}(B_0) \subseteq D_{l+nk}(p)$$

for $n \geq 1$. Note that B_n is in η_{r+q+nk}. Then

$$P_{c_0}^{\circ(q+nk)} : B_n \to C_r$$

is a homeomorphism.

By the structure stability theorem, the points α and p, the union Γ of topological curves, the topological curve γ, and the sets C_r, D_n, and B_n, for $n \geq 0$, are preserved by a small perturbation c of c_0. Therefore they can be constructed for P_c as long as c close to c_0. Let U_0 be a neighborhood about c_0 with $\operatorname{diam}(U_0) \leq 1$ such that the corresponding points $\alpha(c)$ and $p(c)$ and the corresponding union $\Gamma(c)$ of topological curves, corresponding topological curve $\gamma(c)$, and corresponding sets $C_r(c)$, $D_n(c)$, and $B_n(c)$, for $n \geq 0$, are all preserved for $c \in U_0$. As n goes to infinity, the diameter $\operatorname{diam}(B_n(c))$ tends to zero uniformly on U_0 and the set $B_n(c)$ approaches to $p(c)$ uniformly on U_0. Let

$$W_n = W_n(c_0) = \{c \in \mathbf{C} \mid P_c^m(0) \in B_n(c)\}.$$

Since the equation $P_c^{\circ m}(0) - p(c) = 0$ has a unique solution c_0 in U_0 (see [DH2]), the Rouché Theorem implies that the equation $P_c^{\circ m}(0) - x = 0$ has a unique solution for any x in $B_n(c)$ and large n and that the solution is close to c_0. Therefore, there is an integer $N_0 = N_0(c_0) > 0$ such that $W_n \subset U_0$. Thus $\mathrm{diam}(W_n) \leq 1$ for all $n \geq N_0$. A similar argument to the proof of Lemma 5.20 implies that, for $n \geq N_0$,

$$\tilde{\mathcal{M}}_n = \mathcal{M} \cap W_n$$

is connected (see Fig. 5.16).

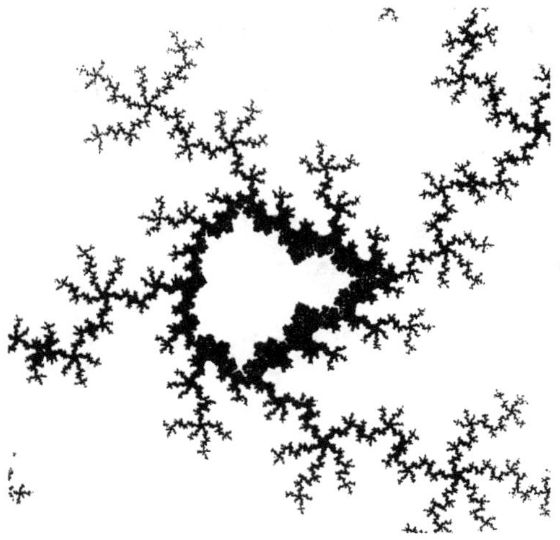

A small copy of the Mandelbrot set and hairs around it.

Fig. 5.16

For any $c \in W_n$, $n \geq N_0$, let $R_n(c)$ be the pre-image of $B_n(c)$ under the map

$$P_{c_0}^{\circ(m-1)} : E_{l+m-1}(c) \to D_l(p, c)$$

and let $C_{m+r+q+nk}(c) = P_c^{-1}(R_n(c))$. Then $C_{m+r+q+nk}(c)$ is the closure of a domain which contains 0 and which is in $\eta_{m+r+q+nk}$. Hence

$$F_{n,c} = P_c^{\circ(q+nk+m)} : \mathring{C}_{m+r+q+nk}(c) \to \mathring{C}_r(c)$$

is a quadratic-like map. Moreover,

$$\{F_{n,c} : \mathring{C}_{m+r+q+nk}(c) \to \mathring{C}_r(c) \mid c \in W_n\}$$

is a full family. Thus W_n contains a copy $\mathcal{M}_n = \mathcal{M}_n(c_0)$ of the Mandelbrot set $\mathcal{M}$ where $\mathcal{M}_n \subset \tilde{\mathcal{M}}_n$. For any

$$c \in \Upsilon_1(c_0) = \cup_{n \geq N_0} \mathcal{M}_n,$$

P_c is once renormalizable.

Now repeat the induction we did for -2, we can construct a subset $\Upsilon(c_0)$ such that every $c \in \Upsilon(c_0)$ is infinitely renormalizable, such that $\mathcal{M}$ is locally connected at every $c \in \Upsilon(c_0)$, and such that c_0 is a limit point of $\Upsilon(c_0)$.

Proof of Theorem 5.32. Let $\Upsilon = \cup_{c_0}\Upsilon(c_0)$ where c_0 runs over all preperiodic points in $\mathcal{M}$. Then every $c \in \Upsilon$ is infinitely renormalizable and $\mathcal{M}$ is locally connected at every $c \in \Upsilon$. From Lemma 5.19, the set Υ is dense in $\partial\mathcal{M}$. ■

Remark 5.9. Eckmann and Epstein [EE1] and Douady and Hubbard [DH3] estimated the size of $\mathcal{M}_n$ in the construction. Since $\tilde{\mathcal{M}}_n \setminus \mathcal{M}_n$ contains hairs (see Fig. 5.16) and may destroy the local connectivity of $\mathcal{M}$, we must estimate the size of $\tilde{\mathcal{M}}_n$.

Remark 5.10. Lyubich [LYU] studied Conjecture 5.2 by combinatorial rigidity which proposes another way to approach this problem. Shishikura [SHI] proved that the Hausdorff dimension of the boundary of the Mandelbrot set is two. Tan Lei [TA1,TA2] proved a certain self-similarity between the Mandelbrot set at a preperiodic point c_0 and the Julia set of $P_{c_0}(z) = z^2 + c_0$.

5.11. Certain Complex Infinitely Renormalizable Quadratic Polynomials

For a parameter c in the set Υ which is constructed in §5.10, we did not show that P_c is unbranched and has the *a priori* complex bounds. We can not conclude from Theorem 5.28 that the filled-in Julia set J_c is locally connected (although we can prove that K_c is locally connected at the critical point 0). However, in this section, we modify the construction in §5.10 and prove that

Theorem 5.33 [JI16]. *There is a subset $\tilde{\Upsilon}$ in $\mathcal{M}$ such that* **(1)** *$\tilde{\Upsilon}$ is dense in the boundary $\partial\mathcal{M}$ of the Mandelbrot set $\mathcal{M}$,* **(2)** *for every c in $\tilde{\Upsilon}$, P_c is unbranched, infinitely renormalizable and has the a priori complex bounds.*

From Theorem 5.28, we have

Corollary 5.3 [JI16]. *The filled-in Julia set K_c of P_c is locally connected for every c in $\tilde{\Upsilon}$.*

Proof of Theorem 5.33. We use the same notation as in the proof of Theorem 5.32. Suppose c_0 is a preperiodic (or Misiurewicz) point in $\mathcal{M}$. Then there is an integer $m > 1$ such that $p = P_{c_0}^{\circ m}(0)$ is a repelling periodic point of period $k \geq 1$. Let α be the separate fixed point of P_{c_0}. Without loss of generality, we assume that P_{c_0} is non-renormalizable. (If P_{c_0} is renormalizable, it must be finitely renormalizable. We would then take α as the separate fixed point of the last renormalization of P_{c_0} (see §5.7).) Let Γ be the union of a cycle of external rays landing at α. Let γ be a fixed equipotential curve of P_{c_0}. Using Γ and γ, we construct the two-dimensional Yoccoz puzzle as follows (see §5.6). Let C_{-1} be the domain bounded by γ. The set Γ cuts C_{-1} into a finite number of closed domains. Let η_0 denote the set of these domains. Let $\eta_n = P_{c_0}^{-n}(\eta_0)$. Let C_n be the member of η_n containing 0.

Let

$$p \in \cdots \subseteq D_n(p) \subseteq D_{n-1}(p) \subseteq \cdots \subseteq D_1(p) \subseteq D_0(p)$$

be a p-end, where $D_n(p) \in \eta_n$. Let

$$c_0 \in \cdots \subseteq E_n(c_0) \subseteq E_{n-1}(c_0) \subseteq \cdots \subseteq E_1(c_0) \subseteq E_0(c_0)$$

be a c_0-end, where $E_n(c_0) \in \eta_n$. We have $P_{c_0}^{\circ(m-1)}(E_{n+m-1}(c_0)) = D_n(p)$.

Since the diameter $\mathrm{diam}(D_n(p))$ tends to zero as $n \to \infty$ and since p is a repelling periodic point, we can find an integer $l \geq m$ such that $|(P_{c_0}^{\circ k})'(x)| \geq \lambda > 1$ for all $x \in D_l(p)$ and such that

$$P_{c_0}^{\circ(m-1)} : E_{l+m-1}(c_0) \to D_l(p)$$

is a homeomorphism. Let $q > 0$ be the smallest integer such that $P_{c_0}^{\circ q}(D_l(p))$ contains 0, i.e., it is C_{r_0} in η_{r_0} where $r_0 \geq 0$. Then

$$f = P_{c_0}^{\circ q} : D_l(p) \to C_{r_0}$$

is a homeomorphism. Let $r > r_0$ be an integer such that $B_0 = f^{-1}(C_r) \subset D_l(p)$ does not contain p where C_r is the member of η_r containing 0. Then

$$P_{c_0}^{\circ q} : B_0 \to C_r$$

is a homeomorphism. The domain B_0 is a member of η_{r+q}. Let $B_n \subseteq D_{l+nk}$ be the pre-image of B_0 under $P_{c_0}^{\circ nk}|D_{l+nk}(p)$ for $n \geq 1$. The domain B_n is a member of η_{r+q+nk} and

$$P_{c_0}^{\circ(q+nk)} : B_n \to C_r$$

is a homeomorphism.

From the structural stability theorem (see [PRZ,SHU]), the points α and p and the sets Γ, C_r, D_n and B_n for $n \geq 1$ are all preserved by a small perturbation c of c_0 (refer to §5.10). Therefore they can be constructed for P_c as long as c near c_0. Let U_0 be a neighborhood about c_0 such that the corresponding points $\alpha(c)$ and $p(c)$ and the corresponding sets $\Gamma(c)$, $C_r(c)$, $D_n(c)$, and $B_n(c)$ for $n \geq 1$ are all preserved for $c \in U_0$. Moreover, as n goes to infinity, the diameter $\mathrm{diam}(B_n(c))$ tends to zero and the set $B_n(c)$ approaches to $p(c)$ uniformly on U_0. Let

$$W_n = W_n(c_0) = \{c \in \mathbf{C} \mid P_c^m(0) \in B_n(c)\}.$$

Then from §5.10, $W_n \subseteq U_0$ for n large enough.

For any $c \in W_n$, Let $R_n(c)$ be the preimage of $B_n(c)$ under the map

$$P_{c_0}^{\circ(m-1)} : E_{l+m-1}(c) \to D_l(p, c)$$

and let $C_{r+q+nk+m}(c) = P_c^{-1}(R_n(c))$. The domain $C_{r+q+nk+m}(c)$ is the member containing 0 in $\eta_{r+q+m+nk}$ and

$$F_{n,c} = P_c^{\circ(q+nk+m)} : X_n(c) = \mathring{C}_{r+q+nk+m}(c) \to Y_n(c) = \mathring{C}_r(c)$$

is a quadratic-like map. Let $A_n(c) = \mathring{C}_r(c) \setminus C_{r+q+nk+m}(c)$. Since the diameter

$$\mathrm{diam}(C_{r+q+nk+m}(c))$$

tends to zero as n goes to infinity uniformly in U_0, there is an integer $N_0 > 0$ such that

$$\mathrm{mod}(A_n(c)) \geq 1$$

for all $n \geq N_0$ and all $c \in W_n$. Since

$$\{F_{n,c} : X_n \to Y_n \mid c \in W_n\}$$

is a full family of quadratic-like maps, W_n contains a copy $\mathcal{M}_n = \mathcal{M}_n(c_0)$ of the Mandelbrot set $\mathcal{M}$ (see [DH3]). For any $c \in \mathcal{M}_n$, P_c is once renormalizable and

$$CO(c) \cap C_{r+q+nk+m}(c) = \{c(j(q+nk+m))\}_{j=0}^{\infty}$$

where $CO(c) = \{c(i) = P_c^{\circ i}(0)\}_{i=0}^{\infty}$. Let

$$\tilde{\Upsilon}_1(c_0) = \cup_{n=N_0}^{\infty} \mathcal{M}_n.$$

We use the induction to complete the construction of the subset $\tilde{\Upsilon}(c_0)$ around c_0. Suppose we have constructed W_w where $w = i_0 i_1 \ldots i_{k-1}$ and $i_0 \geq N_0$, $i_1 \geq N_{i_0}$, ..., $i_{k-1} \geq N_{i_0 i_1 \ldots i_{k-2}}$. There is a parameter $c_w \in \mathcal{M}_w$ such that

$$F_w = F_{w,c_w} : X_w = X_w(c_w) \to Y_w = Y_w(c_w)$$

is hybrid equivalent (see §5.3) to $P(z) = z^2 - 2$. For F_w, let β_w and α_w be its non-separate and separate fixed points. Let $\tilde{\beta}_w$ be another preimage of β_w under F_w. Let Γ_w be the external rays of P_{c_w} landing at α_w. Let Y_{w0} be the domain containing 0 and bounded by ∂X_w and $F_w^{-1}(\Gamma_w)$. Let $\tilde{\beta}_w \in E_{w0}$ and $\beta_w \in E_{w1}$ be the components of the closure of $X_w \setminus Y_{w0}$. Let G_{w0} and G_{w1} be the inverses of $F_w|E_{w0}$ and $F_w|E_{w1}$. Let

$$D_{wn} = G_{w1}^{\circ n}(D_{w0})$$

and

$$B_{wn} = G_{w0}(D_{w(n-1)})$$

for $n \geq 1$. From the structural stability theorem, the points β_w and α_w and the sets Γ_w are all preserved by a small perturbation c of c_w. Therefore we can find a small neighborhood U_w about c_w such that the corresponding domains $D_{wn}(c)$ and $B_{wn}(c)$ can be constructed for P_c, $c \in U_w$ (refer to §5.10). Let

$$W_{wn} = \{c \in \mathbf{C} \mid F_{w,c}(0) \in B_{wn}(c)\}.$$

The diameter $\operatorname{diam}(B_{wn}(c)) \to 0$ as $n \to \infty$ uniformly on U_w. From §5.10, $W_{wn} \subseteq U_w$ for n large.

For each c in W_{wn}, $n \geq N_w$, let $X_{wn}(c) = F_{w,c}^{-1}(\mathring{B}_{wn}(c))$ and $Y_{wn}(c) = \mathring{Y}_{w0}(c)$. Then

$$F_{wn,c} = F_{w,c}^{\circ(n+1)} : X_{wn}(c) \to Y_{wn}(c)$$

is a quadratic-like map. Let

$$A_{wn}(c) = X_{wn}(c) \setminus \overline{Y}_{wn}(c).$$

Since the diameter $\operatorname{diam}(Y_{wn}(c))$ tends to zero as n goes to infinity uniformly in U_w. There is an integer $N_w > 0$ such that

$$\operatorname{mod}(A_{wn}(c)) \geq 1$$

for all $n \geq N_w$ and all $c \in W_{wn}$. Since

$$\{F_{wn,c} : X_{wn} \to Y_{wn} \mid c \in W_{wn}\}$$

is a full family of quadratic-like maps, W_{wn} contains a copy $\mathcal{M}_{wn} = \mathcal{M}_{wn}(c_0)$ of the Mandelbrot set $\mathcal{M}$ (see [DH3]). For any $c \in \mathcal{M}_{wn}$, P_c is *k-times* renormalizable and $A_{wn}(c)$ contains no critical values of P_c. Let

$$\tilde{\Upsilon}_k(c_0) = \cup_w \cup_{n=N_w}^{\infty} \mathcal{M}_{wn}$$

where w runs over all sequences of integers of length k in the induction.

We have thus constructed a subset $\tilde{\Upsilon}(c_0) = \cap_{k=1}^{\infty} \tilde{\Upsilon}_k(c_0)$ such that every $c \in \tilde{\Upsilon}(c_0)$ is infinitely renormalizable and such that c_0 is a limit point of $\tilde{\Upsilon}(c_0)$. From the above construction, for every $c \in \tilde{\Upsilon}(c_0)$, P_c is unbranched and has the a *priori* complex bounds.

Let $\tilde{\Upsilon} = \cup_{c_0} \tilde{\Upsilon}(c_0)$ where c_0 runs over all preperiodic points in $\mathcal{M}$. Then for every $c \in \tilde{\Upsilon}$, P_c is unbranched infinitely renormalizable and has the a *priori* complex bounds. Since the set of all preperiodic is dense in $\partial\mathcal{M}$ (see Lemma 5.19), the set $\tilde{\Upsilon}$ is dense in $\partial\mathcal{M}$. It completes the proof of the theorem. ∎

Remark 5.11. Douady (see [MI3]) constructed an example of an infinitely renormalizable quadratic polynomial whose Julia set is not locally connected. Perez-Marco [PER] modified this method and found a set of infinitely renormalizable quadratic polynomials whose Julia sets are not locally connected. A robust condition was defined by McMullen in [MC1] for a complex infinitely renormalizable quadratic polynomial. A real infinitely renormalizable quadratic polynomial is robust. In [MC1,MC2], McMullen gave a nice discussion about the unbranched condition, the a *priori* complex bounds condition, the robust condition, and many other topics. It is conjectured that the Julia set of a robust infinitely renormalizable quadratic polynomial is locally connected.

Remark 5.12. Douady and Sullivan (see [SU5,MI2]) proved that if the Julia set J_c of a quadratic polynomial P_c is locally connected, then every periodic point in J_c is either repelling or parabolic. Thus if P_c has a Cremer point (see [MI2] for the definition), then its Julia set is not locally connected. More recently, Sørensen [SOR] found a set of quadratic polynomials such that each of them has a Cremer point. Petersen [PET] did some work on the local connectivity of the Julia set of certain quadratic polynomial with an irrational indifferent periodic point.

Chapter Six

Thermodynamical Formalism and the Renormalization Operator

This chapter introduces thermodynamical formalism and its application to the study of the spectrum of the renormalization operator.

Let $n \geq 1$ be an integer and let $\mathbf{R}^n$ (respectively, $\mathbf{C}^n$) be the n-dimensional real (respectively, complex) Euclidean space. Let $A = (a_{ij})_{n\times n}$ be an $n \times n$ matrix having complex entries. An eigenvalue λ of A is a complex number such that $A\mathbf{v} = \lambda\mathbf{v}$ for some non-zero vector $\mathbf{v} \in \mathbf{C}^n$; such a $\mathbf{v}$ is called an eigenvector. let $E_\lambda = \{\mathbf{v} \in \mathbf{C}^n \mid A\mathbf{v} = \lambda\mathbf{v}\}$ be the corresponding eigenspace of A. An eigenvector $\mathbf{v} = (v_1, \cdots, v_n)$ is called positive if $v_i > 0$ for all $1 \leq i \leq n$. An eigenvalue λ is called simple if $\dim(E_\lambda) = 1$. A matrix A is called positive if $a_{ij} > 0$ for all $1 \leq i, j \leq n$. For a positive $n \times n$ matrix A, we have the Perron-Frobenius theorem: A has a unique simple, positive, maximal eigenvalue λ with a positive eigenvector $\mathbf{v}$.

Perron-Frobenius type results have been developed in thermodynamical formalism. Let M^n be a compact, connected, oriented and smooth Riemannian manifold and let $\Omega \subset M^n$ be an open set. Let $\sigma : \Omega \to M^n$ be a C^2-expanding mapping, i.e., $||(D\sigma^{\circ n}(x))\mathbf{v}|| \geq C\lambda^n||\mathbf{v}||$ for all $n > 0$ and all x such that x, $\sigma(x)$, ..., $\sigma^{\circ(n-1)}(x)$ are in Ω, where $C > 0$ and $\lambda > 1$ are constants and where D means the derivative. Let

$$\mathcal{LV} = \{v \mid v \text{ is a complex Lipschitz continuous vector field on } \Omega\}.$$

A vector v is called positive (denoted as $v > 0$) if $v(x) > 0$ for all x in Ω. A linear operator $\mathcal{L} : \mathcal{LV} \to \mathcal{LV}$ is called positive if $\mathcal{L}v > 0$ whenever $v > 0$. An eigenvalue of a linear operator $\mathcal{L}$ is a complex number λ such that $\mathcal{L}v = \lambda v$

for a non-zero vector field $v \in \mathcal{LV}$. Let ϕ be a real Lipschitz function on Ω. A Ruelle's Perron-Frobenius operator $\mathcal{L}_\phi : \mathcal{LV} \to \mathcal{LV}$ is defined as

$$\mathcal{L}_\phi v(x) = \sum_{y \in \sigma^{-1}(x)} e^{\phi(y)} v(y) \tag{6.1}$$

for $v \in \mathcal{LV}$. It is a positive linear operator. Ruelle [RU1,RU2] proved that the operator $\mathcal{L}_\phi$ has a positive, simple, maximal eigenvalue λ with positive eigenvector h, and that the remainder of the spectra of $\mathcal{L}_\phi$ is contained in a disk of radius strictly less than λ. Moreover, if σ' and ϕ are C^k for $k = 1, 2, \ldots, \infty$, ω, then h is a C^k-vector field. Recently, Pollicott [PO1], Tangerman [TAG], and Ruelle [RU5] allowed the function ϕ in Eq. (6.1) to be a complex function. In this setting the operator $\mathcal{L}_\phi$ is not necessarily positive. They were still able to estimate the spectral radius and the essential spectral radius of $\mathcal{L}_\phi$ in the $C^{k+\alpha}$-setting for $k = 1, 2, \ldots, \omega$, and $0 \leq \alpha \leq 1$. An operator $\mathcal{L}_\phi$ with complex function ϕ is called a transfer operator. In [BJL], an exact formula for the essential spectral radius is obtained for a transfer operator acting on C^{1+Z} space and on $C^{1+\alpha}$ space, where Z means Zygmund and where $0 < \alpha \leq 1$. The interesting point of a transfer operator is that it connects with the study of invariant measures, entropy, dynamical zeta functions and many other topics. The reader who is interested in thermodynamical formalism may refer to the article of Sinai [SI2] and the books of Bowen [BO1], Ruelle [RU3], and Mayer [MA1]. The reader who is interested in some recent developments in this field may refer to articles of, among others, Parry and Pollicott [PAP], Mayer [MA2,MA3], Baladi [BAL], Ruelle [RU6], and Rugh [RUG]. Ruelle's monograph [RU3] gives useful background material and connection with statistical mechanics. Sinai's article [SI2] and Bowen's monograph [BO1] give discussions about Gibbs measures, ergodic theory, and hyperbolic dynamical systems. Parry and Pollicott's monograph [PAP] and Ruelle's monograph [RU6] discuss thermodynamical formalism, zeta functions for hyperbolic dynamical systems, and an extension of the theory of Fredholm determinants from the point of view of Markov partitions. Baladi's article [BAL] gives a very brief and clear review to the subject. An interesting research problem is to study thermodynamical formalism, zeta functions, and Fredholm determinants for non-renormalizable quadratic polynomials from the point of view of Yoccoz puzzles (see §5.6). Another interesting topic is to relate the study of thermodynamical formalism and dynamical systems with the study of Teichmüller theory (refer to papers of, among others, Cawley [CAW], Cui [CUI], Pinto and Sullivan [PNS], Sullivan [SU4], and Baladi, Jiang, and Lanford [BJL]).

In this chapter, we use a transfer operator $\mathcal{L}_\Phi$ to study the linearization $T_g\mathcal{R}$ of the renormalization operator $\mathcal{R}$ at the Feigenbaum fixed point

g. We discuss the conceptual proof of the existence of the Feigenbaum universal number δ worked out by the author together with Morita and Sullivan in [JMS]. We note that Eckmann and Epstein [EE2], Christiansen, Cvitanović, and Rugh [CCR], and Pollicott [PO2] have done similar work in this direction.

In §6.1, we introduce some basic results in thermodynamical formalism and prove Ruelle's Perron-Frobenius Theorem. In §6.2, we present some results on estimating the spectral radius and the essential spectral radius of a transfer operator. In §6.3, we use a transfer operator $\mathcal{L}_\Phi$ to study the linearization $T_g\mathcal{R}$ of the renormalization operator $\mathcal{R}$ at the Feigenbaum fixed point g. We prove that the eigenvalues of $\mathcal{L}_\Phi$ are the eigenvalues of the linearization $T_g\mathcal{R}$ except for the value 1. We use the linear operator $\mathcal{L}_n$ with the finite rank 2^{n-1} to approximate $\mathcal{L}_\Phi$ in the C^b-setting (C^b is the space of bounded vector fields). We prove the following statements: **(1)** Each $\mathcal{L}_n$ has an eigenvalue $\lambda_n > 1$ with a positive eigenvector v_n, which means that each component of v_n is positive. **(2)** There is a sequence $\{n_i\}_{i=0}^{\infty}$ of integers such that the limit $\delta = \lim_{i\to\infty} \lambda_{n_i} > 1$ is an eigenvalue of $\mathcal{L}_\Phi$ with an eigenvector $v = \lim_{n\to+\infty} v_{n_i}$ in C^b. **(3)** The limit δ is an eigenvalue of $\mathcal{L}_\Phi$ in the C^{0+1}-setting (C^{0+1} is the space of Lipschitz continuous vector fields). **(4)** The limit δ is an eigenvalue of $\mathcal{L}_\Phi$ in the C^ω-setting (C^ω is the space of analytic vector fields on $g(I)$). The proofs are constructive. By using $\mathcal{L}_n$, one can calculate the approximating expanding manifolds and the rate of expansion of $\mathcal{R}$.

6.1. Gibbs Measures, Pressures, and Ruelle's Perron-Frobenius Operators

Let $I = [-1,1]$, let $-1 < a < 0 < b < 1$, and let $I_0 = [-1,a]$ and $I_1 = [b,1]$. Let $\sigma : I_0 \cup I_1 \to I$ be a C^1 degree two expanding map (see §1.2). Let $g_0 : I \to I_0$ and $g_1 : I \to I_1$ be the two inverse branches of σ.

As in §1.2, we use $w_n = i_0 i_1 \ldots i_{n-1}$ to denote a finite string of 0's and 1's of length n and use $g_{w_n} = g_{i_0} \circ g_{i_1} \circ \cdots \circ g_{i_{n-1}}$ and $I_{w_n} = g_{w_n}(I)$ to denote the composition from g_{i_0} to $g_{i_{n-1}}$ and the image of I under g_{w_n}. Let η_n be the set of all intervals I_{w_n} for every $n \geq 1$ and $\Lambda = \cap_{n\geq 1} \cup_{w_n} I_{w_n}$ be the non-escaping set of σ. Let $\Lambda_{w_n} = \Lambda \cap I_{w_n}$ for all w_n.

Let $\mathcal{C}(\Lambda, \mathbf{R})$ be the Banach space of all continuous functions $\psi : \Lambda \to \mathbf{R}$ with the norm

$$||\psi|| = \max_{x\in\Lambda} |\psi(x)|.$$

Let $\mathcal{C}^*(\Lambda, \mathbf{R})$ be the space of all linear functionals $T : \mathcal{C}(\Lambda, \mathbf{R}) \to \mathbf{R}$. The weak $*$-topology on $\mathcal{C}^*(\Lambda, \mathbf{R})$ is generated by sets of the form

$$U(\psi, \epsilon, T_0) = \{T \in \mathcal{C}^*(\Lambda, \mathbf{R}) \,\Big|\, |T(\psi) - T_0(\psi)| < \epsilon\}$$

with ψ in $\mathcal{C}(\Lambda, \mathbf{R})$, $\epsilon > 0$, and T_0 in $\mathcal{C}^*(\Lambda, \mathbf{R})$. Let

$$\mathcal{C}_0^*(\Lambda, \mathbf{R}) = \{T \in \mathcal{C}^*(\Lambda, \mathbf{R}) \,\Big|\, T(1) = 1 \text{ and } T(\psi) \geq 0 \text{ whenever } \psi \geq 0\}$$

be a subspace of $\mathcal{C}^*(\Lambda, \mathbf{R})$, where $\psi \geq 0$ means that $\psi(x) \geq 0$ for all $x \in \Lambda$.

Let $\mathcal{PM}$ be the space of all Borel probability measures on Λ and let $\mathcal{PM}_\sigma$ be the subset of all invariant ones under σ in $\mathcal{PM}$, that is, $\mu(\sigma^{-1}(A)) = \mu(A)$ for all μ in $\mathcal{PM}_\sigma$ and all Borel sets in Λ. Let $\sigma^*(\mu) = \mu(\sigma^{-1}(A))$.

Lemma 6.1 (Riesz representation). *For each μ in $\mathcal{PM}$ define $T_\mu(\psi) = \int \psi \, d\mu$. Then $\mathcal{Q}(\mu) = T_\mu$ is a bijection between $\mathcal{PM}$ and $\mathcal{C}_0^*(\Lambda, \mathbf{R})$.*

We identify T_μ with μ and often write μ when we mean T_μ. Under this identification, $\mathcal{PM}$ is a compact convex metrizable space. A metric on $\mathcal{PM}$ we use is

$$d(\mu, \mu') = \sum_{n=1}^{\infty} 2^{-n} ||\psi_n|| |\mu(\psi_n) - \mu'(\psi_n)|$$

where $\{\psi_n\}_{n=1}^{\infty}$ is a dense subset of $\mathcal{C}(\Lambda, \mathbf{R})$.

Lemma 6.2. *The subspace $\mathcal{PM}_\sigma$ is a non-empty closed convex set.*

Proof. The map σ^* is a homeomorphism from $\mathcal{PM}$ onto $\mathcal{PM}$. And $\mathcal{PM}_\sigma = \{\mu \in \mathcal{PM} \mid \sigma^*(\mu) = \mu\}$. Pick μ in $\mathcal{PM}$, let

$$\mu_n = \frac{\mu + \sigma^*(\mu) + (\sigma^*)^2(\mu) + \cdots + (\sigma^*)^{n-1}(\mu)}{n}.$$

Choose a subsequence $\{\mu_{n_k}\}_{k=1}^{\infty}$ converging to μ_0. Then μ_0 is in $\mathcal{PM}_\sigma$. So $\mathcal{PM}_\sigma$ is non-empty.

Suppose μ_0 and μ_1 are in $\mathcal{PM}_\sigma$. Let $\mu_t = t\mu_1 + (1-t)\mu_0$ for $0 \leq t \leq 1$. Then

$$\sigma^*(\mu_t) = t\sigma^*(\mu_1) + (1-t)\sigma^*(\mu_0) = t\mu_1 + (1-t)\mu_0.$$

Thus μ_t is in $\mathcal{PM}_\sigma$. This implies that $\mathcal{PM}_\sigma$ is a convex set.

Suppose μ_n is a sequence in $\mathcal{PM}_\sigma$ converging to μ in $\mathcal{PM}$. Then $\sigma^*(\mu_n) = \mu_n$ converges to $\sigma^*(\mu) = \mu$. This says that $\mathcal{PM}_\sigma$ is closed. ∎

Remark 6.1. One can also check that μ is in $\mathcal{PM}_\sigma$ if and only if

$$\int (\psi \circ \sigma)\, d\mu = \int \psi\, d\mu,$$

that is, $\mu(\psi \circ \sigma) = \mu(\psi)$ for all ψ in $\mathcal{C}(\Lambda, \mathbf{R})$.

A function ψ in $\mathcal{C}(\Lambda, \mathbf{R})$ is said to be positive (denoted as $\psi > 0$) if $\psi(x) > 0$ for all x in Λ. A function ψ in $\mathcal{C}(\Lambda, \mathbf{R})$ is said to be α-Hölder for $0 < \alpha \leq 1$ if there is a constant $C > 0$ such that

$$|\psi(x) - \psi(y)| \leq C|x-y|^\alpha$$

for all x and y in Λ. Let $\mathcal{C}^\alpha(\Lambda, \mathbf{R})$ be the set of all α-Hölder continuous functions $\psi : \Lambda \to \mathbf{R}$ and

$$\mathcal{C}^+(\Lambda, \mathbf{R}) = \cup_{0<\alpha\leq 1}\mathcal{C}^\alpha(\Lambda, \mathbf{R})$$

be the union of $\mathcal{C}^\alpha(\Lambda, \mathbf{R})$ for all $0 < \alpha \leq 1$.

Theorem 6.1 (Gibbs measure). *Suppose ϕ is in $\mathcal{C}^+(\Lambda, \mathbf{R})$. Then there are a unique measure μ in $\mathcal{PM}_\sigma$, a unique constant P, and a constant $C > 0$ such that*

$$C^{-1} \leq \frac{\mu(\Lambda_{w_n})}{\exp\Big(-Pn + \sum_{k=0}^{n-1}\phi(\sigma^k(x))\Big)} \leq C$$

for all w_n, all x in Λ_{w_n}, and all $n \geq 0$.

Remark 6.2. The measure $\mu = \mu_\phi$ and $P = P(\phi)$ in the theorem are called the Gibbs measure and the pressure of ϕ.

Two functions ϕ_1 and ϕ_2 in $\mathcal{C}(\Lambda, \mathbf{R})$ are said to be homologous if there is a function u in $\mathcal{C}(\Lambda, \mathbf{R})$ such that

$$\phi_1(x) = \phi_2(x) + u\big(\sigma(x)\big) - u(x)$$

for all x in Λ.

Corollary 6.1. *Suppose ϕ_1 and ϕ_2 in $\mathcal{C}^+(\Lambda, \mathbf{R})$ are homologous. Then they have the same Gibbs measure and pressure.*

Proof. For every x in Λ and $n \geq 0$

$$\begin{aligned}\Big|\sum_{k=0}^{n-1} \phi_1(\sigma^k(x)) - \sum_{k=0}^{n-1} \phi_1(\sigma^k(x))\Big| &= \Big|\sum_{k=0}^{n-1} \Big(u\big(\sigma^{k+1}(x)\big) - u\big(\sigma^k(x)\big)\Big)\Big| \\ &= \big|u\big(\sigma^n(x)\big) - u(x)\big| \leq 2||u||.\end{aligned}$$

Suppose $\mu = \mu_{\phi_1}$ and $P = P(\phi_1)$ are the Gibbs measure and the pressure of ϕ_1, that is,

$$C^{-1} \leq \frac{\mu(\Lambda_{w_n})}{\exp\Big(-Pn + \sum_{k=0}^{n-1} \phi_1\big(\sigma^k(x)\big)\Big)} \leq C$$

for all Λ_{w_n} and all x in Λ_{w_n}. Then

$$C^{-1}\exp(-2||u||) \leq \frac{\mu(\Lambda_{w_n})}{\exp\Big(-Pn + \sum_{k=0}^{n-1} \phi_2\big(\sigma^k(x)\big)\Big)} \leq C\exp(2||u||)$$

for all Λ_{w_n} and all x in Λ_{w_n}. By the uniqueness of the Gibbs measure, $\mu_{\phi_2} = \mu$ and $P(\phi_2) = P$. ∎

For any ϕ in $\mathcal{C}^+(\Lambda, \mathbf{R})$ we define an operator $\mathcal{L}_\phi : \mathcal{C}(\Lambda, \mathbf{R}) \to \mathcal{C}(\Lambda, \mathbf{R})$ by

$$\mathcal{L}_\phi \psi(x) = \sum_{y \in \sigma^{-1}(x)} e^{\phi(y)} \psi(y) \tag{6.2}$$

where ψ is in $\mathcal{C}(\Lambda, \mathbf{R})$ and x is in Λ. It is called a Ruelle's Perron-Frobenius operator. It is a positive operator; this means that $\mathcal{L}_\phi \psi > 0$ if $\psi > 0$.

For a Ruelle's Perron-Frobenius operator $\mathcal{L} = \mathcal{L}_\phi$, let $\mathcal{L}^*$ be its dual operator defined as

$$\mathcal{L}^* \mu(\psi) = \mu(\mathcal{L}\psi)$$

for all μ in $\mathcal{PM}$ and all ψ in $\mathcal{C}(\Lambda, \mathbf{R})$.

Theorem 6.2 [RU1,RU2]. *Suppose ϕ is in $\mathcal{C}^+(\Lambda, \mathbf{R})$ and $\mathcal{L} = \mathcal{L}_\phi$. There are a constant $\lambda > 0$, a positive function h in $\mathcal{C}(\Lambda, \mathbf{R})$, and a measure ν in $\mathcal{PM}$ such that $\mathcal{L}h = \lambda h$, $\mathcal{L}^*\nu = \lambda\nu$, $\nu(h) = 1$, and*

$$\lim_{n\to\infty} ||\lambda^{-n}\mathcal{L}^n\psi - \nu(\psi)h|| = 0$$

for all ψ in $\mathcal{C}(\Lambda, \mathbf{R})$.

Remark 6.3. Then λ is a simple eigenvalue of $\mathcal{L}$ and h is a corresponding eigenvector.

Let us first see how Theorem 6.2 implies Theorem 6.1.

Proof of Theorem 6.1. Suppose h and ν are the function and the measure in Theorem 6.2. Let $\mu = h\nu$. Let us show that μ is invariant under σ.

For any ψ in $\mathcal{C}(\Lambda, \mathbf{R})$

$$\begin{aligned}\Big(\mathcal{L}h \cdot \psi\Big)(x) &= \sum_{y\in\sigma^{-1}(x)} e^{\phi(y)}h(y)\psi(x)\\ &= \sum_{y\in\sigma^{-1}(x)} e^{\phi(y)}h(y)\psi(\sigma(y))\\ &= \Big(\mathcal{L}\big(h\cdot(\psi\circ\sigma)\big)\Big)(x).\end{aligned}$$

Using this, we have that

$$\begin{aligned}\mu(\psi) = \nu(h\psi) &= \nu\Big(\lambda^{-1}\cdot\mathcal{L}h\cdot\psi\Big)\\ &= \lambda^{-1}\nu\Big(\mathcal{L}\big(h\cdot(\psi\circ\sigma)\big)\Big) = \lambda^{-1}\Big(\mathcal{L}^*\nu\big(h\cdot(\psi\circ\sigma)\big)\Big)\\ &= \nu\Big(h\cdot(\psi\circ\sigma)\Big) = \mu(\psi\circ\sigma)\,.\end{aligned}$$

Now let us show that μ is a Gibbs measure. Let $S_n(x) = \sum_{k=0}^{n-1}\phi(\sigma^{\circ k}(x))$. It is easy to check that

$$\mathcal{L}^{\circ n}\psi(x) = \sum_{y\in\sigma^{-n}(x)} e^{S_n(y)}\psi(y).$$

Since σ is an expanding map and ϕ is an α-Hölder continuous function for some $0 < \alpha \le 1$, there is a constant $C_0 > 0$ such that

$$|S_n(x) - S_n(x')| \le C_0 \tag{6.3}$$

for all Λ_{w_n} and all x and x' in Λ_{w_n} (see §1.2). Let $\chi = \chi_{\Lambda_{w_n}}$ be the function taking value 1 in Λ_{w_n} and 0 elsewhere. Since for any z in Λ, there is at most one y' in $\sigma^{-n}(z) \cap \Lambda_{w_n}$

$$\mathcal{L}^{\circ n}(h\chi)(z) = \sum_{y\in\sigma^{-n}(z)} e^{S_n(y)}h(y)\chi(y) = e^{S_n(y')}h(y') \le e^{S_n(x)}e^{C_0}||h||$$

for all x in Λ_{w_n} and all z in Λ. So

$$\mu(\Lambda_{w_n}) = \nu(h\chi) = \lambda^{-n}\nu\Big(\mathcal{L}^{\circ n}(h\chi)\Big) \le \lambda^{-n}e^{S_n(x)}e^{C_0}||h||.$$

On the other hand, for any z in Λ, there is at least one y' in $\sigma^{-n}(z) \cap \Lambda_{w_n}$. Then

$$\mathcal{L}^{\circ n}(h\chi)(z) \ge e^{S_n(y')}h(y') \ge e^{-C_0} \inf_{x\in\Lambda}\{h(x)\}e^{S_n(x)}$$

and

$$\mu(\Lambda_{w_n}) = \lambda^{-n}\nu\Big(\mathcal{L}^{\circ n}(h\chi)\Big) \ge \lambda^{-n}e^{S_n(x)}e^{-C_0} \inf_{x\in\Lambda}\{h(x)\}.$$

Take $C = \max\{e^{C_0}||h||,\ e^{C_0}(\inf_{x\in\Lambda}\{h(x)\})^{-1}\}$. Then we have

$$C^{-1}\lambda^{-n}e^{S_n(x)} \le \mu(\Lambda_{w_n}) \le C\lambda^{-n}e^{S_n(x)}$$

for all Λ_{w_n} and all x in Λ_{w_n}. Hence μ is the Gibbs measure and $P = \log\lambda$. From

$$\lim_{n\to\infty} ||\lambda^{-n}\mathcal{L}^n(\psi) - \nu(\psi)h|| = 0$$

for all ψ in $\mathcal{C}(\Lambda, \mathbf{R})$, one can see that μ and λ are unique. ∎

We prove Theorem 6.2 via several lemmas.

Lemma 6.3. *There is a ν in $\mathcal{PM}$ such that $\mathcal{L}^*(\nu) = \lambda\nu$ for a constant $\lambda > 0$.*

Proof. It is a corollary of the Schauder Tychonoff theorem (see [DUS]) which says that for a non-empty compact convex subset E of a locally convex topological vector space, any continuous map G form E into E has a fixed point. In our case, let $E = \mathcal{PM}$ and $G(\mu) = \Big(\mathcal{L}^*\mu(1)\Big)^{-1}\mathcal{L}^*\mu$. Then ν is a fixed point of G in $\mathcal{PM}$ and $\lambda = \mathcal{L}^*\nu(1)$. ∎

Since σ is expanding and ϕ is α-Hölder for some $0 < \alpha \le 1$, there are constants $A > 0$ and $0 < \tau < 1$ such that

$$\max_{x,x'\in\Lambda_{w_n}} |\phi(x) - \phi(x')| \le A\tau^n$$

for all Λ_{w_n}. Let $B_n = \exp(\sum_{k=n+1}^{\infty} 2A\tau^k)$ and let

$$\Omega = \{\psi \in \mathcal{C}(\Lambda, \mathbf{R}) \mid \psi \ge 0,\ \nu(\psi) = 1,\ \psi(x) \le B_n\psi(x') \text{ whenever } x, x' \in \Lambda_{w_n}\}.$$

Lemma 6.4. *There is an h in Ω with $\mathcal{L}h = \lambda h$ and $h > 0$.*

Proof. Let us check that $\lambda^{-1}\mathcal{L}\psi$ is in Ω when ψ is in Ω. Clearly, $\lambda^{-1}\mathcal{L}\psi \geq 0$ and

$$\nu\Big(\lambda^{-1}\mathcal{L}\psi\Big) = \lambda^{-1}\mathcal{L}^*\nu(\psi) = \nu(\psi) = 1.$$

Suppose x and x' are in Λ_{w_n}. For any y in $\sigma^{-1}(x) \cap \Lambda_{jw_n}$, let y' be the point in $\sigma^{-1}(x') \cap \Lambda_{jw_n}$, where j is 0 or 1. Then

$$e^{\phi(y)}\psi(y) \leq e^{\phi(y')}e^{A\tau^{n+1}}B_{n+1}\psi(y') \leq B_n e^{\phi(y')}\psi(y')$$

and so $\mathcal{L}\psi(x) \leq B_n\mathcal{L}\psi(x')$.

For any x and z in Λ, there is a y' in $\sigma^{-1}(x)$ such that y' and z are in the same Λ_{w_1}. From this fact, we have that for any ψ in Ω,

$$\mathcal{L}\psi(x) = \sum_{y\in\sigma^{-1}(x)} e^{\phi(y)}\psi(y) \geq e^{-||\phi||}\psi(y') \geq e^{-||\phi||}B_1^{-1}\psi(z).$$

Then for $K = \lambda e^{||\phi||}B_1$, we have $K^{-1}\psi(z) \leq \nu(\lambda^{-1}\mathcal{L}\psi) = 1$. This implies that $||\psi|| \leq K$ and

$$|\psi(x) - \psi(x')| \leq (B_n - B_n^{-1})K$$

for any x and x' in Λ_{w_n}. Thus Ω is equicontinuous and thus compact by the Ascoli-Arzelà theorem. The set Ω is not empty because 1 is in it. It is also convex. Applying the Schauder-Tychonoff theorem to $G = \lambda^{-1}\mathcal{L}$ and $E = \Omega$, we get a fixed point h of G in E with $\mathcal{L}h = \lambda h$. Since for any ψ in Ω, there is some z such that $\psi(z) \geq 1$, $\lambda^{-1}\mathcal{L}\psi(x) \geq K^{-1}$ for all x in Λ. In particular, $\inf_{x\in\Lambda}\{h(x)\} = \inf_{x\in\Lambda}\{\lambda^{-1}\mathcal{L}h(x)\} \geq K^{-1}$. But $||h|| \leq K$. So for all x in Λ,

$$K^{-1} \leq h(x) \leq K.$$

■

Lemma 6.5. *There is an η in $(0, 1)$ such that for any ψ in Ω, $\lambda^{-1}\mathcal{L}\psi = \eta h + (1 - \eta)\psi'$ for some ψ' in Ω.*

Proof. Let $\theta = \lambda^{-1}\mathcal{L}\psi - \eta h$ where η is to be determined. If $\eta||h|| \leq K^{-1}$, then $\theta \geq 0$. We must pick $0 < \eta \leq (K||h||)^{-1}$ such that for all x and x' in Λ_{w_n},

$$\eta\Big(B_n h(x') - h(x)\Big) \leq B_n\lambda^{-1}\mathcal{L}\psi(x') - \lambda^{-1}\mathcal{L}\psi(x). \tag{6.4}$$

This gives that $\theta(x) \leq B_n\theta(x')$ for all x and x' in Λ_{w_n}. From the proof of Lemma 6.4, we can get $\mathcal{L}\psi(x) \leq B_{n+1}e^{A\tau^{n+1}}\mathcal{L}\psi(x')$ for any x and x' in Λ_{w_n}. To get Inequality (6.4) it is enough to have

$$\eta(B_n - B_n^{-1})h(x') \leq (B_n - B_{n+1}e^{A\tau^{n+1}})\lambda^{-1}\mathcal{L}\psi(x')$$

or

$$\eta(B_n - B_n^{-1})||h|| \leq (B_n - B_{n+1}e^{A\tau^{n+1}})K^{-1}$$

because $\lambda^{-1}\mathcal{L}\psi(x) \geq K^{-1}$ for all x in Λ (see the proof of Lemma 6.4). Let $L_0 > 0$ be a constant such that $|\log B_n| \leq L_0$ for all $n \geq 0$ and let $L_1 > 0$ and $L_2 > 0$ be two constants such that

$$L_1(x-y) \leq e^x - e^y \leq L_2(x-y)$$

for all $y < x$ in $[-L_0, L_0]$. For Inequality (6.4) to be hold it is enough for $0 < \eta \leq (K||h||)^{-1}$ to satisfy

$$\eta||h||L_2\frac{4A\tau^{n+1}}{1-\tau} \leq K^{-1}L_1A\tau^{n+1}$$

or

$$\eta \leq \frac{L_1(1-\tau)}{4L_2||h||K}.$$

So $\psi' = (1-\eta)\theta$ is in Ω. ∎

Lemma 6.6. *There are constants $D > 0$ and $0 < \beta < 1$ such that*

$$||\lambda^{-n}\mathcal{L}\psi - h|| \leq D\beta^n$$

for all ψ in Ω and all $n \geq 0$.

Proof. Inductively from Lemma 6.5,

$$\lambda^{-n}\mathcal{L}^{\circ n}\psi = (1-(1-\eta)^n)h + (1-\eta)^n\psi_n'$$

where ψ_n' is in Ω. As $||\psi_n'|| \leq K$, one has

$$||\lambda^{-n}\mathcal{L}^{\circ n}\psi - h|| \leq D\beta^n$$

where $D = ||h|| + K$ and $\beta = 1-\eta$. ∎

Let $\mathcal{C}_n(\Lambda, \mathbf{R})$ be the subset of functions ψ in $\mathcal{C}(\Lambda, \mathbf{R})$ such that $\psi|\Lambda_{w_n}$ are constant functions for all Λ_{w_n}.

Lemma 6.7. *If ψ is in Ω and Ψ is in $C_n(\Lambda, \mathbf{R})$ with $\Psi \geq 0$ and $\psi\Psi \not\equiv 0$, then we have that $\left(\nu(\psi\Psi)\right)^{-1} \lambda^{-n}\mathcal{L}^{\circ n}(\psi\Psi)$ is in Ω.*

Proof. Assume x and x' are in Λ_{w_m} and $y \in \sigma^{-n}(x)$ and $y' \in \sigma^{-n}(x')$ are in the same $\Lambda_{w_{n+m}}$. Then $\Psi(y) = \Psi(y')$ and $\psi(y) \leq B_{m+n}\psi(y')$. Since

$$\phi(\sigma^{\circ k}(y)) \leq \phi(\sigma^{\circ k}(y')) + A\tau^{n+m-k},$$

we have that, for $S_n(z) = \sum_{k=0}^{n-1} \phi(\sigma^{\circ k}(z))$,

$$B_{n+m}e^{S_n(y)} \leq B_m e^{S_n(y')}.$$

Thus

$$\mathcal{L}^{\circ n}(\psi\Psi)(x) = \sum_{y\in\sigma^{-n}(x)} e^{S_n(y)}\psi(y)\Psi(y) \leq B_m\mathcal{L}^{\circ n}(\psi\Psi)(x').$$

Using a similar argument as in the proof of Lemma 6.4 (with $\mathcal{L}^{\circ n}(\psi\Psi)$ in place $\mathcal{L}(\psi\Psi)$), we get

$$\lambda^n\nu(\Psi\psi) = \nu\left(\lambda^{-1}\mathcal{L}^{n+1}(\Psi\psi)\right) \geq K^{-1}\mathcal{L}^{\circ n}(\psi\Psi)(z)$$

for any z in Λ. But $(\Psi\psi)(w) > 0$ for some w in Λ gives that $\mathcal{L}^{\circ n}(\psi\Psi)(z) > 0$ for some z in Λ. So $\nu(\psi\Psi) > 0$. ∎

Lemma 6.8. *For ψ in Ω and Ψ in $C_n(\Lambda, \mathbf{R})$ and $m \geq 0$,*

$$||\lambda^{-(n+m)}\mathcal{L}^{\circ(n+m)}(\psi\Psi) - \nu(\psi\Psi)h|| \leq D\nu(\psi\Psi)\beta^m.$$

Proof. Write $\Psi = \Psi^+ - \Psi^-$ with $\Psi^+ > 0$ and $\Psi^- > 0$ and Ψ^+, Ψ^- in $C_n(\Lambda, \mathbf{R})$. Then

$$||\lambda^{-(n+m)}\mathcal{L}^{\circ(n+m)}(\psi\Psi^\pm) - \nu(\psi\Psi^\pm)h|| \leq D\nu(\psi\Psi^\pm)\beta^m.$$

For $\psi\Psi^\pm \equiv 0$, it is trivial; for $\psi\Psi^\pm \not\equiv 0$, we can apply Lemma 6.6. These inequalities add up to give us the lemma. ∎

Proof of Theorem 6.2. For any ψ in $C^+(\Lambda, \mathbf{R})$ and any $\epsilon > 0$, we can find Ψ_1 and Ψ_2 in $C_n(\Lambda, \mathbf{R})$ for n large such that $\Psi_1 \leq \psi \leq \Psi_2$ and

$0 \leq \Psi_2(x) - \Psi_1(x) \leq \epsilon$ for all x in Λ. Therefore, $|\nu(\Psi_1) - \nu(\Psi_2)| < \epsilon$. Let $\psi_0 = 1$. From Lemma 6.8,

$$||\lambda^{-m}\mathcal{L}^{\circ m}\Psi_i - \nu(\psi)h|| \leq ||\lambda^{-m}\mathcal{L}^{\circ m}(\Psi_i\psi_0) - \nu(\Psi_i\psi_0)h|| + ||\nu(\Psi_i\psi_0)h - \nu(\psi)h|| \leq \epsilon(1 + ||h||)$$

for large $m > n$. Since $\lambda^{-m}\mathcal{L}^{\circ m}\Psi_1 \leq \lambda^{-m}\mathcal{L}^{\circ m}\psi \leq \lambda^{-m}\mathcal{L}^{\circ m}\Psi_2$,

$$||\lambda^{-m}\mathcal{L}^{\circ m}\psi - \nu(\psi)h|| \leq 2\epsilon(1 + ||h||)$$

for large $m > n$. Therefore, $\lim_{m\to\infty} ||\lambda^{-m}\mathcal{L}^{\circ m}\psi - \nu(\psi)h|| = 0$. ■

Remark 6.4. The Gibbs measure is also called an equilibrium state. The reason is the following (see [BO1,RU2,RU3] for more details) : we use $h_\mu(\sigma)$ to denote the measure-theoretic entropy of a measure μ in $\mathcal{PM}_\sigma$ and consider the variation principle

$$P(\phi) = \sup_\mu \Big(h_\mu(\sigma) + \int_\Lambda \phi \, d\mu\Big)$$

for $\phi \in \mathcal{C}^+(\Lambda)$ where μ runs over $\mathcal{PM}_\sigma$. Then the Gibbs measure is the unique one in $\mathcal{PM}_\sigma$ such that

$$P(\phi) = h_\mu(\sigma) + \int_\Lambda \phi \, d\mu.$$

Now let us see how can we calculate the pressure P for a function ϕ in $\mathcal{C}^+(\Lambda)$. Let

$$\mathrm{Fix}(\sigma^{\circ n}) = \{x \in \Lambda \mid \sigma^{\circ n}(x) = x\}$$

be the set of all fixed points of $\sigma^{\circ n}$. Since σ is a degree two expanding map, for every string w_n of 0's and 1's of length n, there is a unique point x_{w_n} in $\mathrm{Fix}(\sigma^{\circ n}) \cap \Lambda_{w_n}$.

Theorem 6.3. *Suppose ϕ is in $\mathcal{C}^+(\Lambda, \mathbf{R})$ and $P = P(\phi)$ is the pressure of ϕ. Then P can be calculated as*

$$P = \lim_{n\to\infty} \frac{1}{n} \log\Big(\sum_{x\in Fix(\sigma^{\circ n})} \exp\Big(\sum_{k=0}^{n-1} \phi(\sigma^{\circ k}(x))\Big)\Big).$$

Proof. Let λ and h be the number and the function in Theorem 6.2. From $\mathcal{L}h = \lambda h$, we have that for x in Λ,

$$\mathcal{L}^{\circ n}h(x) = \sum_{y\in\sigma^{-n}(x)} e^{S_n(y)}h(y) = \lambda^n h(x)$$

where $S_n(y) = \sum_{k=0}^{n-1} \phi(\sigma^{\circ k}(y))$. Therefore

$$\lambda^n = \sum_{y \in \sigma^{-n}(x)} e^{S_n(y)} \frac{h(y)}{h(x)}$$

for x in Λ. From Inequality (6.3), there is a constant $C > 0$ such that

$$C^{-1} \sum_{x \in \mathrm{Fix}(\sigma^{\circ n})} e^{S_n(x)} \leq \lambda^n \leq C \sum_{x \in \mathrm{Fix}(\sigma^{\circ n})} e^{S_n(x)}.$$

So

$$P = \log \lambda = \lim_{n \to \infty} \frac{1}{n} \log \Big(\sum_{x \in \mathrm{Fix}(\sigma^{\circ n})} \exp \Big(\sum_{k=0}^{n-1} \phi(\sigma^{\circ k}(x)) \Big) \Big).$$

∎

6.2. A Transfer Operator and Its Spectral Radius and Essential Spectral Radius

Let Λ be a subset of the interval $I = [-1,1]$. A continuous function $\psi : \Lambda \to \mathbf{C}$ is said to be $C^{n+\alpha}$ for an integer $n \geq 0$ and for a real number $0 \leq \alpha \leq 1$ (or C^{∞} or C^{ω}) if ϕ can be extended to a function $\tilde{\psi}$ defined on an open set $U \supset \Lambda$ such that $\tilde{\psi}$ is *n-times* differentiable and its nth derivative is α-Hölder continuous (or such that $\tilde{\psi}$ is *n-times* differentiable for all $n > 0$ or such that $\tilde{\psi}$ is analytic on U). (Here 1-Hölder means Lipschitz.) Let $r = n+\alpha$ or ∞ or ω. A vector field $v(x) = \psi(x)\partial/\partial x$ on Λ is said to be C^r if $\psi : \Lambda \to \mathbf{C}$ is C^r. We call $v = \psi\partial/\partial x$ positive if $\psi > 0$. Let $\mathcal{C}^r(\Lambda, \mathbf{C})$ be the space of all C^r-functions $\psi : \Lambda \to \mathbf{C}$ and let $\mathcal{V}^r(\Lambda)$ be the space of all C^r-vector fields v on Λ. Conceptually, $\mathcal{C}^r(\Lambda, \mathbf{C})$ and $\mathcal{V}^r(\Lambda)$ are different. But since our underline space is just $I = [-1,1]$, we can identify them. Also we can identify the space of linear functionals on $\mathcal{V}^r(\Lambda)$ as $\mathcal{PM}$ and the subspace of invariant ones under σ as $\mathcal{PM}_\sigma$.

Let $\sigma : I_0 \cup I_1 \to I$ be a C^r degree two expanding map and let

$$\tau^{-1} = \min_{x \in I_0 \cup I_1} |\sigma'(x)| > 1$$

be the expansion constant. Let Λ now be the non-escaping Cantor set of σ. For any $\Phi \in \mathcal{C}^r(\Lambda, \mathbf{C})$, we define an operator $\mathcal{L}_\Phi : \mathcal{C}^r(\Lambda, \mathbf{C}) \to \mathcal{C}^r(\Lambda, \mathbf{C})$ as

$$\mathcal{L}_\Phi \psi(x) = \sum_{y \in \sigma^{-1}(x)} \Phi(y)\psi(y). \tag{6.5}$$

It is called a transfer operator. If Φ is a positive function in $\mathcal{C}^r(\Lambda, \mathbf{R})$, we can restate Theorem 6.2 for $\mathcal{L}_\Phi$ (by considering $\phi = \log \Phi$).

Theorem 6.2′. *Let Φ be a positive function in $\mathcal{C}^r(\Lambda, \mathbf{R})$ and $\mathcal{L} = \mathcal{L}_\Phi$. There are a unique constant $\lambda > 0$, a positive vector field h in $\mathcal{C}^r(\Lambda, \mathbf{R})$, and an element ν in $\mathcal{PM}_\sigma$ such that $\mathcal{L}h = \lambda h$, $\mathcal{L}^*\nu = \lambda\nu$, $\nu(h) = 1$, and*

$$\lim_{n \to \infty} ||\lambda^{-n}\mathcal{L}^n v - \nu(v)h|| = 0$$

for all v in $\mathcal{C}^r(\Lambda, \mathbf{C})$.

Remark 6.5. Suppose σ and Φ are analytic in a domain $U \supset I_0 \cup I_1$ and can be extended continuously to $\overline{U}$. Suppose $\Phi|\Lambda$ is positive. Let $\mathcal{C}^\omega(U, \mathbf{C})$ be the set of all vector fields on I which can be extended analytically to U and continuously to $\overline{U}$. Then $\mathcal{L}_\Phi : \mathcal{C}^\omega(U, \mathbf{C}) \to \mathcal{C}^\omega(U, \mathbf{C})$ defined by Eq. (6.5) is a compact operator and λ is the leading eigenvalue of $\mathcal{L}_\Phi$ with an positive eigenvector h in $\mathcal{C}^\omega(U, \mathbf{C})$ (see [RU5]).

Similarly, we can calculate the eigenvalue λ as in Theorem 6.3.

Theorem 6.3′. *Let Φ be a positive function in $C^r(\Lambda, \mathbf{R})$ and let λ be the number in Theorem 6.2′ for Φ. Then*

$$\log \lambda = \lim_{n \to \infty} \frac{1}{n} \log \Big(\sum_{x \in Fix(\sigma^{\circ n})} \prod_{k=0}^{n-1} \Phi(\sigma^{\circ k}(x)) \Big)$$

where $Fix(\sigma^{\circ n}) = \{x \in \Lambda \mid \sigma^{\circ n}(x) = x\}$.

For a transfer operator

$$\mathcal{L} = \mathcal{L}_{\Phi} : C^r(\Lambda, \mathbf{C}) \to C^r(\Lambda, \mathbf{C}),$$

let $\varrho(\mathcal{L})$ be the spectrum of $\mathcal{L}$. The spectral radius $\rho(\mathcal{L})$ of $\mathcal{L}$ is $\rho(\mathcal{L}) = \sup_{z \in \varrho(\mathcal{L})} |z|$ and the essential spectral radius $\rho_{\mathrm{ess}}(\mathcal{L})$ of $\mathcal{L}$ can be calculated by the result of Nussbaum [NUS], which holds for any bounded linear operator on a Banach space,

$$\rho_{\mathrm{ess}}(\mathcal{L}) =$$

$$\lim_{n \to \infty} \Big(\inf\{ \|\mathcal{L}^n - \mathcal{K}\| \mid \mathcal{K} : C^r(\Lambda, \mathbf{C}) \to C^r(\Lambda, \mathbf{C}) \text{ is a compact operator } \} \Big)^{\frac{1}{n}}$$

Let Φ be a fixed function in $C^r(\Lambda, \mathbf{C})$ such that $\Phi(x) \neq 0$ for all x in Λ. Then $|\Phi|$ is a positive function in $C^r(\Lambda, \mathbf{R})$. Let

$$\mathcal{L} = \mathcal{L}_{\Phi} : C^r(\Lambda, \mathbf{C}) \to C^r(\Lambda, \mathbf{C}) \tag{6.6}$$

and let

$$\mathcal{L}^{+} = \mathcal{L}_{|\Phi|} : C^r(\Lambda, \mathbf{C}) \to C^r(\Lambda, \mathbf{C}). \tag{6.7}$$

Let $\lambda = \lambda(|\Phi|)$ be the number in Theorems 6.2′ and 6.3′ for $\mathcal{L}^{+}$.

Theorem 6.4 [RU5]. *The spectral radius of $\mathcal{L}$ is less than or equal to λ and the essential spectral radius of $\mathcal{L}$ is less than or equal to $\tau^r \lambda$.*

Theorem 6.5 [RU5]. *Let λ_0 be a number in the spectrum of $\mathcal{L}$. If $\lambda_0 > \tau^r \lambda$, then it is an isolated eigenvalue of finite multiplicity. Moreover, if σ and Φ are C^s for $s > r$, then every eigenvector h_0 corresponding to λ_0 is a vector in $C^s(\Lambda, \mathbf{C})$. Furthermore, if σ and Φ can be extended analytically to a domain*

$U \supset I$ and continuously to $\overline{U}$, then h_0 can be extended analytically to U and continuously to $\overline{U}$.

Remark 6.6. In other words, if σ and Φ can be extended analytically to a domain $U \supset I$ and continuously to $\overline{U}$, then h_0 is an eigenvector and λ_0 is the corresponding eigenvalue of

$$\mathcal{L} = \mathcal{L}_\Phi : \mathcal{C}^\omega(U, \mathbf{C}) \to \mathcal{C}^\omega(U, \mathbf{C}).$$

Remark 6.7. If Φ is a positive function in $\mathcal{C}^r(\Lambda, \mathbf{R})$, then Theorems 6.2′ and 6.4 imply that λ is a simple eigenvalue of $\mathcal{L}$ and is the spectral radius of $\mathcal{L}_\Phi$.

Let $I = [-1, 1]$. The Zygmund function space $\mathcal{Z}(I, \mathbf{C})$ (refer to §2.4) is the complex vector space of continuous functions $\varphi : I \to \mathbf{C}$ such that

$$\mathcal{Z}(\varphi) = \sup_{x \in I, t > 0, x \pm t \in I} |\mathcal{Z}(\varphi, x, t)| < \infty\,,$$

where

$$\mathcal{Z}(\varphi, x, t) = \frac{\varphi(x+t) + \varphi(x-t) - 2\varphi(x)}{t}.$$

The vector space $\mathcal{Z}(I, \mathbf{C})$ becomes a Banach space when endowed with the norm

$$||\varphi||_{\mathcal{Z}} = \max(\sup_I |\varphi|, \mathcal{Z}(\varphi)).$$

For $0 < \alpha \le 1$, let $\mathcal{C}^\alpha(I, \mathbf{C})$ denote the space of α-Hölder functions, i.e. functions $\varphi : I \to \mathbf{C}$ satisfying

$$|\varphi|_\alpha = \sup_{x \neq y \in I} \frac{|\varphi(x) - \varphi(y)|}{|x - y|^\alpha} < \infty\,.$$

In particular, $\mathcal{C}^1(I, \mathbf{C})(= \mathcal{C}^{0+1})$ is the space of Lipschitz functions. Each $\mathcal{C}^\alpha(I, \mathbf{C})$ is a Banach space for the norm $||\varphi||_\alpha = \max(\sup_I |\varphi|, |\varphi|_\alpha)$. We would like to note that $\mathcal{C}^1(I, \mathbf{C}) \subset \mathcal{Z}(I, \mathbf{C}) \subset \mathcal{C}^\alpha(I, \mathbf{C})$ for $0 < \alpha < 1$, but $\mathcal{C}^1(I, \mathbf{C})$ as well as $\mathcal{Z}(I, \mathbf{C})$ is not dense in $\mathcal{C}^\alpha(I, \mathbf{C})$ for $0 < \alpha < 1$. We would also like to note that the norms

$$||\varphi||_{\mathcal{Z},\alpha} = \max(\sup_I |\varphi|, \mathcal{Z}(\varphi), |\varphi|_\alpha)$$

for $0 < \alpha < 1$ on $\mathcal{Z}(I, \mathbf{C})$ are all equivalent with the norm $|| \cdot ||_{\mathcal{Z}}$. (Indeed, for each $0 \le \alpha < 1$ the space $\mathcal{Z}(I, \mathbf{C})$ is a Banach space for the norm $|| \cdot ||_{\mathcal{Z},\alpha}$, open mapping theorem may then be applied to the identity maps $(\mathcal{Z}(I,\mathbf{C}), \|\cdot\|_{\mathcal{Z},\alpha}) \to (\mathcal{Z}(I,\mathbf{C}), \|\cdot\|_{\mathcal{Z}})$.) In other words, for each $0 \leq \alpha < 1$, there is a constant $K = K(\alpha)$ such that

$$|\varphi|_\alpha \leq K(\alpha)\Big(\sup|\varphi| + \mathcal{Z}(\varphi)\Big), \quad \forall \varphi \in \mathcal{Z}(I,\mathbf{C}).$$

The following lemma can be proved by direct computation:

Lemma 6.9 (Zygmund derivation of a product). *For all φ, ψ in $\mathcal{Z}(I,\mathbf{C})$, all $x \in I$, and all $t > 0$,*

$$\begin{aligned}\mathcal{Z}(\varphi\psi, x, t) &= \varphi(x)\mathcal{Z}(\psi, x, t) + \psi(x)\mathcal{Z}(\varphi, x, t)\\ &\quad + t\Delta_+(\varphi, x, t)\Delta_+(\psi, x, t) + t\Delta_-(\varphi, x, t)\Delta_-(\psi, x, t), \end{aligned} \tag{6.8}$$

where $\Delta_+(v,x,t) = (v(x+t) - v(x))/t$ and $\Delta_-(v,x,t) = (v(x) - v(x-t))/t$.

Now let $\sigma : I_0 \cup I_1 \to I$ be a $C^{1+\beta}$ degree two expanding map where $0 < \beta \leq 1$ and let Λ be the non-escaping Cantor set of σ. For a function $\Phi \in \mathcal{Z}(I,\mathbf{C})$, we define the transfer operator

$$\mathcal{L}_{\Phi,\mathcal{Z}}\psi(x) = \sum_{y\in\sigma^{-1}(x)} \Phi(y)\psi(y) : \mathcal{Z}(I,\mathbf{C}) \to \mathcal{Z}(I,\mathbf{C}). \tag{6.9}$$

For a function $\Phi \in C^\alpha(I,\mathbf{C})$, $0 < \alpha \leq 1$, we define the transfer operator

$$\mathcal{L}_{\Phi,C^\alpha}\psi(x) = \sum_{y\in\sigma^{-1}(x)} \Phi(y)\psi(y) : C^\alpha(I,\mathbf{C}) \to C^\alpha(I,\mathbf{C}). \tag{6.10}$$

The following lemma is proved in [BJL] by applying Lemma 6.9. (A more general setting is proved in [BJL].)

Lemma 6.10. *The linear operator $\mathcal{L}_{\Phi,C^\alpha}$ for $0 < \alpha \leq 1$ is bounded; the linear operator $\mathcal{L}_{\Phi,\mathcal{Z}}$ is bounded.*

Remark 6.8. We would like to point out that the transfer operator $\mathcal{L}_{\Phi,\mathcal{Z}}$ may be unbounded if σ is just a C^1 map (even for a constant function Φ). Indeed, it is known that there exist a Zygmund function φ and a C^1 diffeomorphism f such that $\varphi\circ f$ is not Zygmund (but $f\circ\varphi$ is Zygmund) (see, for example, [BJL]).

For a function Φ and an integer $n \geq 1$, let

$$\Phi^{(n)}(y) = \prod_{i=0}^{n-1} \Phi(\sigma^{\circ i}(y))$$

if $\sigma^{\circ i}(y) \in I$ for all $0 \leq i \leq n-1$. The following exact formula for the essential spectral radius is proved in [BJL]. (A more general setting is proved in [BJL].)

Theorem 6.6 [BJL].

1. *The essential spectral radius* $\rho_{\text{ess}}(\mathcal{L}_{\Phi,\mathcal{Z}})$ *of the operator* $\mathcal{L}_{\Phi,\mathcal{Z}}$ *is equal to*

$$\rho_{\text{ess}}(\mathcal{L}_{\Phi,\mathcal{Z}}) = \lim_{n\to\infty}\left(\sup_{x\in I}\sum_{y\in\sigma^{-n}(x)}\frac{|\Phi^{(n)}(y)|}{|(\sigma^{\circ n})'(y)|}\right)^{1/n}$$

(in particular, the limit on the right exists).

2. *The essential spectral radius* $\rho_{\text{ess}}(\mathcal{L}_{\Phi,\mathcal{C}^\alpha})$ *of the operator* $\mathcal{L}_{\Phi,\mathcal{C}^\alpha}$ *for* $0 < \alpha \leq 1$ *is equal to*

$$\rho_{\text{ess}}(\mathcal{L}_{\Phi,\mathcal{C}^\alpha}) = \lim_{n\to\infty}\left(\sup_{x\in I}\sum_{y\in\sigma^{-n}(x)}\frac{|\Phi^{(n)}(y)|}{|(\sigma^{\circ n})'(y)|^\alpha}\right)^{1/n}.$$

Remark 6.9. The Zygmund function space $\mathcal{Z}$, which has been much used in dynamical systems in recent years, notably in Sullivan's analysis of renormalization (see §4.4, §4.5, and §4.6), is interesting not only because $\mathcal{C}^1 \subset \mathcal{Z} \subset \mathcal{C}^\alpha$ for all $0 < \alpha < 1$ but also because it arises in the study of quasiconformal mappings and Teichmüller theory: let $\mathbf{T}^1$ denote the circle $\mathbf{R}/\mathbf{Z}$, and choose three points $p_1 < p_2 < p_3$ in $\mathbf{T}^1$. A homeomorphism h of $\mathbf{T}^1$ fixing p_i for $i = 1$, 2, and 3 is called quasisymmetric if

$$||h||_{qs} = \sup_{x\in\mathbf{T}^1;x+t,x-t\in\mathbf{T}^1}\frac{|h(x+t)-h(x)|}{|h(x)-h(x-t)|} < \infty.$$

Let $\mathcal{U}$ be the set of all orientation-preserving quasisymmetric homeomorphisms of $\mathbf{T}^1$ which fix p_i for $i = 1, 2, 3$, endowed with the distance $d(h_1,h_2) = \log||h_1\circ h_2^{-1}||_{qs}$. The set $\mathcal{U}$ with distance d is a model for the universal Teichmüller space (see [AH1,GAR,LEH]). For a fixed quasisymmetric homeomorphism h_0 in $\mathcal{U}$, the right composition $R_{h_0}(h) = h\circ h_0$ acting on $\mathcal{U}$ is a continuous map (but the left composition is not continuous) (see Remark 6.8), and sends a neighborhood of the identity to a neighborhood of h_0. This makes $\mathcal{U}$ into a homogeneous space. It is also known that $\mathcal{U}$ is a complex manifold (see [GAR,LEH]). Thus $\mathcal{U}$ has a tangent space at the identity, which is also the tangent space at any point h_0. This tangent space is a Banach space of certain continuous vector fields $\phi(x)\partial/\partial x$ defined on $\mathbf{T}^1$, and, when factored by the two-dimensional subspace of affine functions, can be identified with the Zygmund function space $\mathcal{Z}$ (see [REI]). Therefore a transfer operator $\mathcal{L}$ acting on the Zygmund function space $\mathcal{Z}$ can be viewed as acting on the tangent space of the universal Teichmüller space. It is hoped that the knowledge of a transfer operator may be applied to the study of Teichmüller theory and vice versa. An especially interesting case is when $\mathcal{L}$ is the tangent map $\mathcal{DR}$ to some nonlinear operator $\mathcal{R}$ acting on the universal Teichmüller space. In the next section, we calculate an example which can be used to illustrate this research problem.

6.3. The Renormalization Operator and Transfer Operators

Let I be the interval $[-1,1]$ and $U \subset \mathbf{C}$ be a connected open subset containing I. Let $\mathcal{B} = FM^{\omega}(2,U)$ be the space of folding mappings f from I into I with a unique non-degenerate critical point 0 such that f can be extended to U analytically and to $\overline{U}$ continuously. Let $\mathcal{B}_0 = RFM^{\omega}(2,U)$ be the subspace of mappings in $\mathcal{B}$ satisfying the conditions $f(0)=1$ and

$$f^{\circ 3}(0) > -f^{\circ 2}(0) > f^{\circ 4}(0) > 0 > f^{\circ 2}(0).$$

The period doubling operator $\mathcal{R}: \mathcal{B}_0 \to \mathcal{B}$ is defined by (compare to Chapter Four)

$$\mathcal{R}(f)(x) = -\alpha_f f \circ f(-\alpha_f^{-1}x), \quad x \in I,$$

for $f \in \mathcal{B}_0$, where $\alpha_f = -1/f(1)$.

Suppose g is the Feigenbaum fixed point of $\mathcal{R}$ and suppose U is a domain contained in the domain of g. Let $\alpha = -1/g(1)$, $J = g(I)$ and $\Omega = g(U)$. Then g satisfies the Cvitanović-Feigenbaum functional equation,

$$g(x) = -\alpha g \circ g(-\alpha^{-1}x).$$

Suppose $\mathcal{C}^{\omega}(J,\Omega)$ is the space of real vector fields v on J such that v can be extended to Ω analytically and to $\overline{\Omega}$ continuously. This space equipped with the supremum norm is a Banach space. The number $\alpha = 2.5\ldots$ is universal and can be calculated by the Cvitanović-Feigenbaum functional equation (see also the beginning of Chapter Four).

Suppose J_0 and J_1 are the intervals $[g(1), g^{\circ 3}(1)]$ and $[g^{\circ 2}(1), 1]$. We define σ from $J_0 \cup J_1$ onto J by

$$\sigma(x) = \begin{cases} -\alpha x, & x \in J_0; \\ -\alpha g(x), & x \in J_1. \end{cases}$$

Then σ is a degree two expanding mapping with the expansion constant

$$\tau^{-1} = \inf_{x \in J_0 \cup J_1} |\sigma'(x)| = \alpha$$

because $|g'(x)| > 1$ for $x \in J_1$. The mapping has an analytic extension, which we still denote as σ, to $\Omega_0 \cup \Omega_1 \supset J_0 \cup J_1$ with also the expansion constant α. Here Ω_0 and Ω_1 are disjoint subdomains of Ω and contain J_0 and J_1, respectively. Moreover, the restrictions $\sigma|\Omega_0$ and $\sigma|\Omega_1$ of σ to Ω_0 and Ω_1

are bijective from Ω_0 and Ω_1 to Ω and can be extended continuously to the boundaries $\partial\Omega_0$ and $\partial\Omega_1$, respectively (see Fig. 6.1).

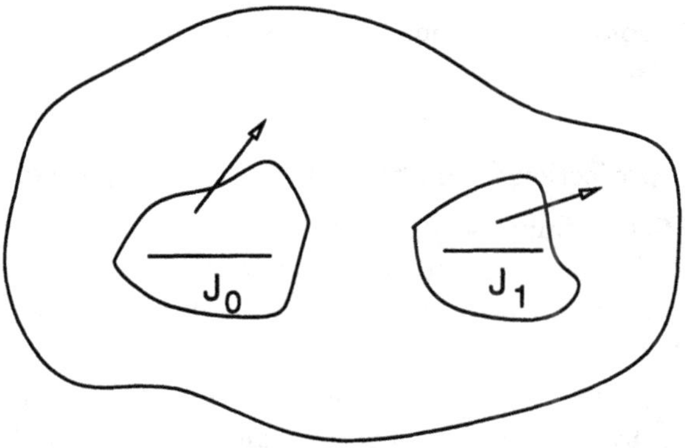

Fig. 6.1

Suppose $\mathcal{A}$ is the attractor of g (see §4.6) and Λ is the non-escaping set of σ (see §1.2).

Lemma 6.11. *The set Λ and the set $\mathcal{A}$ are the same.*

Proof. One can check it by the equation $g(x) = -\alpha g \circ g(-\alpha^{-1}x)$. ∎

Let $\Phi = \sigma'$ on $\Omega_0 \cup \Omega_1$. Then we have a transfer operator $\mathcal{L}_\Phi : \mathcal{C}^\omega(J, \Omega) \to \mathcal{C}^\omega(J, \Omega)$ defined as

$$\mathcal{L}_\Phi v(z) = \sum_{w \in \sigma^{-1}(z)} \Phi(w) v(w). \tag{6.11}$$

This operator is bounded and compact (by Montel's theorem).

Suppose $T_g\mathcal{B}_0$ is the tangent space of $\mathcal{B}_0$ at g and

$$T_g\mathcal{R} : T_g\mathcal{B}_0 \to T_g\mathcal{B}_0 = T_g\mathcal{B}$$

is the tangent map of $\mathcal{R}$ at g. The following lemma is easy to prove.

Lemma 6.12. *The mapping g_* from $\mathcal{C}^\omega(J, \Omega)$ into $T_g\mathcal{B}_0$ defined by*

$$\Big(g_* v\Big)(x) = v(g(x))$$

for $x \in \Omega$ and $v \in \mathcal{C}^\omega(J, \Omega)$ is an isomorphism.

Lemma 6.13. *The operators $\mathcal{L}_\Phi$ and $T_g\mathcal{R}$ have the same eigenvalues (counted with multiplicity) except for the value* 1.

Proof. By some calculations, we can show that

$$\mathcal{L}_\Phi = g_*^{-1} \circ T_g\mathcal{R} \circ g_* + e_1,$$

where e_1 is the projection from $\mathcal{C}^\omega(J,\Omega)$ to the eigenspace of eigenvalue one.

Remark 6.10. Suppose $V_{2m-1}(x) = g'(x)x^{2m-1} - (g(x))^{2m-1} \in T_g\mathcal{B}_0$ and $v_{2m-1} = g_*^{-1}(V_{2m-1}) \in \mathcal{C}^\omega(J,\Omega)$. The vector v_{2m-1} is an eigenvector of $\mathcal{L}_\varphi$ with the eigenvalue $\lambda_{2m-1} = \alpha^{-(2m-2)}$ for $m = 1, 2, \ldots$. They are trivial eigenvalues (see [VSK]).

Lemma 6.13 tells us that we can use the transfer operator $\mathcal{L}_\Phi$ which has an explicit form to study the eigenvectors and eigenvalues of $T_g\mathcal{R}$ except for the value 1. We use it to find the expanding direction and the rate of $\mathcal{R}$ at the Feigenbaum fixed point g.

Suppose v is a real vector field on Λ. We say it is a Lipschitz continuous if there is a constant $L > 0$ such that $|v(x) - v(y)| \leq L|x - y|$ for any x and y in Λ. We say it is bounded if there is a constant $B > 0$ such that $|v(x)| \leq B$ for any x in Λ. Let $\mathcal{C}^1(\Lambda, \mathbf{R})$ be the space of real Lipschitz continuous vector fields on Λ and let $\mathcal{C}^b(\Lambda, \mathbf{R})$ be the space of bounded vector fields on Λ. Suppose $\Phi(x)$ is the derivative $\sigma'(x)$ on Λ. We define two linear operator by the same formula. One is $\mathcal{L}_{\Phi,L} : \mathcal{C}^1(\Lambda, \mathbf{R}) \to \mathcal{C}^1(\Lambda, \mathbf{R})$ defined by

$$\mathcal{L}_{\Phi,L}v(x) = \sum_{y \in \sigma^{-1}(x)} \Phi(y)v(y)$$

and the other is $\mathcal{L}_{\Phi,B} : \mathcal{C}^b(\Lambda, \mathbf{R}) \to \mathcal{C}^b(\Lambda, \mathbf{R})$ defined by

$$\mathcal{L}_{\Phi,B}v(x) = \sum_{y \in \sigma^{-1}(x)} \varphi(y)v(y).$$

They are bounded but not compact.

Let $\mathcal{L}_{|\Phi|,L} : \mathcal{C}^1(\Lambda, \mathbf{R}) \to \mathcal{C}^1(\Lambda, \mathbf{R})$ defined by

$$\mathcal{L}_{|\Phi|,L}v(x) = \sum_{y \in \sigma^{-1}(x)} |\Phi(y)|v(y).$$

Then it is a positive transfer operator. Let $\lambda = \lambda(|\Phi|)$ be the number in Theorems 6.2 and 6.2′ for $|\Phi|$.

Lemma 6.14. *Suppose λ_0 is an eigenvalue of $\mathcal{L}_{\Phi,B}$ and $\lambda_0 > \alpha+1$. Then it is an eigenvalue of $\mathcal{L}_{\Phi,L}$.*

Proof. There is a non-zero vector field v_0 in $\mathcal{C}^b(\Lambda, \mathbf{R})$ such that

$$\mathcal{L}_{\Phi,B} v_0 = \lambda_0 v_0.$$

This is

$$-\alpha v_0(-\alpha^{-1}x) - \alpha g'(g^{-1}(-\alpha^{-1}x))v_0(g^{-1}(-\alpha^{-1}x)) = \lambda_0 v_0(x) \quad (6.12)$$

for any x in Λ. From this we can have an inequality

$$\sup_{x\neq y\in\Lambda}\left(\frac{|v(x)-v(y)|}{|x-y|}\right) \leq \frac{D}{\lambda_0-\alpha-1}$$

where D is a positive constant. In other words, v_0 is Lipschitz continuous on Λ and is an eigenvector of $\mathcal{L}_{\Phi,L}$ with the eigenvalue λ_0. ∎

Lemma 6.15. *Suppose λ_0 is an eigenvalue of $\mathcal{L}_{\Phi,L}$ and $\lambda_0 > \alpha+1$. Then λ_0 is an eigenvalue of $\mathcal{L}_{\Phi}$.*

Proof. Remember that σ is the expanding map induced from the period doubling operator $\mathcal{R}$ and Φ is the derivative σ'. By some combinatorial arguments (see Chapter Four), we have that

$$\sum_{x\in Fix(\sigma^{\circ n})}\prod_{i=0}^{n-1}|\Phi(\sigma^{\circ i}(x))| \leq (\alpha^2+\alpha)^n,$$

Applying Theorem 6.3′,

$$\log\lambda = \lim_{n\to\infty}\frac{1}{n}\log\Big(\sum_{x\in Fix(\sigma^{\circ n})}\prod_{i=0}^{n-1}|\Phi(\sigma^{\circ i}(x))|\Big) \leq \log\Big(\alpha(\alpha+1)\Big).$$

Therefore $\tau\lambda \leq \alpha+1$. Now Theorem 6.5 (also Theorem 6.6) implies the lemma. ∎

From Lemmas 6.13, 6.14, and 6.15, if λ_0 is an eigenvalue of $\mathcal{L}_{\Phi,B}$ and $\lambda_0 > \alpha+1$, then it is an eigenvalue of $T_g\mathcal{R}$.

6.4. The Expanding Direction of the Renormalization Operator

We prove that the transfer operator $\mathcal{L}_\Phi : \mathcal{C}^\omega(J,\Omega) \to \mathcal{C}^\omega(J,\Omega)$ defined in Eq. (6.11) has an expanding direction. We also construct this direction. The transformation of this direction under g_* is the expanding direction of the period doubling operator $\mathcal{R}$.

Let $\sigma : J_0 \cup J_1 \to J$ be the degree two expanding mapping in §6.3 induced from the Feigenbaum fixed point g. Let Λ be the non-escaping set of σ. Let $\alpha = 2.5\ldots$ be the universal number in the Feigenbaum-Cvitanović functional equation (see §6.3). We note that the derivative of σ on J_1 is strictly greater than one and the derivative of σ at the right end point of J is α^2.

Suppose Φ_1 is the function defined by

$$\Phi_1 = \begin{cases} -\alpha, & x \in J_0 \cap \Lambda; \\ \alpha^2, & x \in J_1 \cap \Lambda \end{cases}$$

and $\mathcal{L}_1 : \mathcal{C}^b(\Lambda, \mathbf{R}) \to \mathcal{C}^b(\Lambda, \mathbf{R})$ is the corresponding operator defined by

$$\mathcal{L}_1 v(x) = \sum_{y \in \sigma^{-1}(x)} \Phi_1(y) v(y). \tag{6.13}$$

It is easy to check that the number $\lambda_1 = \alpha(\alpha - 1)$ is an eigenvalue of $\mathcal{L}_1$ with an eigenvector $v_1 = 1$ on Λ.

Suppose $\sigma^{-2}(J) = J_{21} \cup J_{22} \cup J_{23} \cup J_{24}$ and $J_{23} = [a_{21}, b_{21}]$, $J_{24} = [a_{22}, b_{22}]$ (see Fig. 6.2). Let $\beta_{21} = |g'(b_{21})|$ and $\beta_{22} = |g'(b_{22})| = |g'(1)| = \alpha$. Because g is a concave function (see [LA1,LA2]) (this also follows from the recent work of Sullivan [SU4] and McMullen [MC2], which gives a conceptual proof of the existence of g, and from the work of Lanford [LA1,LA2], which calculates g from the Feigenbaum-Cvitanović functional equation), we have that $\beta_{21} \leq \beta_{22}$. Suppose Φ_2 is the function defined by

$$\Phi_2(x) = \begin{cases} -\alpha, & x \in J_0 \cap \Lambda; \\ \alpha\beta_{21}, & x \in J_{23} \cap \Lambda\ ; \\ \alpha\beta_{22}, & x \in J_{24} \cap \Lambda \end{cases}$$

and $\mathcal{L}_2 : \mathcal{C}^b(\Lambda, \mathbf{R}) \to \mathcal{C}^b(\Lambda, \mathbf{R})$ is the corresponding operator defined by

$$\mathcal{L}_2 v(x) = \sum_{y \in \sigma^{-1}(x)} \Phi_2(y) v(y).$$

Let k_{21} be the vector field on Λ defined by

$$k_{21}(x) = \begin{cases} 1, & x \in J_0 \cap \Lambda; \\ 0, & x \in J_1 \cap \Lambda \end{cases}$$

and $k_{22} = 1 - k_{21}$. The space $\mathbf{R}^2 = \text{span}\{k_{21}, k_{22}\}$ is a subspace of $\mathcal{C}^b(\Lambda, \mathbf{R})$. For any $v = x_{21}k_{21} + x_{22}k_{22}$,

$$\mathcal{L}_2 v(x) = (k_{21}, k_{22}) \begin{pmatrix} -\alpha & \alpha\beta_{21} \\ -\alpha & \alpha\beta_{22} \end{pmatrix} \begin{pmatrix} x_{21} \\ x_{22} \end{pmatrix}.$$

Let A_2 be the matrix

$$\begin{pmatrix} -\alpha & \alpha\beta_{21} \\ -\alpha & \alpha\beta_{22} \end{pmatrix}.$$

Lemma 6.16. *The maximal eigenvalue of A_2 is*

$$\lambda_2 = \alpha \frac{(\beta_{22} - 1) + \sqrt{(\beta_{22} - 1)^2 + 4(\beta_{22} - \beta_{21})}}{2},$$

with an eigenvector $v_2 = (t_{21}, 1)$, $t_{21} < 1$.

Proof. The proof is to use linear algebra. ∎

Furthermore, suppose $\sigma^{-n}(J) = J_{n1} \cup J_{n2} \cup \cdots \cup J_{n2^{n-1}} \cup J_{n(2^{n-1}+1)} \cup \cdots \cup J_{n2^n}$ and $J_{n(2^{n-1}+i)} = [a_{ni}, b_{ni}]$ (see Fig. 6.2).

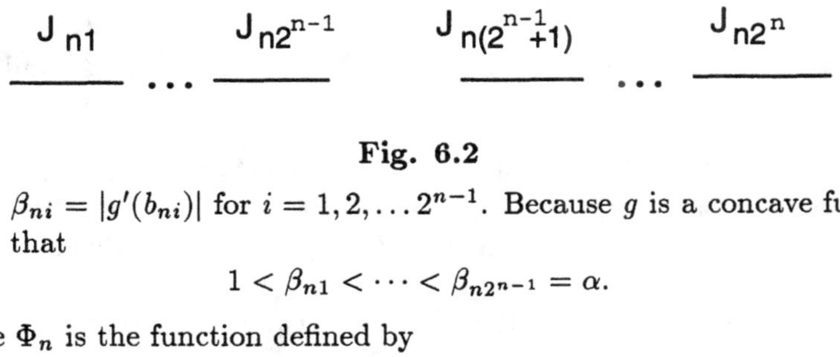

Fig. 6.2

Let $\beta_{ni} = |g'(b_{ni})|$ for $i = 1, 2, \ldots 2^{n-1}$. Because g is a concave function, we have that

$$1 < \beta_{n1} < \cdots < \beta_{n2^{n-1}} = \alpha.$$

Suppose Φ_n is the function defined by

$$\Phi_n(x) = \begin{cases} -\alpha, & x \in J_0 \cap \Lambda; \\ -\alpha\beta_{ni}, & x \in J_{n(2^{n-1}+i)} \cap \Lambda, \quad i = 1, 2, \ldots 2^{n-1} \end{cases},$$

and $\mathcal{L}_n : \mathcal{C}^b(\Lambda, \mathbf{R}) \to \mathcal{C}^b(\Lambda, \mathbf{R})$ is the corresponding operator defined by

$$\mathcal{L}_n v(x) = \sum_{y \in \sigma^{-1}(x)} \Phi_n(y) v(y).$$

Let k_{ni} be the vector field on Λ defined by

$$k_{ni}(x) = \begin{cases} 1, & x \in (J_{n(2i-1)} \cup J_{n(2i)}) \cap \Lambda; \\ 0, & x \in \Lambda \setminus ((J_{n(2i-1)} \cup J_{n(2i)}) \cap \Lambda) \end{cases}$$

for $i = 1, 2, \ldots, 2^{n-1}$. The space $\mathbf{R}^{2^{n-1}} = \mathrm{span}\{k_{n1}, \ldots, k_{n2^{n-1}}\}$ is a subspace of $\mathcal{C}^b(\Lambda, \mathbf{R})$. For any $v = x_{n1}k_{n1} + \cdots + x_{n2^{n-1}}k_{n2^{n-1}}$, we have that

$$\mathcal{L}_n v = K_n A_n X_n^t$$

where $K_n = (k_{n1}, \ldots, k_{n2^{n-1}})$ and $X_n = (x_{n1}, \ldots, x_{n2^n})$ and A_n stands for the $2^{n-1} \times 2^{n-1}$-matrix

$$\begin{pmatrix}
0 & 0 & \cdots & 0 & -\alpha & \alpha\beta_{n1} & 0 & \cdots & 0 & 0 \\
0 & 0 & \cdots & 0 & -\alpha & \alpha\beta_{n2} & 0 & \cdots & 0 & 0 \\
0 & 0 & \cdots & -\alpha & 0 & 0 & \alpha\beta_{n3} & \cdots & 0 & 0 \\
0 & 0 & \cdots & -\alpha & 0 & 0 & \alpha\beta_{n4} & \cdots & 0 & 0 \\
\vdots & \vdots & \cdots & \vdots & \vdots & \vdots & \vdots & \cdots & \vdots & \vdots \\
0 & -\alpha & \cdots & 0 & 0 & 0 & 0 & \cdots & \alpha\beta_{n(2^{n-1}-3)} & 0 \\
0 & -\alpha & \cdots & 0 & 0 & 0 & 0 & \cdots & \alpha\beta_{n(2^{n-1}-2)} & 0 \\
-\alpha & 0 & \cdots & 0 & 0 & 0 & {} & \cdots & 0 & \alpha\beta_{n(2^{n-1}-1)} \\
-\alpha & 0 & \cdots & 0 & 0 & 0 & 0 & \cdots & 0 & \alpha\beta_{n2^{n-1}}
\end{pmatrix}.$$

Lemma 6.17. *The matrix A_n has an eigenvalue λ_n which is greater than $\alpha(\alpha - 1)$.*

Proof. Let CN_n be the set

$$\{(x_{n1}, \ldots, x_{n2^{n-1}}) \in \mathbf{R}^{2^{n-1}} \mid x_{ni} \geq 0 \text{ for } i = 1, \ldots, 2^{n-1}, \text{ and } x_{n1} \leq x_{n2} \leq \cdots \leq x_{n2^{n-1}}\}.$$

It is easy to check that CN_n is a convex cone and A_n maps this cone into the interior of this cone and zero vector. By the Brouwer fixed point theorem, we conclude that there is a unique direction $\mathbf{R}^+ v_n$ in this cone which is preserved by A_n. Suppose $v_n = (t_{n1}, \ldots, t_{n2^{n-1}})$ with $t_{n2^{n-1}} = 1$ is an eigenvector with the eigenvalue λ_n. By the equation $A_n v_n = \lambda_n v_n$, we have that $-\alpha t_{n1} + \alpha^2 = \lambda_n$. Because $t_{n1} < 1$, we get $\lambda_n > \alpha(\alpha - 1)$. ∎

Remark 6.11. We can prove more that $\{\lambda_n\}_{n=1}^{\infty}$ is an increasing sequence. But we do not use this fact because we would like the following arguments to be more general.

Lemma 6.18. *There is a subsequence $\{n_i\}$ of the positive integers such that the continuous extension of the limit $v = \lim_{i\to\infty} v_{n_i}$ on the critical orbit $Or(g) = \{g^{\circ n}(0)\}_{n=1}^{\infty}$ is an eigenvector of $\mathcal{L}_{\Phi,B}$ with the eigenvalue $\delta = \lim_{i\to\infty} \lambda_{n_i}$.*

Proof. Because $Or(g)$ is a countable set, we can find a subsequence $\{n_i\}_{i=0}^{\infty}$ such that for every $a \in Or(g)$, the limit $v_{n_i}(a)$ exists as i goes to infinity. We denote this limit as $v(a)$. For the sequence $\{\lambda_{n_i}\}_{i=0}^{\infty}$, we can find convergent subsequence. Let δ be the limit of this subsequence. Then we have that

$$\mathcal{L}_{\Phi,B}v(a) = \delta v(a)$$

for any $a \in Or(g)$. Now by using Eq. (6.12) and the fact $\alpha(\alpha-1) > \alpha+1$ which can be implied by $\alpha > 1+\sqrt{2}$, we can show that v has a continuous extension on Λ which is the closure of $Or(g)$. ∎

Lemmas 6.13 to 6.18 and the work of Sullivan [SU4] and McMullen [MC2] give a conceptual proof of the following theorem (see [JMS]):

Theorem 6.7. *Suppose $\mathcal{R}$ is the period doubling operator on $\mathcal{B}_0$. Then its tangent map $T_g\mathcal{R}$ at its unique fixed point g has a unique expanding eigenvalue δ with one-dimensional eigenspace $< V >$ in $T_g\mathcal{B}_0$.*

The expanding direction $V = g_*(v)$ of $\mathcal{R}$ can be obtained by a subsequence of $\{v_n\}_{n=0}^{\infty}$. We can say more about the sequence $\{v_n\}_{n=0}^{\infty}$ and the corresponding eigenvalues $\{\lambda_n\}_{n=0}^{\infty}$. For example, $\{\lambda_n\}$ is an increasing sequence and for every $a \in \Lambda$, $\{v_n(a)\}$ is a monotone sequence. In practice, we can use these good properties to give an effective program to find the expanding direction V and the rate of the period doubling operator as follows: Suppose u is a vector in $\mathbf{R}^k$. We use $(u)_i$ to denote its ith-coordinate.

Program. **(1)** *Start from the constant function $v_1 = 1$. Consider it as a vector in $\mathbf{R}^2$ and compute the limiting vector*

$$v_2 = \lim_{l\to\infty} \frac{A_2^l v_1}{(A_2^l v_1)_2}$$

and the corresponding eigenvalue $\lambda_2 = \alpha(\alpha - (A_2 v_2)_1)$.

$$\vdots$$

(n) Let $v_{n-1} \in \mathbf{R}^{2^{n-2}}$ be the eigenvector of A_{n-1} with the eigenvalue λ_{n-1}. Consider v_{n-1} as a vector in $\mathbf{R}^{2^{n-1}}$ and compute the limiting vector

$$v_n = \lim_{l\to\infty} \frac{A_n^l v_{n-1}}{(A_n^l v_{n-1})_{2^{n-1}}}$$

and the corresponding eigenvalue $\lambda_n = \alpha(\alpha - (A_n v_n)_1)$.

(∞) The limiting vector

$$V = \lim_{n \to \infty} g_*(v_n)$$

is the expanding direction and the limiting value

$$\delta = \lim_{n \to \infty} \lambda_n$$

is the rate of expansion of the period doubling operator $\mathcal{R}$ at the fixed point g.

Bibliography

[AH1] L. V. Ahlfors, *Lectures on Quasiconformal Mappings.* D. Van Nostrand-Reinhold Company, Inc., Princeton, New Jersey, 1966.

[AH2] L. V. Ahlfors, *Conformal Invariants: Topics in Geometric Function Theory.* McGraw-Hill Book Co., New York, 1973.

[AH3] L. V. Ahlfors, *Complex Analysis.* McGraw-Hill Book Co., New York, 1979.

[ARN] V. I. Arnold, *Ordinary Differential Equations.* M.I.T. Press: Cambridge, MA., 1973. (Russian original, Moscow, 1971).

[BAL] V. Baladi, *Dynamical zeta functions. In Real and Complex Dynamical Systems* (B. Branner and P. Hjorth, eds). Kluwer Academic Publishers, 1995.

[BJL] V. Baladi, Y. Jiang, and O. E. Lanford III, *Transfer operators acting on Zygmund functions.* FIM/ETH, Zürich Preprint, March 1995 and Trans. of the Amer. Math. Soc., to appear.

[BEF] T. Bedford and A. Fisher, *Ratio geometry, rigidity and the scenery process for hyperbolic Cantor sets.* IMS preprint 1994/9, SUNY at Stony Brook.

[BIE] L. Bieberbach, *Conformal Mapping.* Chelsea Publishing Company, New York, 1953.

[BSTV] B. Bielefeld, S. Sutherland, F. Tangerman, and J. J. P. Veerman, *Dynamics of certain non-conformal degree two maps of the plane.* Experimental Mathematics, **2**, 1993, pp. 281-300.

[BLA] P. Blanchard, *Complex analytic dynamics on the Riemann sphere.* Bull. of the Amer. Math. Soc. **11**, 1984, pp. 85-141.

[BO1] R. Bowen, *Equilibrium States and the Ergodic Theory of Anosov Diffeomorphisms.* Lecture Notes in Mathematics, Vol. **470**, Springer-Verlag, New York, Berlin, 1975.

[BO2] R. Bowen, *Hausdorff dimension of quasi-circles.* Publ. Math. Inst. des Hautes Études Scientif., No. **50**, 1979, pp. 11-25.

[BO3] R. Bowen, *A horseshoe with positive measure.* Invent. Math., **29** (1975), 203-204.

[BO4] R. Bowen, *Markov partitions for Axiom A diffeomorphisms.* Amer. Journal of Math. **92**, 1970, pp. 907-918.

[BRA] L. de Branges, *A proof of the Bieberbach conjecture.* Acta Mathematica, **154**, 1985, pp. 137-152.

[BRH] B. Branner and J. Hubbard, *The iteration of cubic polynomials, Part I : The global topology of parameter space & Part II : Patterns and parapatterns.* Acta Math **160**, 1988, pp. 143-206 & Acta Math, 169, 1992, pp. 229-325.

[CAR] C. Carathéodory, *Über die Begrenzung einfach zusammenhängender Gebiete.* Math. Ann. **73**, 1913, pp. 323-370. (Gesam. Math. Schr., v. **4**).

[CAG] L. Carleson and T. Gamelin, *Complex Dynamics.* Springer-Verlag, Berlin, Heidelberg, 1993.

[CAW] E. Cawley, *The Teichmüller space of an Anosov diffeomorphism of T^2.* IMS preprint 1991/9, SUNY at Stony Brook.

[CAE] M. Campanino and H. Epstein, *On the existence of Feigenbaum's fixed point.* Commun. Math. Phys., **79**, 1981, pp. 261-302.

[CER] M. Campanino, H. Epstein, and D. Ruelle, *On Feigenbaum's functional equation $g \circ g(\lambda x) + \lambda g(x) = 0$.* Topology, **21**, 1982, pp. 125-129.

[CCR] F. Christiansen, P. Cvitanović, and H. Rugh, *The spectrum of the period doubling operator in terms of cycles.* J. Phys. A, **23**, 1990, pp. L713-L717.

[COE] P. Collet and J.-P. Eckmann, *Iterated Maps on the Interval as Dynamical Systems.* Progress in Physics, **Vol. 1**, Birkhäuser, Boston, 1980.

[COT] P. Coullet and C. Tresser, *Itération d'endomorphismes et groupe de renormalisation.* C. R. Acad. Sci. Paris Ser., A-B **287**, 1978, pp. A577-A580. (J. Phys. Coll. 39:C5 (1978), 25-28; supplément au 39:8).

[CUI] G. Cui, *Circle expanding maps and symmetric structures.* Preprint.

[CVI] P. Cvitanović, *Universality in Chaos.* Adam Hilger Ltd., Bristol, 1984.

[DEN] A. Denjoy, *Sur les courbes définies par les équations différentielles à la surface du tore.* J. Math. Pure et Appl. **11** (IV), 1932, pp. 333-375.

[DH1] A. Douady and J. H. Hubbard, *Itération des polynômes quadratiques complexes.* C.R. Acad. Sci. Paris, **294**, 1982, pp. 123-126.

[DH2] A. Douady and J. H. Hubbard, *Étude dynamique des polynômes complexes I & II.* Publ. Math. d'Orsay, 1984 & 1985.

[DH3] A. Douady and J. H. Hubbard, *On the dynamics of polynomial-like mappings*. Ann. Sci. Éc. Norm. Sup., Paris, **18**, 1985, pp. 287-343.

[DH4] A. Douady and J. H. Hubbard, *A proof of Thurston's topological characterization of rational maps*. Preprint, Institute Mittag-Leffler 1984.

[DUS] N. Dunford and J. Shwartz, *Linear Operators*. Vol. I, II, and III, Interscience, 1985.

[EE1] J.-P. Eckmann and H. Epstein, *Scaling of Mandelbrot sets generated by critical point preperiodicity*. Commun. Math. Phys., **101**, 1985, pp. 283-289.

[EE2] J.-P. Eckmann and H. Epstein, *Bounds on the unstable eigenvalue for period doubling*. Commun. Math. Phys., **128**, 1990, pp. 427-435.

[ECW] J.-P. Eckmann and P. Wittwer, *A complete proof of Feigenbaum conjectures*. J. of Stat. Phys., **46**, 1987, pp. 455-475.

[EPS] H. Epstein, *New proofs of the existence of the Feigenbaum functions*. Commun. Math. Phys., **106**, 1986, pp. 395-426.

[FAL] K. J. Falconer, *The Geometry of Fractal Sets*. Cambridge Univ. Press, 1985.

[FAR] E. de Faria, *Proof of universality for critical circle mappings*, Thesis, Graduate Center of CUNY, 1992.

[FAM] E. de Faria and W. de Melo, *Proof of universality for critical circle mappings*. Preprint in preparation.

[FE1] M. Feigenbaum, *Quantitative universality for a class of non-linear transformations*. J. Stat. Phys., **19**, 1978, pp. 25-52.

[FE2] M. Feigenbaum, *The universal metric properties of non-linear transformations*. J. Stat. Phys., **21**, 1979, pp. 669-706.

[FLE] M. Flexor, *Théoréme du secteur d'aprés D. Sullivan*, Preprint, 92-26. Université de Paris-Sud, Orsay, 1993.

[GAR] F. Gardiner, *Teichmüller Theory and Quadratic Differentials*. A Wiley-Interscience Publication, John Wiley & Sons, New York, 1987.

[GS1] F. Gardiner and D. Sullivan, *Lacunary series as quadratic differentials in conformal dynamics*. Contemporary Mathematics, **169**, 1994, pp. 307-330.

[GS2] F. Gardiner and D. Sullivan, *Symmetric and quasisymmetric structures on a closed curve*. Amer. J. of Math., 114, no. **4**, (1992), pp. 683-736.

[GRS] J. Graczyk and G. Świątek, *Private talk*.

[GU1] J. Guckenheimer, *Limit sets of S-unimodal maps with zero entropy*. Commun. Math. Phys., **110**, 1987, pp. 655-659.

[GU2] J. Guckenheimer, *Sensitive dependence on initial conditions for one-dimensional maps*. Commun. Math. Phys., **70**, 1979, pp. 133-160.

[GUJ] J. Guckenheimer and S. Johnson, *Distortion of S-unimodal maps.* Annals. Math., **132**, 1990, pp. 71-130.

[HER] M. R. Herman, *Sur la conjugaison différentiable des difféomorphismes du cercle á des rotations.* Publ. Math. Inst. des Hautes Études Scientif., No. **49**, 1979, pp. 5-234.

[HPS] M. W. Hirsh, and C. C. Pugh, and M. Shub, *Invariant Manifolds.* Springer Lectures Notes in Mathematics, Vol. **583**, Springer-Verlag: New York, Heidelberg, Berlin, 1977.

[HUS] J. Hu and D. Sullivan, *Topological conjugacy of circle diffeomorphisms.* IMS preprint 1995/7, SUNY at Stony Brook.

[HUB] J. H. Hubbard, *Local connectivity of Julia sets and bifurcation loci: three theorems of J. -C. Yoccoz.* Topological Methods in Modern Mathematics, A Symposium in Honor of John Milnor's Sixtieth Birthday, Publish or Perish, Inc., 1993, pp. 467-512.

[JAK] M. Jakobson, *Quasisymmetric conjugacy for some one-dimensional maps inducing expansion.* Preprint.

[JI1] Y. Jiang, *Geometry of Cantor systems.* Preprint. (Also *Leading gap determines the geometry of Cantor sets.* Preprint of IHES, June/1989.)

[JI2] Y. Jiang, *Local normalization of one-dimensional maps.* Preprint of IHES, June/1989.

[JI3] Y. Jiang, *Generalized Ulam-von Neumann transformations.* Thesis, 1990, Graduate School of CUNY and UMI dissertation service.

[JI4] Y. Jiang, *Dynamics of certain one-dimensional mappings: I. $C^{1+\alpha}$-Denjoy-Koebe distortion lemma.* IMS preprint 1991/1, SUNY at Stony Brook.

[JI5] Y. Jiang, *On quasisymmetrical classification of infinitely renormalizable maps – I. Maps with Feigenbaum topology, and II. Remarks on maps with a bounded type topology.* IMS preprint 1991/19, SUNY at Stony Brook.

[JMS] Y. Jiang, T. Morita, and D. Sullivan, *Expanding direction of the period doubling operator.* Commun. Math. Phys., **144**, 1992, pp. 509-520.

[JI6] Y. Jiang, *Asymptotic differentiable structure on Cantor set.* Commun. Math. Phys., **155**, 1993, pp. 503-509.

[JI7] Y. Jiang, *Geometry of geometrically finite one-dimensional maps.* Commun. Math. Phys., **156**, 1993, pp. 639-647.

[JI8] Y. Jiang, *Dynamics of certain non-conformal semi-groups.* Complex variables, Vol. **22**, 1993, pp. 27-34. (also [JI8ims] IMS preprint 1992/5, SUNY at Stony Brook under the same title.)

[JI9] Y. Jiang, *On Ulam-von Neumann transformations.* Commun. Math. Phys., **172**, 1995, pp. 449-459.

[JI10] Y. Jiang, *Markov partitions and Feigenbaum-like maps.* Commun. in Math. Phys., **171**, 1995, pp. 351-363.

[JI11] Y. Jiang, *Smooth classification of geometrically finite one-dimensional maps.* Trans. of the Amer. Math. Soc., 1996, to appear.

[JI12] Y. Jiang, *On rigidity of one-dimensional maps.* Preprint.

[JI13] Y. Jiang, *Infinitely renormalizable quadratic Julia sets.* Preprint.

[JIH] Y. Jiang and J. Hu, *The Julia set of the Feigenbaum polynomial is locally connected.* Preprint.

[JI14] Y. Jiang, On Sullivan's sector theorem. Preprint.

[JI15] Y. Jiang, *Renormalization on one-dimensional folding maps.* Proceedings of the International Conference on Dynamical Systems and Chaos, Volume **1**, World Scientific Publishing Co. Pte. Ltd., 1995, pp. 116-125.

[JI16] Y. Jiang, *The Renormalization method and quadratic-like maps.* MSRI preprint No. 081-95.

[JI17] Y. Jiang, *Local connectivity of the Mandelbrot set at certain infinitely renormalizable points.* MSRI preprint No. 063-95, Berkeley.

[KAH] J. Kahn, *Unpublished work.*

[KAO] Y. Katznelson and D. Ornstein, *The differentiability of conjugation of certain diffeomorphisms of the circle.* Ergod. Th. & Dynam. Sys. **9**, 1989, pp. 643-680.

[KHS] K. M. Khanin and Ya. G. Sinai, *A new proof of M. Herman's theorem.* Commun. in Math. Phys., **112**, 1987, pp. 89-101.

[LA1] O. E. Lanford III, *A computer-assistant proof of the Feigenbaum conjecture.* Bull. of the Amer. Math. Soc., **6**, 1982, pp. 427-434.

[LA2] O. E. Lanford III, *A shorter proof of the existence of Feigenbaum fixed point.* Commun. in Math. Phys., **96**, 1984, pp. 521-538.

[LEV] G. Levin and S. van Strien, *Local connectivity of the Julia sets of real polynomials.* IMS preprint 1995/5.

[LEH] O. Lehto, *Univalent Functions and Teichmüller Spaces.* Springer-Verlag, New York, Berlin, 1987.

[LMM] R. de la Llave, J. M. Marco and R. Moriyon, *Canonical perturbation theory of Anosov systems and regularity results for the Livsic cohomology equation.* Annals of Mathematics, **123**, (1986), pp. 537-611.

[LL1] R. de la Llave, *Invariants for smooth conjugacy of hyperbolic dynamical systems I.* Commun. Math. Phys. **109**, (1987), pp. 681-689.

[LL2] R. de la Llave, *Invariants for smooth conjugacy of hyperbolic dynamical systems II.* Commun. Math. Phys. **109**, (1987), pp. 369-378.

[LYM] M. Lyubich and J. Milnor, *The Fibonacci unimodal map.* J. Amer. Math. Soc. **6**(1993), pp. 425-457.

[LYU] M. Lyubich, *Geometry of quadratic polynomials: moduli, rigidity, and local connectivity.* IMS preprint 1993/9, SUNY at Stony Brook. And later developments.

[LYY] M. Lyubich and M. Y. Yampolsky, *Dynamics of quadratic polynomials: complex bounds for real maps.* MSRI preprint No. 034-95.

[MAC] R. MacKay, *A simple proof of Denjoy's theorem.* Math. Proc. Camb. Phil. Soc., **103**, 1988, pp. 299-303.

[MAM] J. M. Marco and R. Moriyon, *Invariants for smooth conjugacy of hyperbolic dynamical systems III.* Commun. Math. Phys. **112**, (1987), pp. 317-333.

[MA1] D. Mayer, *The Ruelle-Araki Transfer Operator in Classical Statistical Mechanics.* Lecture Notes in Physics, vol. **123**, Berlin, Heidelberg, New York: Springer, 1980.

[MA2] D. Mayer, *On the thermodynamics formalism for the Gauss map.* Commun. Math. Phys. **130**, 1990, pp. 311-333.

[MA3] D. Mayer, *The thermodynamic formalism approach to Selberg's zeta function for* $PSL(2, \mathbf{Z})$. Bull. of the Amer. Math. Soc., Vol. **25**, No. 1, 1991, pp. 55-60.

[MC1] C. McMullen, *Complex Dynamics and Renormalization.* Ann. of Math. Stud., vol **135**, Princeton Univ. Press, Princeton, NJ, 1994.

[MC2] C. McMullen, *Renormalization and 3-manifolds which fiber over the circle.* Preprint.

[MV1] W. de Melo and S. van Strien, *A structure theorem in one-dimensional dynamics.* Annals of Math., **129**, 1989, pp. 519-546.

[MV2] W. de Melo and S. van Strien, *One-Dimensional Dynamics.* Springer-Verlag, Berlin, Heidelberg, 1993.

[MI1] J. Milnor, *On the concept of attractor.* Commun. in Math. Phys., **99**, 1985, pp. 177-195.

[MI2] J. Milnor, *Dynamics in one complex variable: Introductory lectures.* IMS preprint 1990/5, SUNY at Stony Brook.

[MI3] J. Milnor, *Local connectivity of Julia sets: expository lectures.* IMS preprint 1992/11, SUNY at Stony Brook.

[MIT] J. Milnor and W. Thurston, *On iterated maps of the interval: I and II.* In Lecture Notes in Mathematics, **Vol. 1342**, pp. 465-563, Springer, New York, Berlin, 1988, pp. 465-563.

[MOS] D. Mostow, *Strong Rigidity of Locally Symmetric Spaces.* Ann. of Math. Stud., vol **78**, Princeton Univ. Press, Princeton, NJ, 1972.

[NEW] S. E. Newhouse, *Nondensity of Axiom A (a) on* S^2. Proc. Symp. Pure Math., **14**, 1970, pp. 191-202.

[NUS] R. D. Nussbaum, *The radius of the essential spectrum.* Duke Math. J. vol. **37**, 1970, pp. 473–478.

[PAL] W. Pałuba, *Talks in Seminars* and Thesis, 1992, the Graduate school of CUNY.

[PAP] W. Parry and M. Pollicott, *Zeta Functions and the Periodic Orbit Structure of Hyperbolic Dynamics.* Société Mathématique de France, Astérisque, vol. **187-188**, Paris, 1990.

[PAM] J. Palis and W. de Melo, *Geometric Theory of Dynamical Systems: An Introduction.* Springer-Verlag: New York, Heidelberg, Berlin, 1982.

[PER] R. Perez-Marco, Private conversation with a third party.

[PET] C. Petersen, *Local connectivity of some Julia sets containing a circle with an irrational rotation.* IHES preprint, April, IHES/M/1994/26.

[PIS] T. Pignataro and D. Sullivan, *Ground state and lowest eigenvalue of the Laplacian for non-compact hyperbolic surfaces.* Commun. Math. Phys., **104**, pp. 529–535.

[PIR] A. Pinto and D. Rand, *Global phase space universality, smooth conjugacies and renormalization: 2. The $C^{k+\alpha}$ case using rapid convergence of Markov families.* Nonlinearity, **4**, (1992), pp. 49-79.

[PNS] A. Pinto and D. Sullivan, *The circle and solenoid.* Preprint.

[PO1] M. Pollicott, *A complex Ruelle-Perron-Frobenius theorem and two counterexamples.* Ergod. Th. and Dynam. Sys. **4**, 1984, pp. 135-146.

[PO2] M. Pollicott, *A note on the Artuso-Aurell-Cvitanović approach to the Feigenbaum tangent operator.* J. Stat. Phys., **60**, 1991, pp. 257-267.

[PRZ] F. Przytycki, *On U-stability and structural stability of endomorphisms satisfying Axiom A.* Studia-Math., **60**, no. 1, 1977, pp. 61–77.

[RAN] D. Rand, *Global phase space universality, smooth conjugacies and renormalization: 1. The $C^{1+\alpha}$ case.* Nonlinearity, **1**, (1988), pp. 181-202.

[REI] M. Reimann, *Ordinary differential equations and quasiconformal mappings.* Invent. Math., vol. **33**, 1976, pp. 247–270.

[RIC] S. Rickman, *Removability theorem for quasiconformal mappings.* Ann. Ac. Scient. Fenn, **499**, 1969, pp. 1-8.

[RU1] D. Ruelle, *Statistical mechanics of a one-dimensional lattice gas.* Commun. in Math. Phys., **9**, 1968, pp. 267-278.

[RU2] D. Ruelle, *A measure associated with Axiom A attractors.* Amer. J. Math., **98**, 1976, pp. 619-654.

[RU3] D. Ruelle, *Thermodynamic Formalism : The Mathematical Structures of Classical Equilibrium Statistical Mechanics.* **5**, Addison-Wesley: Reading, Mass., 1978.

[RU4] D. Ruelle, *Repellers for real analytic maps*. Ergod. Th. & Dynam. Sys., **2**, 1982, pp. 99-107.

[RU5] D. Ruelle, *The thermodynamical formalism for expanding maps*. Commun. Math. Phys. **125**, 1989, pp. 239-262.

[RU6] D. Ruelle, *Dynamical Zeta Functions for Piecewise Monotone Maps of the Interval*. CRM monograph series, Volume **4**, AMS, Providence, RI, 1994.

[RUG] H. H. Rugh, *Generalized Fredholm determinants and Selberg zeta functions for Axiom A dynamical systems*. Preprint.

[SHI] M. Shishikura, *The Hausdorff dimension of the boundary of the Mandelbrot set and Julia sets*. IMS preprint 1991/7, SUNY at Stony Brook

[SHU] M. Shub, *Endomorphisms of compact differentiable manifolds*. Amer. J. Math., **91**, 1969, pp. 129-155.

[SHS] M. Shub and D. Sullivan, *Expanding endomorphisms of the circle revisited*. Ergod. Th & Dynam. Sys., **5**, 1985, pp. 285-289.

[SI1] Ya. G. Sinai, *Markov partitions and C-diffeomorphisms*. Func. Anal. and its Appl., **2**, no. 1, 1968, pp. 64-89.

[SI2] Ya. G. Sinai, *Gibbs measures in ergodic theory*. Russian Math. Surveys, 27, **no. 4** 1972, pp. 21-69.

[SIN] D. Singer, *Stable orbits and bifurcations of maps of the interval*. SIAM J. - Appl. Math., **35**, 1978, pp. 260-267.

[SOR] D. E. K. Sørensen, *Complex Dynamical Systems: Rays and non-local connectivity*. Ph. D. Thesis, Technical University of Denmark, 1995.

[SW1] G. Świątek, *Rational rotation numbers for maps of the circle*. Commun. in Math. Phys., **119**, 1988, pp. 109-128.

[SW2] G. Świątek, *Hyperbolic is dense in the real quadratic polynomials*. IMS preprint 1992/10, SUNY at Stony Brook. And later developments with J. Graczyk.

[STA] M. Stark, *Smooth conjugacy and renormalization for diffeomorphisms of the circle*. Nonlinearity, **1**, 1988, pp. 541-575.

[STR] S. van Strien, *On the bifurcations creating horseshoes in Dynamical Systems and Turbulance*. Lecture Notes in Mathematics, **898**, 1981, Springer, Berlin, New York, pp. 316-351.

[SU1] D. Sullivan, *Seminar on conformal and hyperbolic geometry*. IHES Preprint, March/1982.

[SU2] D. Sullivan, *Quasiconformal homeomorphisms and dynamics I, solution of the Fatou-Julia problem on wandering domains*. Ann. Math., **122**, 1985, pp. 401-418.

[SU3] D. Sullivan, *Differentiable structure on fractal like sets determined by intrinsic scaling functions on dual Cantor sets.* The Proceedings of Symposia in Pure Mathematics, Vol. **48**, 1988.

[SU4] D. Sullivan, *Bounds, quadratic differentials, and renormalization conjectures.* American Mathematical Society Centennial Publications, Volume **2**: Mathematics into the Twenty-First Century, AMS, Providence, RI, 1992, pp. 417-466.

[SU5] D. Sullivan, *Conformal dynamical systems.* In Geometric Dynamics edited by J. Palis. Lecture Notes in Math. **1007**, Springer, 1983, pp. 725-752.

[SU6] D. Sullivan, *Quasiconformal homeomorphisms in dynamics, topology, and geometry.* Proceedings of ICM, 1986, Berkeley, pp. 1216-1228.

[TA1] L. Tan, *Similarity between the Mandelbrot set and Julia sets.* Commun. in Math. Phys., **134**, 1990, pp. 587-617.

[TA2] L. Tan, *Voisinages connexes des points de Misiurewicz.* Ann. Inst. Fourier, Grenoble 124, 1992.

[TAG] F. M. Tangerman, *Meromorphic continuation of Ruelle zeta functions.* Thesis, 1986, Boston University.

[THU] W. Thurston, *Geometry and topology of three-manifolds.* Preprint, Princeton University, 1979.

[TUK] P. Tukia, *Differentiability and rigidity of Möbius groups.* Invent. Math., **82**, 1985, pp. 557-578.

[VET] J. J. P. Veerman and F. M. Tangerman, *A remark on Herman's theorem for circle diffeomorphisms.* IMS preprint 1990/13, SUNY at Stony Brook.

[VSK] E. B. Vul, Ya. G. Sinai, and K. M. Khanin, *Feigenbaum universality and the thermodynamic formalism.* Russian Math. Surveys, Volume **39**, 1984, pp. 1-40.

[YO1] J.-C. Yoccoz, *Conjugaison différentiable des difféomorphismes du cercle dont le nombre de rotation vérifie une condition Diophantienne.* Ann. Sci. Ec. Norm. Sup., **17**, 1984, pp. 333-361.

[YO2] J.-C. Yoccoz, (1984), *Il n'y a pas de contre-exemple de Denjoy analytique.* C.R. Acad. Sci. Paris, **298** , série I, no. 7, 1984, pp. 141-144.

[YO3] J.-C. Yoccoz, *Centralisateurs et conjugaison différentiable des diffeomorphisms du cercle.* Thesis, 1985, Univ. Paris-Sud, Orsay.

[YO4] J.-C. Yoccoz, *unpublished work.*

[ZHL] M. Zhang and W. Li, *A rigidity phenomenon in the conjugacy for family of diffeomorphisms.* Preprint.

[ZYG] A. Zygmund, *Smooth functions.* Duke Math. J., vol. **12**, 1945, pp. 47-76.

Notation Index

Subject Index

www.ingramcontent.com/pod-product-compliance
Lightning Source LLC
Chambersburg PA
CBHW050635190326
41458CB00008B/2276